D1496187

Transforming Rural Life

Revisiting Rural America

SERIES EDITORS

Pete Daniel
Deborah K. Fitzgerald

Mary Neth • *Preserving the Family Farm: Women, Community, and the Foundations of Agribusiness in the Midwest, 1900–1940*

Sally McMurry • *Transforming Rural Life: Dairying Families and Agricultural Change, 1820–1885*

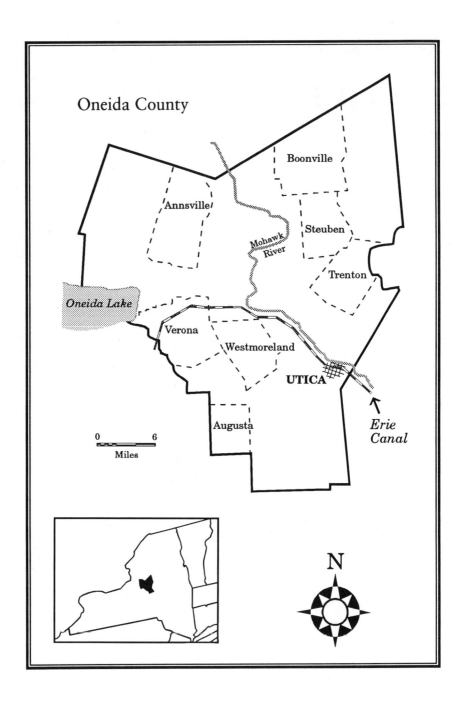

Transforming Rural Life

Dairying Families and Agricultural Change, 1820–1885

Sally McMurry

The Johns Hopkins University Press
BALTIMORE AND LONDON

© 1995 The Johns Hopkins University Press
All rights reserved. Published 1995
Printed in the United States of America
on acid-free paper
04 03 02 01 00 99 98 97 96 95   5 4 3 2 1

The Johns Hopkins University Press
2715 North Charles Street
Baltimore, Maryland 21218-4319
The Johns Hopkins Press Ltd., London

SF
274
.U6
M38
1995

ISBN 0-8018-4889-X

Library of Congress Cataloging-in-Publication Data
will be found at the end of this book.

A catalog record for this book is available
from the British Library.

Frontispiece: Oneida County, with the seven
towns, Oneida Lake, the Mohawk River,
and the Erie Canal. Map by George Skoch

For my family

Contents

Acknowledgments . . . . *xi*

Introduction *1*

1 Agriculture in the Dairy Zone: The Farming System of Cheese Dairying, 1820–1860 . . . . *6*

2 Sharp Dealings: Cheesemakers and the Market *43*

3 The Social Organization of Cheesemaking Households *62*

4 "Intense Interest and Anxiety": The Women's Work of Household Cheesemaking *72*

5 The Social Dynamics of Household Dairying: Forces for Change *100*

6 The Rise of the Factory *123*

7 The Social Organization of Factory Work: Cheesemaking Becomes Men's Work *148*

8 The Dairy Zone Embattled: Factory-Era Agriculture *172*

9 Fragmentation and Reorientation: Dairying Households in the Factory Era *194*

10 Rural Communities Transformed *212*

Conclusion *233*

Notes *237*

Essay on Sources *281*

Index *283*

Acknowledgments

This project has been a long time in the making. Like an old-time cheesemaker, I could not, at the outset of the process, foresee what shape the final product would take or what capricious transmutations would occur along the way. But at every step I was fortunate in being able to count on other people's help, generously given.

Institutional and collegial support from the Pennsylvania State University has been instrumental throughout. Gary Gallagher, head of Penn State's history department, has offered constant and much-appreciated encouragement. Seed money from Penn State's Institute for the Arts and Humanistic Studies and from the university's Research Initiation Grant program funded the travel and purchases to lay the groundwork. Supported by these grants in the summer of 1985, able research assistants Ellyn Bartges, Betty Koblitz, and Carol Lee performed yeoman service. A grant-in-aid from the American Council of Learned Societies provided further support at the outset.

As I traveled in Oneida and Herkimer Counties in 1985 and on later occasions, local historians and residents shared resources and memories. I especially thank the Boonville Public library staff, the Clinton Historical Society staff, David Maldwyn Ellis, Raymond Ernenwein, the Herkimer County Historical Society staff, Dorothea Ives, Lorena Jerson, Mary Reynolds, the Rome Historical Society staff, Edith Pangburn, Lauren Poese, Douglas Preston, the Utica Public Library staff, and Steve Wright. Special thanks to the Pellman family—Sam, Colleen, Emily, and (later) John—for putting me up during several research trips. It was a delight to return from a dusty day in the archives to share good company and pizza (with double cheese topping, of course). Further afield, the staffs of the New York State Historical Association, the Bentley Library in Michigan, and the State Historical Society of Wisconsin also helped in many ways.

As the research took shape, valuable comments from colleagues raised questions and prodded me to think hard about the arguments. I thank Katherine Boardman, Joan Brumberg, Faye Dudden, Steve Fales, Lori Ginzberg, Peter Gottlieb, Robert Gross, Ben Hudson, Carol Lee, Judith McGaw, Debra Reid, Robert Proctor, Lena Sommestad, Laurel Ulrich, and Nan Woodruff.

Acknowledgments

Two anonymous readers for the Johns Hopkins University Press also made very helpful suggestions, as did editors Robert J. Brugger and Deborah K. Fitzgerald. I would also like to thank Bob Brugger for his constant support and encouragement.

At a crucial point in the work a Rhodes Postdoctoral Fellowship from St. Hilda's College, Oxford, offered me the time, facilities, and resources to think carefully about the project in its comparative context. To the Fellows of St. Hilda's I extend my deepest gratitude for their generosity and Senior Common Room hospitality.

My last debt to mention is also my greatest. My husband, Barry Kernfeld, has helped in innumerable ways that I know about and probably in many of which I am mercifully unaware. And Paul and Eric waited cheerfully and patiently for the book to be finished. All my deepest thanks to you.

Transforming Rural Life

Introduction

This book concerns the fundamental cultural, social, and intellectual changes that transformed the nineteenth-century rural North. It began more modestly, as a brief, article-length survey of cheese dairying in central New York, undertaken in order to inquire why centralized cheesemaking plants so rapidly replaced home manufacture there between 1860 and 1875, causing in the process a major transfer of dairying labor from women to men. I soon found that in order to understand the disappearance of home-based cheesemaking, it was necessary to know something about the household system in its antebellum prime. As the structure and workings of cheesemaking households unfolded, a complex social dynamic emerged. An attempt to explain one of the most striking changes in nineteenth-century agriculture—its "defeminization"— is now also an effort to trace many of the epoch-making transformations of the era: How and why did northern society move so rapidly toward a new social order in which national markets, urban centers, and large-scale production and consumption assumed central importance? How did rural people—a majority through much of the period—shape and respond to these changes?

Dairying offers a window on these dramatic shifts. During most of the nineteenth century, dairying encompassed far more tasks and purposes than do today's dairy farms, which are devoted exclusively to the production of milk to be sold off the farm and either consumed in liquid form or processed into other products such as yogurt, cheese, butter, cream, and ice cream. In antebellum America family members themselves processed milk on the farm, into either cheese or butter. They consumed some of the product and usually bartered or sold some.

During the nineteenth century more and more people practiced these skills as dairying assumed a prominent place in the American economy. By midcentury it had developed into a major enterprise in New England, New York, and Ohio's Western Reserve. Later in the century farmers gave over large expanses of the Midwest, especially Wisconsin, Illinois, Minnesota, and Iowa, to dairying. Improved sanitation and refrigeration, urbanization, and the rise of efficient rail transport made fluid-milk production more important. The number of milch cows rose steadily. In the half century after 1830, production of all types increased 300 percent. By 1880 prominent American dairyman H. E. Alvord declared that dairying had become "the greatest single agricultural interest in America."[1]

The years 1820 and 1885 frame a coherent period in the history of American dairying. Home dairying in America had always been common, of course, but it was seldom practiced on a substantial scale and did not expand geographically or include large numbers of farms until the Erie Canal was built between 1817 and 1825. The antebellum period witnessed the establishment of a flourishing home-based dairy production, then during the Civil War era, cheesemaking quickly became centralized. By 1885 more than 90 percent of cheese was factory-made, bringing to a close this first great transformation in dairying. Soon technical innovations (the silo, the Babcock test for butterfat content, and the cream separator) and business consolidation (national corporations' challenge to local control)[2] would usher in yet another era.

The analysis offered here concentrates on one branch of dairying, that devoted to producing cheese. Though relatively few families carried on cheese dairying, it shared so many characteristics with the other major nineteenth-century dairying pursuit, buttermaking, that in almost all respects cheesemaking is representative of dairying more generally. At first glance, some differences appear. Butter, unlike cheese, was made on virtually every northern farm, often chiefly for home consumption or for small surpluses over which the farm woman had control, while farm-based cheese was destined almost exclusively for market, and women could not always hope to control the income. But this is misleading in terms of "representativeness." Throughout the nineteenth century, cheesemaking households almost always also produced significant quantities of butter, and cheese dairying families raised the same animals and plants as other farm families—hay, oats, potatoes, small amounts of grain, forest products, hogs, beef cattle—in slightly different proportions. The difference between cheese dairying farms and other farms was more a matter of emphasis and proportion than a question of a fundamental difference in type.

It is also likely that cheese dairying farms shared many social characteristics in common with butter dairying farms, particularly among those that produced large surpluses. Any dairying on a substantial scale, no matter what the end product, was tied closely to the family's life cycle, not only because of labor demands, but also because it took time to accumulate the resources necessary to own and maintain a significant-sized herd of milch cows. Conditions on these butter dairying farms likely resembled the conditions on cheesemaking farms, because while buttermaking was not as complicated a skill as cheesemaking, it demanded from adult laborers physical strength and stamina, skill, undivided attention, and considerable investment of time. Of course, the primary workers in both cases were women.

Buttermaking and cheesemaking underwent the same process of centralization, though at different times. Families curtailed home buttermaking when they began to send fluid milk to the city or to send cream to the creamery for processing into butter. These transitions took place later and more slowly than the disappearance of home cheesemaking.[3] Not until the cream separator appeared in the 1880s did creameries become economically feasible. Even after the separator appeared, creameries were apparently much slower to take hold than cheese factories, probably because butter held a much more prominent role than cheese in family subsistence and local markets.[4] But the three-quarters of buttermaking farms that sent large quantities to outside markets converted to creamery production in the 1880s and 1890s, echoing the shift in home cheesemaking a decade earlier.[5] The switch to fluid milk may also have served the same function for buttermaking families as the decision to become factory patrons had served for cheesemaking families. In short, cheesemaking anticipated the direction of change in eastern agriculture more generally: toward a heavier emphasis upon dairying, especially upon milk production; and toward a redefinition of women's work.

Finally, a common set of customs—a more or less consistent dairying culture—prevailed across the entire North, regardless of what specific product farm families manufactured. These customs arose from a strong New England influence, perpetuated through steady migration, first as Yankees poured into New York State and Ohio's Western Reserve, later once removed as Yorkers and Buckeyes moved further westward. Shared ways of selecting animals, organizing household labor, operating feeding and cultivation systems, and arranging fields united the families who took up dairying.

The analysis offered here approaches the subject from both the "bottom up"—that is, from the individual farm family's viewpoint—and on the level of national developments. To gain an appreciation of local conditions in rural communities and on specific farms, Oneida County, New York, presents an ideal setting. Oneida was a dairying county early on, ranking among the state leaders in home cheese production, and the first cheese "factories" were built there.[6] Practical matters enter into consideration here, for Oneida County brings with its experience excellent historical resources, both primary and secondary.[7] Individual census data, local newspapers, and family papers and records allow one to draw a detailed portrait of dairying household structure and farming practices.

Transforming Rural Life

Yet because cheese dairying so closely parallels dairying in general, material from a wider geographical range, and from butter dairying farms, also bears on the analysis. Material from national and regional agricultural journals, archival sources concerning cheese factories in different regions, farmers' manuals, and artifacts whose origins range from New England to the Midwest all help to integrate developments on individual farms with events that occurred all over the North.

In order to capture the complexity and richness of nineteenth-century agriculture, the study approaches dairying as a *farming system*. A farming system encompasses such basic realities as crop mixes and cultivation methods. Culture—here especially values, ideas, gender relations, household structure, and artifacts—assumes a central place.[8] The intent here is to reconstruct the cultural history of agricultural and rural transformation in this crucial transitional period. The concept of a farming system permits the merging of insights from two usually separate historiographical traditions: the earlier agricultural history, which focused primarily upon economic life, and the "new rural history," which applies to rural life the methods and assumptions of social history.[9]

Applied to dairying, this approach helps to address important questions about the social transformations that accompanied market expansion. How did men, women, sons, daughters, and hired helpers interact to shape the social setting of home cheesemaking? How did the household, in turn, relate to neighboring households and to the market? How did internal household divisions—fractures along gender, generational, or class lines—shape the transformation of dairying? What changes did crossroads cheese factories bring to farming practices, to household organization, and to rural communities? The household, as it remains central to the inquiry, offers a key to shifts in farming practices (for example, mechanization, decreased seasonality, an increased proportion of crops strictly for market); the disappearance of live-in hired labor and its replacement with wage labor; altered relationships among family members and among neighbors; and the emergence of new forms of rural communities, tied more firmly to the national culture and to the cities.

These questions confront the fundamental shift that occurred all over the nation: the transformation from an agrarian economy and culture to an urban, industrial society. The pace of change was uneven, and its reception covered a spectrum from outright rejection to ambivalence to enthusiastic endorsement. But whatever particular local form the change assumed, at the center of the process stood always the people involved in it. In historical analyses of the impact of this "Great Transformation," we have heard from urban workers, from southern yeomen, of course from capitalists and industrialists, and even from the environment.[10] But the voices of northern farm families—equally important actors in the process—have not been fully recovered.[11] The story that follows shows how dairying communities reorganized and reoriented themselves to face—and shape—the emerging modern world.

1

Agriculture in the Dairy Zone
The Farming System of Cheese Dairying, 1820-1860

> Most plants and animals have their natural zone, beyond which they do not live.—William Townsend, 1839

During the waning decade of the eighteenth century, American yeomen and their families rushed westward in unprecedented numbers, impelled by population pressure and land scarcity at home and buoyed by the promise of the new Republic. Many found new homes in New York State, which as yet had few Euro-American inhabitants outside the Hudson Valley. The Mohawk River valley offered a principal corridor for westering settlers, and as population increased, new counties straddling the valley were among the first created from the vast unorganized territory west of the Hudson (see frontispiece). Oneida was one of these counties, carved in 1798 from Herkimer County. Thousands of Yankees and immigrants from Europe and the Hudson Valley flooded into the county. Elkanah Watson was so amazed that he mixed his metaphors: people, he marveled in 1791, were "swarming into these fertile regions in shoals, like the ancient Israelites seeking the land of promise." From a population of less than 2,000 in 1790, the county grew to nearly 23,000 by 1800, 33,800 in 1810, and nearly 60,000 in 1820. These migrants came mostly from New England, often in groups of relatives or neighbors. Some were war veterans. In other cases early Oneida County settlers came from abroad, most notably from Wales.[1]

On the eve of the great migration of Americans and Europeans into the New York State interior, these lands had been very nearly deserted. This had not always been so. As its very name attests, Oneida County was originally occupied by Native Americans, specifically a subgroup of Iroquoian-speaking people living in small villages. In the area that was to become Oneida County lay the four easternmost of about twenty-seven known Oneida village sites.

7
Agriculture in the Dairy Zone

From their settlements—only two to four of which were occupied at any given time—the Oneidas followed the practices central to their culture. They cultivated a variety of plants using slash-and-burn methods, the famous Three Sisters—corn, beans, and squash—forming the nucleus of their agriculture. There is some archaeological evidence that the Oneidas had adopted domesticated cattle and swine, an indication, perhaps, of borrowings from European custom. Women tilled, men hunted for the abundant game that provided sustenance and trade goods. They also fished in the creeks and lake and gathered nuts, fruits, and berries.[2]

The history of the Oneidas after their first contact with Europeans early in the seventeenth century followed the now-familiar dismal pattern, though it was distinctive in its details. The fur trade brought them into close contact with European people and into world commodity markets. More devastating, contact brought disease, as Europeans carried microorganisms to which the Oneidas had no historically derived immunities. From a precontact population estimated at three thousand, the Oneidas lost at least half of their people to virgin soil epidemics. A period of crisis ensued as the Oneidas and the other members of the Iroquois confederacy confronted the challenge of European encroachment. After the War for Independence, when they split from the rest of the confederacy because of their support for the colonies, the Oneidas left the valley. They were rewarded for their loyalty by being defrauded of their lands in treaties in the 1790s, and eventually they were removed to Wisconsin. In 1792 an anonymous commentator traveling from Clinton in the east found no human inhabitants until he reached Oneida Lake.[3]

In terrain and vegetation the land the European newcomers came to occupy bore some resemblance to the environment that colonial-era settlers had encountered as they pushed inland from the coast, and also to the Appalachian landscape simultaneously being traversed all up and down what was then the western United States. Emigrants encountered a "much diversified" topography there, as do visitors today. A central valley, from two to twelve miles wide, divides the county into northern and southern sections. In the eastern portion of this valley flows the celebrated Mohawk, whose headwaters rise in the northern part of the county; in the western portion Wood Creek flows into Oneida Lake—formed during the last glacial epoch, and now constituting part of the county's western boundary. The two parts of the valley come within about a mile of each other near the county's center. To the north and south jut hills dissected by numerous creeks flowing in all directions. Elevations range from about five hundred feet to fifteen hundred feet above sea level. The climate in this part of the state is damp—it now receives the highest annual rainfall, forty-eight inches, in New York State—and winters are cold and snowy, especially in the northeast, where the Adirondacks rise. Summers are correspondingly cool.[4]

The dominant features of the late-eighteenth-century landscape in Oneida County were trees. Contemporary observations show that much of the county was heavily timbered. William Cooper, author of an emigrants' guide intended for prospective settlers from the British Isles, advised his readers that they would likely find white pine, "sugar-tree" (sugar maple), red, white, and black oak, poplar, beech, basswood (linden), hickory, chestnut, hemlock, ash, elm, and birch, among others. The mix of species varied from the valley to the "table lands" above to the still higher regions. These were formidable stands; the hemlocks reached as high as ninety feet, and in many spots the woods were so thick as to be impassable.[5]

The pioneers viewed the forest variably as a resource, an obstacle, and a source of signs that, properly interpreted, would indicate the agricultural potential of the soil beneath. In these ways they followed in the cultural patterns set by previous generations, for they appraised the woodland with an eye to settled, Euro-American agriculture. The resources that emigrants found in the forests are familiar. Cooper, for example, noted that white pine could be used for spars, boards, shingles, and rails; the sugar tree not only yielded maple sugar but made good fuel wood; sturdy farming utensils could be made from hickory wood, and furniture from birch. Hemlock bark was best for tanning. John Lincklaen, a scout for the Holland Land Company, noted that the chief product of early Whitestown, Oneida County (just west of Utica), was pearlash. Two men could manufacture twenty tons a year, seven acres of woodland to the ton. "Wild" fruit trees—Cooper mentioned apples, plums, cherries, and pears—of course were valued for their fruit. (Some of these doubtless were not "wild" but had been cultivated by the Oneidas.) Finally, the timber stands were scrutinized for signs of what the soil beneath would be like for farming. Cooper thought that bass, butternut, maple, ash, elm, and others signified "good soil for both grain and grass" and warned his readers to beware of lands with pitch pine, black or white oak, or birch. Settlers often preferred high ground not only because the auspicious trees grew there, but also because uplands were well drained (presumably healthier) and less heavily forested (therefore more easily cleared).[6]

To obtain these products, the trees must be felled. The settlers' zeal for clearing is legendary. In the Mohawk Valley as elsewhere, the forests disappeared rapidly before their energetic advances. Elkanah Watson witnessed a scene near Whitestown that was replicated thousands of times: "Stumps, half burnt logs, girdled trees, and confusion" dominated the landscape. The mania for clearing was so ferocious that some settlers found it unnerving. Mary Archbald, a Scottish immigrant living in the lower Mohawk Valley, wondered why "might not the ax be at least a little more mercifully used than it is in this part of the country[;] not a single tree nor even a sapling is spared." In a telling comparison she wrote that "the Indians had cleared many places around here with great taste . . . but we are a nation of traders. . . . Money money is everything and there seems not to exist in this neighborhood the smallest idea of observing or relishing the beauties of nature."[7]

The streams were another crucial consideration as pioneers deliberated on where to stake a claim. The supply of water for human and animal consumption was critical. Its purity commanded intense concern in a time when "ague" and "fever" were sometimes attributed to bad drinking water. Creeks, like forests, were supposed to give signs about agricultural promise. William Cooper advised prospective emigrants that they might discern by its color whether a stream had flowed through good farming country. Finally, observers also evaluated streams for their possibilities in the realm of waterpower. Horatio Spafford, surveying the county in 1813, enthusiastically described "an absolute profusion of mill-seats" and predicted an industrial future for the county.[8]

By 1819 the traveler William Darby could see the results of the pioneer generation's work. As he left the little village of Utica and set out northward for Trenton, he climbed into the hills above the town. After a half-dozen miles he stopped to look about him: "Certainly I had more than a thousand farms spread before me, many hundreds could be seen at one glance. I had seen many more sublime views, many more grand, but not one had ever before met my eye, that so completely answered to my conceptions, of the truly soft and beautiful landscape."[9]

Many other residents doubtless shared Darby's pride. While large portions of the county were yet unsettled, in many areas well-bounded fields demarcated by post-and-rail fences, orchards, meadows, and building complexes soon filled the landscape. The Oriskany Creek valley, for example, was described thus by an 1836 gazetteer: "Its flats are exuberantly fertile and are adorned by several beautiful, busy and thriving villages. Its banks forming one of the finest farming countries in the world, are crowned with vestiges of the ancient dense forests, or decked with comfortable farms, houses, and gay villages."[10]

Log barns and houses were the first buildings to be erected; the missionary John Taylor reported in 1802 that "log houses may be found in every direction." Watson found well-filled log barns near Whitestown in 1791. After the rudimentary log barn came the "English barn," usually 30 by 40 feet, with side entry and a tripartite division of stables, drive floor, and hay bays. This type, along with others, arrived with the Yankee migration.[11]

Now that the soils were actually under tillage, farmers and observers differentiated among them more closely. Darby classified soils into four major groups. On the Mohawk flats a fertile "deep black alluvial loam" proved very productive. Away from the riverbanks, "interval" land was more gravelly and somewhat less fertile. Higher still was a "mica slate," as fertile as interval land, but because of elevation, early frosts curtailed productivity there. Ebenezer Emmons, author of a *Natural History of New-York* (1846–54), placed southern Oneida County in what he called the "grain" district, with soils underlain by red and green shales; the rest of the county he classified as a "grazing" district, with varied soils.[12]

Early Oneida County agriculture was diversified; farmers raised grain, grass, fruits, vegetables, and livestock. An anonymous "Description of the Country between Albany and Niagara in 1792" declared that in the vicinity of Utica the "lap of Ceres was full. Most of the land on each side of the Mohawk River, is a rich flat highly cultivated with every species of grain." John Taylor found "crops of corn, oats, and grass, equal if not superior to any I have seen"; in the Black River valley farmers sowed spring wheat and corn. Rye and, where the climate permitted it, flax also appeared in Oneida County fields. Orchards thrived; though some trees, such as peaches, could not survive the cold, apples, pears, and plums could be grown. Livestock included horses, oxen, beef cattle, milch cows, sheep—important in the "age of homespun"— and the ubiquitous swine. Farm families also raised turkeys, chickens, and geese. A host of other plants too were cultivated; Taylor mentioned "hops, grapes, cranberries, . . . strawberries, gooseberries, blackberries, raspberries, currants, plantain, dock, English parsley, French sorrel, peppermint, common mint, and catnip."[13]

Despite this diversity, farmers clearly relied for cash on one crop: wheat. Early on, winter wheat was a staple, but the harsh winters often killed it off, and so most farmers switched over to spring wheat. David Maldwyn Ellis characterizes early Mohawk Valley agriculture as "basically predatory," because farmers took crop after crop of grain without rotating, thereby depleting the soil of fertility, reducing yields, and rendering the remaining plants still more vulnerable to the "rust" and midges that infested them. By the mid-1830s a commentator in the *Albany Cultivator* declared that "Oneida . . . is no longer distinguished as a wheat growing country, the specific food of this grain being . . . exhausted in her soil." A critic of Mohawk Valley farming charged that the lands there had deteriorated by 25 percent in only a decade.

Agriculture in the Dairy Zone

Another agricultural reformer, himself a longtime resident of Oneida County, thought that its Yankee settlers must have regarded the land as an "Eldorado they had no idea could ever be impoverished, by unfruitful, worn-out fields," for they had obliviously plowed and cropped fifteen, even twenty successive years. They seem not to have learned much from the experience of their forefathers, who had encountered the same problems in New England a generation earlier. Other crops also took their toll on soil fertility; early residents of Remsen (Oneida County) found that flax was a demanding crop.[14]

Cattle were treated little better than the land. During the summer they roamed the open ranges and during the winter eked out a scant subsistence on hay. One British sojourner in upstate New York complained to the Board of Agriculture in 1821 that "indifferent grass [was] made into hay at unseasonable times, and [abounded] in the worst weeds that can infest the ground." Even in the 1830s the agricultural reformer Jesse Buel lamented what he regarded as "the slovenly and wasteful practice of feeding at stacks in the field, where the sole of the grass is broken, the fodder waste, and the dung of little effect."[15] Only when deep winter weather was upon them were the cattle sheltered, in "a bad shed, or no shed . . . having a little hay flung to them two or three times a day."[16]

Challenges to this system of farming arose almost as soon as the land was cleared. Ellis has characterized the years 1808–25 as "years of uncertainty." Jefferson's embargo, cutting off exports of wheat, brought economic hardship. Then in 1817 construction on the Erie Canal began, and by 1823 it traversed the county from east to west. Farmers soon felt the impact of competition from wheat growers in the newly opened Genesee County and westward. An 1836 survey of agriculture in the region west of Utica discovered that only about a fifth of the wheat the region's residents consumed was locally grown.[17]

But the canal also presented an opportunity. Oneida County farmers slowly initiated a series of "painful readjustments" that would lead many into dairying. In 1837 an observer there reported that because insect pests had blasted the wheat crop, "cattle and sheep husbandry, and root culture, are annually increasing in extent." Soon the New York State–based agricultural periodicals began to feature discussions of how to convert arable land to permanent meadow or pasture—another indication of rising interest in livestock farming.[18]

A few enterprising farmers in the Mohawk Valley, especially in eastern Oneida and southern Herkimer Counties, had already begun dairying in earnest. The earliest Herkimer County cheese to reach Albany arrived there about 1808, and by the mid-1820s dairying pioneers were providing expertise and leadership for others. Ephraim Perkins of South Trenton (Oneida County) and Alonzo L. Fish of Litchfield (Herkimer County) became well known for their success in cheese dairying. A report from central New York in 1849 also attributed substantial influence to Welsh immigrants, tracing the rise of dairying to "a small settlement" of Welsh in the towns of Remsen and Boonville who began butter dairying in the 1820s. This observer thought that "their success in making good butter . . . has changed the whole farming system in the counties of Oneida, Otsego, Lewis, and Herkimer. The farmers in these counties have changed from the plow to the milk pan."[19] New Yorkers' Yankee background probably contributed too; cheesemaking on a substantial scale was well established in pockets of New England, including Rhode Island and Goshen, Connecticut. An early settler of Herkimer County, Jared Thayer, came from Cheshire County, Massachusetts, equipped with "a practical knowledge of cheese-making."[20] By midcentury thousands of central New York farm families were busy producing tons of butter and cheese, destined for markets nearby and in the West, the South, and the British Isles.[21]

These people shaped the farming system of cheese dairying. The notion of a "farming system" not only encompasses specific production methods and crop mixes but also extends to other components, equally important but sometimes less tangible. Cultural values influence the choices farm families make and the processes they follow. So do ideas, especially ideas about the land. Material culture both shapes and reflects a farming system. Finally, social relationships, especially those revolving around gender and household structure, are crucial dimensions of a farming system.[22]

This chapter will explore the interaction between cultural values, ideas, and material forms in shaping dairying practice in the mid-nineteenth-century North. It draws upon source material from New England, New York State, and Ohio, and where it discusses individual farms, Oneida County occupies center stage.

The Dairy Zone

The demise of arable farming in favor of dairying changed rural New Yorkers' perception of the landscape and its potentialities. Increasingly the terrain and land were evaluated in terms and for purposes different from those the earlier Euro-American residents had employed. Most intriguing was the emergence of the idea of a Dairy Zone, an optimistic reevaluation of the landscape in keeping with new necessities. The notion of a Dairy Zone encompassed not only observations about topography, herbage, and climate but also a benign vision of the entire society and culture.

Agriculture in the Dairy Zone

In 1839 William Townsend, author of *The Dairyman's Manual,* declared that the "*Dairy Zone,* or cheese and butter district . . . is circumscribed between the parallels of 40 and 45 degrees north latitude." This "great dairy district of the Union" lay within the "north lines of Pennsylvania and New Jersey, embracing the northern borders of the Mohawk Valley, and stretch[ed] from Lake Erie in to New England."[23] Others located the Dairy Zone differently, suggesting, for example, that "the true dairy region of the United States is confined mainly to the streams and side hills of the several spurs of the Alleghany mountains that drain into the Atlantic, north of the Chesapeake."[24] Though different people drew its boundaries in different places, the Dairy Zone was conceptually well bounded in the sense that it was believed to possess definite natural limits. As Townsend put it confidently, "Most plants and animals have their natural zone, beyond which they do not live."[25]

A number of characteristics marked out the Dairy Zone. Townsend listed them: "Nature has provided the requisites—a suitable climate, and good air, and good water and good herbage."[26] The climate in the Dairy Zone was regarded as salutary for several reasons. Dairy products would store well in the cool temperatures, and pasture and meadow grasses were not as vulnerable to cold as grains were. Topography and climate seemed to be interrelated: the "broken surface and high hills seem to attract clouds loaded with vapor and rain, which . . . irrigate the whole district."[27]

In combination, these attributes seemed to afford the ideal setting for dairying. Cows, for example, would thrive. An American proponent of dairying argued thus in 1849: "It is, I believe, a well established fact that the cow must have access to pure water, in addition to sweet pasture, to produce the richest and most milk; shade, too, is important. . . . The cow seeks the pasture of white clover, so natural to this county [Oswego], with the dew of the morning on it, with the shade tree and water at noon." Animals could even enjoy improved health in the Dairy Zone: "Perhaps it may not be deemed a fine spun argument," speculated an optimistic Yorker dairyman in 1854,

> to say that the health and vigor of the cow tends to have an appreciable effect upon the quality of the milk. The upland ranges, the principal domain of cheesemaking, are proverbial for the health of man and beast. The atmosphere plays over them in its native purity, nor less pure are the springs and brooks met with there in great abundance. But above all is the quality of the grass the soil produces. It luxuriates as in a chosen spot. . . . [Hay] of equal bulk is heavier from the hills. This marked difference becomes apparent in the quality of the dairy product.[28]

Some carried the notion still further to make claims for particular localities within the Dairy Zone. These were much attenuated versions of ancient

English beliefs associating particular regional cheeses (Cheshire, for instance) with idiosyncratic local geological formations and soils. Claims were made for the "sweet pastures" of Winchester and Goshen, Connecticut, and for Herkimer County. An *American Agriculturist* writer observed that "the cheeses of the granite hills and valleys of New England differ from those of the secondary soils of Herkimer, Oneida, and northern New York, while the latter differ from those produced on the shales of the 'southern tier' and northern Pennsylvania; and they again are a different article from the cheeses made on the slaty clays of the Ohio Western Reserve." These writers tied the characteristics of particular local cheeses to local conditions, impossible to replicate.[29]

The implications of the perceived boundedness of the Dairy Zone extended beyond simple physical characteristics. When they accepted a clearly demarcated Dairy Zone, for all their devotion to "artificial" products, dairymen also acknowledged natural limits to agricultural practice. A well-known Herkimer County dairyman, Alonzo Fish, gave expression to this notion when he advised farmers to "adapt their practice to the provisions of nature." As an example, he urged farmers not to force their cows to calve too early, because they would only lag behind: "*Nature* provides in *Spring time*, a principle of *general progressiveness* in the animal and vegetable kingdom." These limitations encouraged caution and restraint in the farmer's approach to the natural world. In turn, this ethos received reinforcement from the ideology of the Dairy Zone, because if dairying were confined to a specific region, by implication competition would also be limited. This was no small consideration for farmers smarting as they watched western wheat float by on canal barges. The rhetoric of the Dairy Zone resounded with a tone of optimism approaching glee: the dairy business was "safe . . . with regard to product and price," crowed Townsend.[30]

Those who promoted dairying also evaluated the landscape in terms of its meaning for society. Dairying pioneer Ephraim Perkins, who lived and farmed in Trenton (Oneida County) beginning early in the century, believed that during the frontier era lumbering and fishing had diverted people's energies from pursuits he regarded as more productive and more conducive to moral uprightness. For Perkins, *artifice* and *improvement* were synonymous; lumbering and fishing were activities too close to nature to warrant respect. The "improvement" attendant upon the rise of dairying was "salutary" because it was achieved through the "exclusive reliance upon the artificial products of the farm." Perkins's perspective is a reminder that for nineteenth-century New Yorkers, the Dairy Zone symbolized not only a geographical region but also a mentality in which the "artificial" products of people's labor—cheese and butter manufactured from the raw material of milk—assumed paramount importance.[31]

Agriculture in the Dairy Zone

Finally, the ideology of the Dairy Zone possessed political overtones. The Dairy Zone's benign, even positive economic influence was linked directly with contemporary notions of political economy, in a blend of republicanism and anticapitalist sentiment:

> We use the term *profit* emphatically—as we consider the country and the employment calculated to foster and perpetuate those social and republican virtues which are the great ornament and blessing in rural life. If not the richest in dollars, we think the [dairy] district . . . is to become the richest in moral worth, in republican virtue—in the treasures which improve society and render man happy—of any portion of our country.[32]

The *Albany Cultivator* gave voice to similar sentiments in 1838, arguing that the hilly district west of the Hudson and south of the Mohawk furnished the resources for a cattle husbandry that would make possible a more egalitarian social order than under grain cultivation. The writer thought that the dairying populace enjoyed "independence," "competent wealth," and "equality and social enjoyment." The accuracy of these statements is not here at issue; it was doubtless disingenuous to assume that investment in cattle was within every farmer's reach. But the description portrayed society, environment, and economy in happy harmony. The association of the Dairy Zone with republicanism suggests that for dairy enthusiasts, a degree of commercialism was fully compatible with republicanism, as it had been for the Jeffersonian farmers who embraced grain production for the world market as a means of attaining independence.[33] Thus the ideology of the Dairy Zone also successfully combined pursuit of independence with the liberal celebration of material progress.

The grand notion of a Dairy Zone, then, provided a conceptual underpinning and a rationale for the everyday practices and ideas that were to develop over the next few decades. In all of these locales the essential elements of the dairy farm were three: cows, land, and buildings.

What Sort of Cow?

In 1820 local officials took inventory of the estate of one Nicholas Gardiner, of Trenton, Oneida County. Among his possessions were "6 Red Line Back cows 13 yr. old, 4 pied and white cows, 1 2 yr old Red Heifer, and 1 2 yr old Red Line Back Heifer." A few years later, in his manuscript diary, farmer and cheese producer Francis W. Squires recorded the daily milk yields from selected cows in his herd. His names for them—Old Spot, Dun, Brindle, and White Tail—give a sense of what they looked like. And in 1849 a representative from the *Cultivator* visited the Oneida County Fair. He had scant praise for the fair and its exhibits, perhaps least of all for the cattle, which "possessed but few of the characteristics indicative of proper breeding—that is, they had not the uniformity of points which showed that the breeders had any distinct or definite object in view—their occasional good qualities appearing to be rather the result of accident than a well defined system."[34]

As these descriptions attest, the dairy cows roaming over Oneida County's expansive pastures were not the black and white behemoths that comprise today's American dairy herds. Nineteenth-century milch cows made a motley lot, their varied sizes, shapes, and colors reflecting equally varied genetic and geographical origins. These "native" cattle were descended from animals brought by settlers in the seventeenth century. Their ancestors had come from places as varied as the origins of the people who immigrated to the New World. British people imported cattle from disparate regions of that nation, from red Devons to Channel Islands Alderneys to Scottish Ayrshires. Scandinavian immigrants brought their favorite cattle, such as the yellow Danish cattle. In 1840 the prominent Vermont livestock importer Henry Randall wrote that "but a few years since, on the Mohawk and Hudson Rivers, there existed undoubted remnants of the stock imported by the Dutch settlers." In the United States rural people developed detailed taxonomies to describe these varied beasts.[35] Significant importations of cattle ceased very early, as the animals brought to America began to multiply. The stock intermixed continuously from the moment of settlement, either because of intentional breeding or because they roamed, and therefore bred, freely. Geographical mobility also promoted genetic mixing. Most Oneida County farmers, for example, acquired at least some stock by purchase, often from among herds fattened and driven east by enterprising drovers in places such as Ohio's Scioto Valley. When spring rolled around in 1851 herds of cattle from the West thronged local markets. The *Rome Sentinel* noted in March 1851 that "the annual migration of cows, always witnessed at this season of the year, seems to have commenced. A large drove passed through the village on Saturday last. Our dairies are usually replenished by arrivals from the Western and Southern counties [of New York], from Ohio, and from Canada. Central New York absorbs thousands from these droves."[36] No wonder, then, that American native cattle lacked the kind of distinguishing characteristics associated with animals from particular breeds.

Like the visitor to the county fair, agricultural reformers often scorned the natives, regarding them as mongrels unfit for any useful purpose. They vigorously participated in the agricultural journals' debates over the activities of British and American breeders and publicizers. A great deal of discussion revolved around the respective merits for the dairy of the Shorthorn, Ayrshire, Holderness, or Channel Islands cattle.[37]

Agriculture in the Dairy Zone

The agricultural press's passionate interest in "improved" animals represented a strain of thought that ultimately prevailed: today in the United States a small number of specialized breeds has replaced the variety of the past. Black and white Holsteins dominate the dairy scene; white-faced red Herefords are raised for beef. As a result, historians have sometimes tended, in Whiggish fashion, to assume that the journals' coverage of specific, identifiable breeds reflected the progressive thinking of the day.[38]

But the volume of printer's ink devoted to famous breeds of cattle obscures an important reality: natives made up an overwhelming proportion of the cattle in nineteenth-century America. Even in England only a small fraction of the national herd consisted of animals that were recognized as belonging to definite breeds, and there is ample evidence that still fewer found their way to America. In 1841, for example, an observer in New York's Cortland County estimated that of the county's herd of 33,700 cows, a handful were pureblooded and only about 100 were "grade" level—that is, having some blood of animals regarded as distinguished (because of detailed knowledge of ancestry, affiliation with a registry, or the like).[39]

The paucity of purebred animals persisted not only because of the technical and financial difficulties involved in importations, but also because working farmers—and some agricultural reformers as well—frequently questioned the very value of purebreds, instead preferring the natives. The skepticism about famous, recognizable breeds was everywhere apparent. When organizing the local fair in 1845, the Oneida County Agricultural Society did not offer premiums for famous dairy breeds such as the Jersey or Ayrshire; rather, it awarded premiums for "heifers" and "cows." Nor did the New York State Agricultural Society, in its extensive surveys of dairying, even inquire regarding breeds of cattle.[40]

This way of thinking was reflected in a contribution made by one Jonathan Talcott, a well-to-do farmer residing in the Oneida County town of Rome, in the heart of the cheese-producing district. In 1860 he published an article in the *Genesee Farmer* titled "The Breeding and Raising of Farm Stock." Despite the title, Talcott dwelt first and at length upon feeding practice. Though Talcott possessed Shorthorn cattle himself, in his conclusion he wrote, almost casually, "I have said nothing thus far about breeds, or races of farm stock, preferring to let each judge for himself; knowing, that when farmers set themselves to ascertain by actual experiment the growth of an animal, . . . they will learn what is best for them—well knowing, that one breed is not fitted for every farm and purpose so perfectly as to exclude all others."[41]

There were good reasons for the indifference to imported stock. Because farmers in the dairy regions often bought new stock in addition to breeding and raising their own, they were as apt to evaluate a cow's present condition as to inquire about her lineage. Buying cattle assured farmers of getting only females; they did not have to take a chance on breeding useless bull calves. Even dairy farmers turned superannuated milch cows into beef, so that while good milking qualities figured prominently, fattening tendencies were also relevant. Few cattle were outstanding for both beef and dairy purposes, the two qualities being incompatible—another reason the idea of "breed," which usually connoted a single purpose, was not very useful. Those farmers who did want to breed their own stock risked more in dairying than in other types of livestock raising, because a cow's productivity did not show for several years, until she had given birth to a calf. In the meantime a farmer had to invest time and money.[42]

The native cattle were esteemed for a quality that many imported animals lacked: hardiness. Correspondents in the agricultural journals complained that animals imported from the milder British Isles needed special pampering in order to thrive in the harsher North American climate. Lewis Falley Allen, who compiled several well-known American herd books, noted (for example) that the Ayrshire had not yielded as bountifully in the rigorous American climate as in the milder Channel Islands.[43]

Another consideration in determining why farmers were so slow to adopt "improved" breeds is that for families whose primary products were homemade butter and cheese, the properties of milk for processing were at least as important as its sheer quantity. One writer in the 1848 *Cultivator* criticized contemporary methods of evaluating dairy cattle, arguing that these gauges focused on quantity and failed to determine how well a given cow's milk could be processed. If all a buyer desired was a large amount of milk, the author continued, then present systems of evaluating dairy cattle were sufficient, but not if the purchaser wanted to convert the milk into something else. Here again, as with the idea of a Dairy Zone, an acceptance of limits was implied; the goal was not to produce as much as possible, but rather to turn out the best possible product.[44]

Even when quantity was desired, little reliable evidence existed to prove that purebreds consistently outproduced natives. For dairy purposes the natives seemed to compare favorably; the English observer John Fowler remarked that "the native cattle are generally raised, and are considered full as well adapted to the country as the imported ones; the cows are small, but good milkers." Testimony as to the inconclusive results of trials with breeds comes from an unlikely source, one of the most active local agricultural societies in New England. In 1850 the Essex County, Massachusetts, society published a

Agriculture in the Dairy Zone

report on the "Improvement of Stock" that concluded, "Experience, so far has shown that importations from abroad, and the crossing with them, have in no way benefited our milch cows. They may have furnished us better oxen in some respects; but they have not yet helped us to any more butter and cheese. . . . We have no better cows than we did [thirty years ago]."[45] In the same year the Essex Society remarked on the local entrants in the competition on "Dairy Management":

> We are struck with the fact that, but two of the cows were of foreign breeds (so called), viz.: Mr. McNaughton's, Durhams of Byfield, whose produce was the seventh in quantity. . . . Why is it that the farmers of Essex are so slow in introducing these classes of animals. . . . [Many] efforts have been made to make known their superiority; but still the real *hard hands* do not take hold of them. On whose judgment, then, shall we rely, the *gentlemen farmers,* or the *operative farmers?* The *theory* of the one recommends the Durhams and the Ayrshires for the dairy, as being the greatest producers, the *practice* of the other adopts the natives.

That the only purebred in the contest came in seventh surely explains at least partly why local farmers were slow to adopt them despite the publicity. They recognized the implications of failure in the contests: if the cows did poorly under presumably optimum conditions, how would they fare under ordinary treatment?[46]

The unimpressive record of purebred cattle was particularly evident in dairying. In England breeders had slighted breeding for milking capacity; huge beasts for beef commanded the greatest prestige. Livestock authority Randall related the story of a fiasco concerning a milking strain of Durhams; he charged that British breeders had tinkered too much with them, with the result that many developed a hereditary "posterial deformity" that made calving difficult. Americans derisively dubbed them "pumpkin rumps."[47]

The confusion about the respective merits of purebreds and natives shows clearly that the art and science of breeding was a mystery even to the successful. A strong sense of limitation pervaded discussions of breeding. Given the paucity of knowledge about heredity, writers acknowledged the unpredictability of attempts at systematic breeding. Authorities could not even agree about where to start. Charles Flint, author of a number of manuals on dairying, explained in an 1858 piece:

> To work successfully with our common cattle would require great experience, a quick eye for stock, a mind free from prejudice, and a patience, and perseverance quite indefatigable. . . . This mode would require a long series of

years to arrive at any fixed and satisfactory results, owing to the fact that our "native" cattle, made up as they are of so infinite a variety of incongruous elements, do not produce their like, that the defects of an ill-bred ancestry will be continually "cropping out" for several generations.[48]

For these reasons Flint advocated beginning with already "improved" imported breeds.

But the Essex Agricultural Society's Committee on Bulls reached the opposite conclusion: "The only successful mode of improving our stock, is by a judicious, systematic, enlightened attempt, which has for its basis the native stock of the country. . . . [You] may spend money, you may import stock, you may offer premiums, and no more benefit be derived from it, than has been from what had been done by this and other societies for the last thirty years."[49]

It is not surprising that nineteenth-century livestock owners were puzzled by the vagaries of breeding. Even today's sophisticated geneticists have not unraveled heredity in all its complexity, and it is still very difficult to sort out the differential effects of heredity and environment. Thus while nineteenth-century farmers occasionally achieved some desired results through breeding, they readily acknowledged the limitations of breeding as a tool for increasing milk production. Most farmers accepted their animals as "givens" and entertained few thoughts of extensively manipulating their biological makeup over generations. In historical context, then, ideas about breed formed only a part of a larger constellation of ideas about dairy cows, in which concepts other than breed predominated. The concept of breed is best set aside when asking what sorts of cows farm families possessed and valued 150 years ago.[50]

Rejecting breed, farmers employed a range of criteria for selecting milch cows. They judged according to elaborate, traditional notions about how a cow's external features corresponded to her productive capabilities. These ancient criteria included a bewildering variety, both minutely specific and astonishingly vague. No less an authority than Lauren B. Arnold, in the widely distributed manual *American Dairying,* advised farmers to watch for a "feminine appearance, . . . a cowy look" in prospective dairy stock. The Chenango County dairyman John Shattuck believed that only cows with golden-hued skin would produce good butter. "Yankee Drover" Asa Sheldon remembered a neighbor, Mrs. Deacon Parkers, who had a reputation as an infallible judge of a good dairy cow; she looked for a "transparent" horn. Of one promising cow she remarked, "Her horns are all butter." A cow ought to possess good digestive powers, externally indicated by a "large mouth, thick and strong lips." Many believed that soft skin indicated good milking capacity, and some thought it an attribute of good cheese-producing cows. Temperament was an important consideration; a good cow was docile and (of course!) contented.⁵¹

Good dairy cattle were not regarded as beautiful creatures; Allen characterized them as sharp-pointed, loose-limbed beasts, in conformation the opposite of the classic beef animal. A piece in the *New England Farmer* described a good dairy cow as having wide-apart horns, thin head, large dewlap, broad back, full breast, large udder, long teats, broad buttocks, and long tail. With variations this formula appeared often in agricultural literature, epitomized in the oft-repeated form of an old English verse handed down over many generations:

> She's long in her face, she's fine in her horn;
> She'll quickly get fat without cake or corn;
> She's clean in her jaws, and full in her chine;
> She's heavy in flank and wide in her loin;
> She's broad in her ribs, and long in her rump;
> A straight and flat back, without ever a hump;
> She's wide in her hips, and calm in her eyes;
> She's fine in her shoulders, and thin in her thighs;
> She's light in her neck, and small in her tail;
> She's wide in her breast, and good at the pail;
> She's fine in her bone, and silky of skin;
> She's a grazier's without, and a butcher's within.⁵²

At midcentury newer systems of assessing external signs gained popularity; the best known of these was that of François Guenon, who classified cows according to the shape of the "milk mirror" or "escutcheon," the pattern of hair growth along the udder and hindquarters. Guenon's system (figure 1.1) was complicated; it was subdivided into dozens of categories and subcategories. Though many questioned its reliability, Guenon's system did focus attention in a logical direction, that is, on the milk-producing organs.⁵³

In the cultural and intellectual climate of the day, all of these criteria of appearance made sense. They were as reliable as any other way of selecting; moreover, they fit into patterns of nineteenth-century American thinking in other realms. As humans often describe the animal world in anthropocentric terms, perhaps it is not going too far afield to see some conceptual parallels. In democratic rhetoric, for example, farmer citizens were accustomed to hearing that individual ability and accomplishment mattered more than hereditary acquisition. Some partisans of native cattle invoked nationalist sentiments, mistrusting imported "foreign" animals, especially those from aristocratic Britain. One correspondent of the *Cultivator* smugly declared that in his dairy of twelve cows, "not one . . . possess[ed] a drop of royal blood."⁵⁴

Figure 1.1. François Guenon's "milk mirrors." The "escutcheon," or pattern of hair growth around the udder, was supposed to predict milking capacity. *New England Farmer* 5 (May 1853): 209.

In other ways too the taxonomy of dairy cattle reflected contemporary patterns of popular thought. This is especially notable in the realm of thinking about women. The ideal milch cow, docile and contented, with her small head and large udder, reflected a larger pattern of thinking about female capability. In popular thought on human physiology, for example, female intellectual and reproductive activities were often cast as antithetical. Victorian-era theorists, in developing the etiology of "hysteria," posited a direct connection between the womb and the mind. Bookish young women were frequently warned that excessive intellectual activity could damage their reproductive capacity; this argument would culminate in Dr. Edward Clarke's *Sex in Education* (1873). These parallels obviously cannot and should not be pushed too far, but they do suggest that cultural influences were at work as farmers selected and dealt with dairy cows.[55]

Thus when we consider the process by which farmers selected their milch cows, we find that they relied substantially upon what might fairly be described as folk and popular knowledge as means of identifying a good milker, if anything amplifying those traditions. Whether or not they succeeded in the tricky business of detecting productive capacity, farmers who were to one degree or another "commercially" inclined were tapping the old ways as frequently as the newer theories of breeding.

This use of a traditional means toward new ends is significant, for it suggests that in this case, at this time, agricultural tradition and modernity complemented one another. Historians of American agricultural practice have often tended to view "traditional" rural culture as conflicting with a commercial outlook. One group characterizes as precapitalist a particular set of cultural and economic qualities. In this scheme production for family use or maintenance—a value distinct from and opposed to profit—predominates. In this interpretation "traditional" ways of selecting cows would be ranged with the imperatives of family use, and emphasis on "improved" breeds with the profit motive. Other historians show less sympathy for traditional culture but still place it in opposition to "innovation," seeing "innovators"—usually portrayed as progressive—in conflict with hidebound traditionalists. But the evidence introduced here for nineteenth-century dairying tends to support the interpretations of a third group, those who take issue with the strict polarization of "traditionalists" and "capitalists." They have shown that rural Americans have frequently employed traditional means to unconventional ends, or have maintained a compatible mix of traditional and new institutions.[56]

Once a farmer had cobbled together his motley herd, he let them out to pasture and began the annual seasonal cycle of feeding, milking, and making cheese. The cow's productive capacity evoked yet another conceptualization of her that was very much in keeping with the times: she was often compared to a machine, a pervasive analogy characteristic of an age when amazing machines were transforming life everywhere, not least along the canal, where pile drivers and drills had made "Clinton's Ditch" a reality in a few short years. "Yankee Drover" Asa Sheldon wrote in 1862, "I endorse the opinion of Hon. Josiah Quincy, Jr., that 'a cow is a machine; you can get nothing out of her but what you put into her'; but let us remember, the better the machine and the better order it is kept in, the better pay we shall get for running it." This machine analogy appeared with some frequency in the farm journals. In 1853 the *Genesee Farmer* advocated that dairymen learn anatomy and physiology in order to "be able to use the living machines in his possession to the best advantage. A good machine in the hands of an ignorant unskilled operative, uniformly turns out bad work. That is, the net profit is less than it should be." Advocating proper shelter for dairy cows, one correspondent of the *Cultivator* compared cows to "so many furnaces [that] must be kept heated."[57]

This machine analogy, of course, referred not to an animal's genetic makeup but rather to her ability to transform grass into milk, and so farmers' discussions of animal nutrition and physiology frequently had mechanistic overtones. "The natural secretion of milk," one writer explained confidently, "can be vastly augmented by artificial means." Understanding that food was used for respiration, growth, and repair of body tissue as well as for producing milk, farmers sought to manipulate the balance. In a lead article titled "Cows and Carrots," the *Genesee Farmer* argued that "the primary wants of nature being satisfied in a skillful manner, we may proceed to separate milk from the extremities of arteries into veins in the udder. . . . [Every] one should have a clear idea of nature's plan of separating from arterial blood a quart of rich milk an hour. There is no hocus-pocus or black art in this beautiful operation." The language of the "operation," and of the regular, clockwork "secretion," carried implications of a simple process humans could understand and exploit.[58]

What to Feed Them?

> When snows are deep, if you desire
> To *comfort* take, keep a good fire
> Your cattle, too, well fed must be
> If from the *mange* you'd keep them free.[59]

The mechanistic analogy may have been crude, but in one sense it was accurate: the more fuel the cow received, the more milk she would make. The native cow's ability to make use of extra feed probably accounts for most of the rise in yields between 1850 and 1900.[60] Cheese dairy farms, with their expanses of pasture and meadow, epitomized this emphasis upon feeding. They were "grass farms." Various "grasses"—botanically speaking, some actually members of the grass family, some not—raised on pasture for grazing and on meadow for cutting, were the staple of the nineteenth-century dairy cow's diet.

Dairy cow feeding, in Oneida County and in other dairying regions, followed practices that both grew from and diverged from earlier ways. Despite livestock's importance to eighteenth-century settlers, the quality of feed and shelter was minimal. One elderly resident of Oneida County recalled in 1849 that "as in all new settlements, their cattle had in summer to graze in the wide forest for their subsistence." Wherever they encroached onto new lands, Anglo-American livestock quickly decimated the indigenous grasses, making way for more tenacious European clovers, grasses, and "weeds." In the eighteenth century, farmers began systematically to sow certain plants for pasture or hay. But they seldom manured their pastures or practiced crop rotation, so animals had to forage for scanty pasture grasses in the summer, and the yearly harvest of hay often ran short before spring arrived. Often livestock would be slaughtered as winter approached because of an insufficiency of feed for the cold season. (This was another reason farmers often purchased cattle rather than raising them.) Only occasionally did stock receive extra feed in the form of roots or grains. Of the energy they derived from their meager rations, cattle probably expended much just to live and grow, so it is not surprising that milk yields were low.[61]

Nineteenth-century dairying in the Northeast continued these older patterns in part but also diverged from them. Dairy farmers continued to rely upon the traditional mainstays of cattle food. During the warm season cows were turned out, about May 1, to graze on pasture grasses. In the winter they were fed hay that had been cut and cured during the summer months. This was an age-old pattern in its general outlines, yet while earlier farmers had essentially allowed the livestock to do the work of "creating" pastures, thereafter paying minimal attention to maintenance, their successors applied more overt management.

As their New England counterparts had done, the New York dairymen expanded the area of pasture and grassland in proportion to that under cultivation.[62] In Oneida County, dairying brought a noticeable shift in the proportion of land under the plow to that in pasture and meadow. In 1855, for example, the same amount of acreage was planted to crops as in 1845, yet the total amount of improved acreage rose by over fifty thousand; we may assume that most of this went into grassland. Some individual examples from cheesemaking farms demonstrate this pattern. In Annsville in 1855 Nathaniel Churchill plowed only fifteen of his one hundred improved acres; he put forty in pasture and forty-five in meadow. In Trenton Orville Coombs reported 265 improved acres and plowed only 33; he kept over 150 acres of pasture and 80 of meadow. Data for forty selected cheesemaking farms in the towns of Westmoreland and Boonville show that in 1855 they averaged about a quarter of their improved acres in tillage and a third in meadow.[63]

Grazing possessed definite advantages. It required little human labor; the cow did the work of finding a good "bite." It also demanded little capital investment after initial preparation. It reduced the farmer's susceptibility to drought and rain, since he depended less upon field crops for feed, and less upon hay.

There were two ways of procuring a grass crop. One was to maintain "permanent pasture"—land that was seldom or never plowed, and where the farmer relied upon the spontaneous growth of perennial plants, usually European natives. X. A. Willard claimed that some pastures in central New York had not been plowed for sixty years. An observer of cheesemaking in Barre, Massachusetts, reported to the *Country Gentleman* that pastures there consisted mainly of white clover and were rarely plowed. Available evidence indicates that the area of pasture required for one cow's summer keep varied, but in central New York it amounted to just under two acres or as many as three acres, depending on soil, weather, and drainage.[64]

The second, usually complementary, source of grass was land deliberately plowed, then seeded, from which the winter's supply of hay would be cut. In Herkimer County, famous for its cheese, observers noted that farmers sowed meadow grasses "in the same sward" together and then plowed the land when it began to fail. These "reseeded" meadows were often used as pastures for part of their lives, so that no clean distinction should be made between "permanent" pasture and "temporary" meadow. By the mid-nineteenth century in much of the Northeast, meadows of "artificial" grass made up part of a fairly well established rotation.[65] (The term *artificial grass* in the nineteenth century did not refer to the plastic turf reviled by today's purist baseball fans; it meant a field deliberately seeded with selected mixes of grasses and legumes.) The farmer would raise corn, oats, peas, and potatoes in successive years (the order varied), then "seed down" to grass for a longer period, usually about four years. In 1851 Albert Ford of Herkimer County reported that "when any portion of my grass land does not produce well I plow it up, manure it well and plant it with corn, the next season sow it with oats or barley and seed it with timothy and clover." Commonly the farmer would plow up and reseed a small plot every year, reaching the original starting point after ten years or so. Dairy-farming families debated the merits of various crops, often stressing not their profitability in the short term but their function in the rotation. According to Alonzo Fish, for example, potatoes were useful because they pulverized the soil well. Often the rotation called for grass mixed with a "catch-crop" of grain. Again, the "catch-crop" was chosen not only with reference to market considerations but also with regard to how compatibly it would grow with the emerging grass seedlings.[66] Because the new "sward" would remain for several years, "seeding down" was an important operation. The *Rome Sentinel* noted in 1851 that "Re-seeding with Grass" caused "great anxiety" for Oneida County farmers.[67]

The term *artificial* itself bespeaks a conscious intent to control the makeup of pastures, and indeed articles often prescribed specific mixes of seed to plant to grass. Most who were familiar with dairying and with pastoral farming in general recommended a mix of species. As early as 1828 the *New England Farmer* reprinted advice from the Pennsylvania Agricultural Society that listed no fewer than eighteen varieties and recommended amounts of seed per acre, with timothy, white clover, and red clover predominating. Advocates reasoned that a variety would provide a constant supply of fresh grass, since plants

Agriculture in the Dairy Zone

have different growing seasons and flowering times. Variety would also ensure a thick mat of grass, hence discouraging weeds. Native or locally grown plants would thrive best. Agricultural journalists and other writers offered an analogy to nature to justify their argument: "There is no reason," argued the editors of the *Genesee Farmer* in a front-page article, "why we should confine ourselves to [a] single grass in re-seeding. In old pastures there is a great variety of grasses, and this is one reason of their superiority, and it would be well to take a hint from nature in the formation of our pastures." Charles Flint, author of a popular and often reprinted dairying manual, succinctly praised diversity: "This is nature's rule."[68]

But as the editors' comments indicated, dairy farmers preferred a few mainstays. Oneida County papers carried advertisements from merchants for grass seed, raised locally or purchased from Philadelphia-area Quakers, New York State Shakers, and Herkimer County growers. Trade in grass seed flourished; Erie Canal registers showed that in 1836 alone 491,000 pounds traveled on canal barges.[69] The most often mentioned varieties included "white clover"—probably not one but several white clovers, including the ordinary white or Dutch clover (*Trifolium repens*) and Swedish or alsike clover (*Trifolium hybridum*). The former, which grew spontaneously on local soils, was a short legume and served well as a pasture plant, but because of its short stature it was not a good hay plant. Red clover (*Trifolium pratense*) was also a legume. Timothy (*Phleum pratense*) was the prime ingredient of the famous "English hay," a high-yielding, palatable grass.[70] Redtop (*Agrostis vulgaris*) also made good hay. Orchard grass (*Dactylis glomerata*) rounded out the list of preferred pasture and meadow plants.

Diaries and account books from the central New York region show sales and purchases of grass seed mixes, though they seldom detail their specific content. Samuel Hinckley, a storekeeper in Herkimer County, sold "grass and clover seed." On the other side of the ledger Peleg and Cornelia Babcock, of Bridgewater (Oneida County), in 1852 spent $14.25 on clover seed. David Hughes, of Deerfield (Oneida County), in 1849 bought four bushels of timothy seed for $14.00 and a half-bushel of clover seed for $2.50.[71] The proportion of varieties frequently changed, depending upon the purpose of the plantings. Newbury Bronson, of Warsaw (New York), in 1847 grew "white clover and timothy, with a mixture of red top, the clover predominating, for *pasture*, and timothy, red clover, and a small amount of red top for *meadow*." Mr. and Mrs. William Ottley, prizewinning cheesemakers in 1847, believed that they derived their richest cheese from cows grazed on clover pasture. Albert Ford, a cheese dairying farmer from Herkimer County, believed that "timothy and clover are best for dairy purposes."[72]

Transforming Rural Life

In their deliberations about grass, American cheesemakers lacked the sophistication of their English counterparts. Yet they recognized that feeding was a key influence on the cheese's quality. Mr. and Mrs. Ottley, for example, were correct to associate clover pasture with rich cheese, for clover provides the nutrients necessary for conversion to milk with high butterfat content. And while Yorkers valued timothy for its high yields of both hay and milk, this plant did not dominate the most-mentioned mixtures. This is important to note, because it indicates that household cheesemakers placed a high value on quality, downplaying sheer quantity in their plans. This was reflected in output statistics: New York State cheese production climbed during the 1830s, as more families got into the business, but during the 1840s and 1850s it essentially leveled off. To concentrate on producing a limited amount of good-quality cheese not only fit with the prevailing mentality, it also made sense under a system in which family members provided most of the labor and had to spread their efforts over a number of endeavors.

Once the grass was planted and growing, probably the most important influence on yield was the weather. In 1854, for example, a severe drought reduced hay yields by a quarter and even by half in some towns. But another crucial factor was soil fertility. Some Oneida County meadows and pastures, especially those "interval" lands in the valleys, would sustain good yields for years—one editor estimated over two tons per acre—but most needed attentive husbandry to maintain fertility.[73]

Evolving grassland practices helped to preserve or improve soil fertility in several ways. The rotations described above contributed. All plants contain nitrogen, and when they are plowed under, as were crops or grasses in the rotation, the nitrogen is released into the soil, providing nutrients for plants to utilize. Moreover, some of the species in the rotations, notably the clovers, were legumes. These nitrogen-fixing plants, while still living, actually release nitrogen into the soil. Legumes also improve soil conditions and disease resistance. In early modern Europe the land's capacity to yield had been substantially increased as farmers discovered the value of clovers. By midcentury scientists knew that leguminous plants took nitrogen from the air and fixed it in the soil; while neither farmers nor scientists fully understood the workings of this process, they comprehended its beneficial effects.[74]

Of course, the cows made manure. In this respect livestock contributed a crucial element in the maintenance of fertility. Since cows stayed in barns more of the year (see below), manure could be collected and distributed. Farmers wasted less than in earlier days, when some observers had been appalled to see them dumping manure in the river. By the mid-nineteenth century that practice was long past, and a writer in the *Transactions* of the New York

State Agricultural Society summarized the transformation: "Like the lands in the olden time, their worn out fields required to be sown to grass and clover to enjoy their jubilee. The ten year old manure heaps about the barn, had to be put in requisition, and it was found by sad experience, that it would have been far easier to have preserved the land in its fertility, than to renovate that thus early worn out." Numerous diaries show that farmers such as Deerfield's David Hughes were spending long hours procuring and hauling fertilizer; Hughes drew "muck" for his meadow and also traveled into Utica to get stable manure. Another New York State farmer, Francis Squires, also included rotations and systematic manuring as he produced butter and cheese.[75] By the 1850s frequent testimonials appeared from dairy farmers finding, as did Rodney Wilcox of Herkimer County, that their farms were "improving in productiveness." "My land produces now five times the amount of grass annually per acre that it did when I commenced upon it," he reported in 1852.[76]

In Oneida County several products of plaster or lime were also popular.[77] In the New York State manuscript census for 1855, for example, farmers from Westmoreland and Boonville occasionally reported under the category of "special manures used"; most often they had used "plaster." "PLASTER, PLASTER," proclaimed an ad in the *Rome Sentinel* in the spring of 1851; a local merchant had obtained three hundred tons of fresh-ground "Manlius plaster" from further west. Local farmers made "lime" a profitable sideline: William J. Babcock, of Holland Patent, operated a limekiln and offered to "sell in lots to suit purchasers." Indeed, the Herkimer County atlas for 1868 shows limekilns scattered all about the rural areas. This material was put to good use: a "Farmer" writing in the *Black River Herald* (Boonville, Oneida County) in 1857 said that he "sow[ed] plaster on newly seeded meadows" until the grass reached a height of three feet. He had done this for eight years, friends for as many as twenty. They claimed not merely sustained yields but improved ones.[78] Indeed, these locally available fertilizers probably contributed significantly to maintaining fertility. Legumes helped to meet the need for nitrogen; lime would help to maintain a proper soil pH. Contemporary extension agents in Oneida County still recommend applications of lime for the county's dairy farms.[79]

Other, indirect evidence suggests that dairying practice resulted in better-than-average soil fertility. Hay yields ought to serve as a reasonably good indicator. Calculated in several different ways, probable hay yields in Oneida County cheesemaking farms come out well above the statewide averages for the 1850s. Oneida County cheesemaking farms produced from one and a third to two tons per acre, well above the county average of one.[80]

Grass alone, fresh or dried, was a rather monotonous and nutritionally incomplete diet. Agricultural reformers, in both the specialized farm press and local newspapers, encouraged farmers to feed their cows other items. Everyone who wrote about the subject advocated a different mix of feeds. Writers commonly mentioned corn, carrots, potatoes, apples, turnips, whey, sugar beets, beans, oilcake (made from crushed cottonseed), rutabaga, parsnips, grains, pumpkins, and mangel-wurzels (a type of large beet). Judging from comments made in the agricultural journals about the subject, some of these, especially corn, carrots, and potatoes, were used more frequently than others. The New York State manuscript census corroborates this; many dairy farms were recorded as producing significant quantities of carrots, and all grew potatoes (an average of 112 bushels in 1850, as against a state norm of 90). One observer from central New York noted that "large quantities [of potatoes] were fed to our dairy cows at the close of winter." For cheese dairies whey was often a logical feed supplement. Apples, grown on nearly every farm, also frequently ended up in cattle feed.[81]

Farmers and columnists argued endlessly about how best to prepare these foods so that the cow would derive the most benefit. Some recommended steaming grains and sometimes roots, to "render the food more digestible, and easier assimilated by the absorbing vessels, and therefore more economical." One of the better-known feed-cooking devices, Mott's Agricultural Furnace, was available in Rome in the late 1850s. Most farmers probably did not go to such lengths, but fodder cutters seem to have found a place in the farm barn; advertisements for them commonly appeared in the agricultural press, and in local newspapers as well.[82]

Where to Put Them?

Progressive limitations upon cows' movements accompanied and complemented the new feeding regimens. This primarily took the form of new quarters for milking and for winter shelter. Like the other practices, these were significant changes from the era of free grazing.

By midcentury dairy farmers were housing their cattle longer and more carefully than earlier generations had done. Now they had their milch cows "come to the barn in the fall." Once there, they stayed put. One correspondent declared that his "cattle need scarcely go out of the stable in a month," so well did his barn provide feed, water, and comfortable surroundings. Others offered more specific details, explaining that their cows were shut in for sixteen, sometimes as many as twenty-two, hours of every twenty-four. Some even advocated keeping the cows near the barn all year round. In 1858 the *American Agriculturist* published a letter from a Tioga County farmer arguing that cows should be housed beginning in the autumn and kept in the barnyard in summer, "to be handy for milking and to save manure."[83]

In addition to being sheltered and fed indoors in winter, cows were increasingly milked indoors, even when they spent their warm-weather days in pasture. This was a significant change from earlier practice, when milking had been an outdoor task. X. A. Willard recalled in 1877 that in the early days of Herkimer County dairying (the 1820s), the milking had been done in open yards (see figure 4.1). A later era sentimentalized the bucolic scene with milkmaid, her stool, and a pail in the farmyard, but in these early days it was a common enough sight. In the yard, milkers accommodated to the animals' proclivities.[84]

To facilitate indoor milking, cows were tied up. Stanchions—variously called "stanchials" or "stancheals"—were introduced as early as the 1830s in dairy counties. These were pairs of vertical bars arranged so that once the cow inserted her head between the bars, she could move but little either horizontally or vertically. Some dairymen preferred a simple chain. James Whitney of Big Flatts, New York, described his "standards" in 1865: "First, I set my standards four feet apart, have a ring made of . . . iron about six inches across, put over the standard, and then put the bow through the ring and over the animal's neck. . . . The rings will slip up and down to suit the animal's convenience. . . . Thus you can control the most vicious of animals, and make them perfectly submissive." Other arrangements were more elaborate. The 1860 *New England Farmer* described a "Jerseyman's" arrangement thus:

> When milking time comes, a strap with a buckle is passed through [a] ring; the cow's hind foot on the side of the milker drawn back, . . . the strap passed around her ankle and buckled, the neck straps being arranged so as to keep the cow's head up; it is impossible for the most kicking cow to overset the pail, or strike the milker. The most stubborn cows are subdued by this means, and without violence or harm to the cow, or to the temper of the milker. And in the severest fly time, no loss is occasioned by overset pails.[85]

These various ways of confining cows not only kept the animals immobile for milking and feeding but also aided in the systematic collection of manure. Numerous articles in the farm press described barns with manure cellars. A farmer writing under the pseudonym Pine Hill described his arrangement in 1858. He intended to collect manure all year by stabling his cows at night during the summer and all day during the winter. He further processed the manure by adding swamp muck, leaves, and the like. Another farmer also writing under a pseudonym described a manure cellar that not only collected his cows' dung from openings in the stable above but even received household waste from a pipe. He concluded that the farmer's "*fancy fertilizers* should be manufactured by himself, in the cellar of his own barn." New Hampshire's L. L. Pierce agreed, arguing that farmers who made their own manure would escape "the necessity of purchasing commercial manures, at reckless prices."[86]

A pervasive concern with system informed the arrangement of barns. In keeping with the portrayal of cows as machines, barn design facilitated regular, systematic feeding and milking. "The capacity of cows for giving milk," intoned Alonzo Fish, "is varied much by habit." Gurdon Evans insisted that a cow's "lacteal organs" would be "deranged" by erratic feeding. Another advised farmers to feed their cows "systematically—at regular hours each day." The same applied to milking. Numerous dairy farmers testified that they never deviated from a regular milking schedule, arranging all other work around it. William S. Lincoln even observed "the same regularity . . . in the order in which the cows are milked as in the hours of milking." Some made an effort to train cows to occupy the same milking place daily. Even their walks in the yard were sometimes regulated; one herd was let out precisely at the hours of ten in the morning and four in the afternoon on calm winter days.[87]

Barns express a great deal about these systematic aspects of the developing dairy culture in central New York. They certainly indicate greater human control of animals. They facilitated the obsession with cleanliness and system, also questions of control. They betokened a link in a nutrient cycle. Perhaps most profoundly of all, they altered old seasonal patterns. Replacing them was a routine more regular throughout the year, less marked by seasonality than in the past. The *Genesee Farmer* noted that under the new regime "the work can be done there [in the barn], when, on winter mornings, in the old way, it could not." Because the cows not only received better food but burned fewer calories, they gave milk for a longer period than before. Farmer W. H. Yale noticed the impact of good feed and shelter on the length of the milking season:

> The cows['] hair lies smooth, and shines, and their eyes look bright, in place of dead and sunken eyes and bristly hair. Their bags are full, and teats full and smooth, and when I go to milk them, the same as they are in summer, and they will give a quart more at night when tied in the stable, than when they have been in the yard the warmest days during the winter. . . . When I kept my stock at the old barn, my cows would go dry two or three months every winter; now they will give milk till they calve.[88]

The new functions and priorities demanded new designs. By the mid-nineteenth century the traditional "English barn" was coming under criticism, as it failed to answer for dairying purposes. The English barn had followed after the rudimentary pioneer-era log barns. English barns usually measured 30 by 40 feet, with side entry and a tripartite division of stables, drive floor, and hay bays, arranged perpendicular to the roof ridge. Advertisements in the Oneida County papers occasionally mentioned 30- by 40-foot barns as if the reference were commonly understood.[89]

Inadequate space for storage and shelter for stock were the main drawbacks of the English barn. A critic writing in the *New England Farmer* sarcastically described haying time in an English barn: "When a respectable sized load of hay was driven into it . . . the load was tightly pressed on each side and the farthest end, the other being 'out-doors.'" William C. Stuart of Herkimer County explained why he replaced his English barn in 1846: "As a matter of course, you will understand that it is thirty by forty feet, with fourteen feet posts, and done off in the old stereotype plan. . . . Considering that buildings, as well as other things, should be fashioned in that manner which will best serve the purpose for which they were intended, and *my purpose being to keep neat stock,* are my only apology for the innovation"[90] (emphasis added).

Barn designers seeking solutions to these problems began to fashion variations on the old tripartite arrangement. The discussion of barns is necessarily limited here; barns in the dairy regions have been little studied, and so what follows is a general survey based upon agricultural literature. Barn builders' "innovations" in the dairy regions can be roughly categorized into three groups. The first two types of barn, arranged on one level, distributed main functions horizontally. A third type—the one from which the dominant dairy barn design eventually derived—distributed functions vertically, over two, sometimes more, levels, with entrances at many points.

The variations that maintained the old one-story, tripartite division usually reoriented one or more sections. For example, "A.L.W." submitted a plan of a 58- by 28-foot barn to the *Genesee Farmer* in 1853, specifically for housing milch cows. The bay and floor occupied their customary place, but the stables were enlarged and oriented parallel to the roof ridge. Cows secured in stanchions faced inward to a central feeding alley. They ate from mangers and received light from windows. Their manure was dumped through gutters to a cellar below.[91]

Others extended this reorientation to all three sections; this shift has been described by Thomas Hubka. For example, a "Ground Plan of a Convenient Barn" (1856) shows a 30- by 65-foot barn with lengthwise bay, floor, and stables. The cows, restrained with stanchions, stood on platforms. Behind them a trench collected manure and a walkway provided room for milkers to maneuver. Instead of storing the cows' droppings in a cellar beneath, this farmer chose to put manure in an adjoining shed. Windows allowed light in, and movable scaffolds provided increased storage space. Another, even more elaborate example was the "Great Barn" in Massachusetts observed by William D. Brown. This 125- by 54-foot structure contained stables, floor, and bay, all arranged parallel to the long side; now the primary entrance was in the gable end. S. P. Wheeler, the owner, organized his barn space to operate like a well-oiled machine. The cellar stored manure, of course, but also had room for "immense piles of roots." A pump (for water), cutting machine, and mixing trough indicate that feeding the animals now involved more complex and technologically sophisticated processes than simply tossing hay into a manger.[92]

As dairying became more complex, another form appeared: milking barns, also called milking houses. These specialized structures were introduced to dairy regions of New York State in the 1840s. Alonzo Fish described a 30- by 50-foot milking barn in 1843; a timber-framed structure, the building had three side doors for the cows to enter by, in addition to an eight-foot central aisle running lengthwise and eight-foot doors on either end. B. Andrews of Ashtabula County, Ohio, explained to the *Ohio Cultivator* in 1859 that his "cow-barn" was "for storing hay, and milking; it is 32 by 64 feet on the ground, and 44 by 60 feet over the cows, leaving a projection of six feet on each side, to cover manure; it takes forty cows at a time, and when we milk more than forty, we milk a few, turn them out, and drive in those that were left out."[93]

In 1851 Paris Barber of Cortland County published a plan for a milking shed in the *Transactions* of the New York State Agricultural Society (figure 1.2). This long, narrow (24- by 75-foot) structure contained "stantials and feeding boxes for fifty cows." This evidently was a balloon-frame building, and its simple, linear plan permitted efficiency in feeding and milking as well as in carrying the milk to processing points. Above the stable was a "loft for storing hay or corn fodder." The milking shed itself was situated in the cattle yard, which also contained the barn where Barber wintered his cows. Other plans ranged the cows crosswise, tails inward, with a raised platform between the cows. The reason for this plan was that the manure was easier to clear away, and that the milkers could approach the cows and carry out the milk unobstructed.[94]

The specialized milking barn represented a building philosophy that diverged from the advice often promulgated in the agricultural press. The agricultural journals frequently associated old-fashioned farming with a multiplicity of small buildings, each having a specialized function. Agricultural reformers never tired of pointing out that farmers could save time, timber, and money if they would only erect large, centralized barns to replace the clutter of

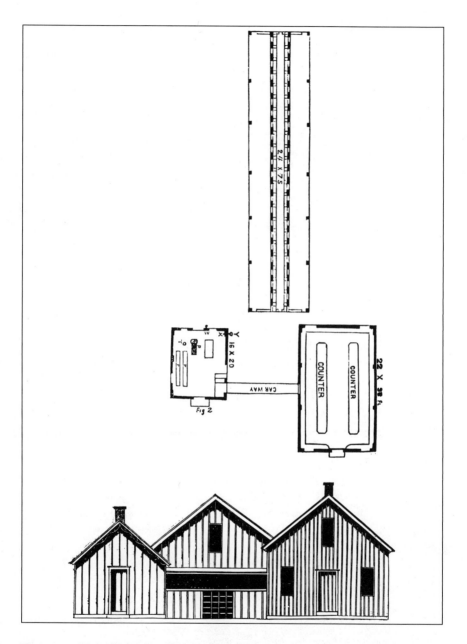

Figure 1.2. Paris Barber's milking barn and cheese house. The milking barn represents the rise of an efficient plan for milking large numbers of cows, and the cheese house shows a typical arrangement for manufacture and storage. New York State Agricultural Society *Transactions* 1851:263, 268.

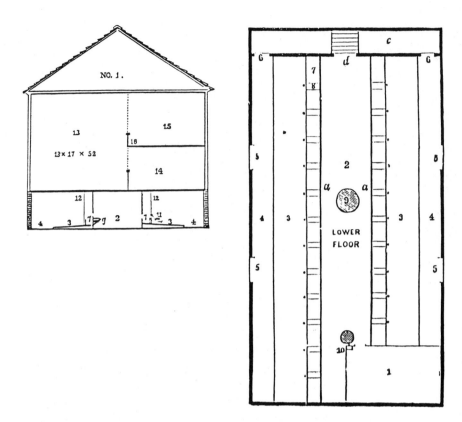

Figure 1.3. Barn plan showing labor-saving arrangements for stabling, milking, and feeding cows. All numbers apply to both section and plan. Section (*left*): 2, feed aisle; 3, floor where cattle stand; 4, manure gutter; 7, mangers; 8, feeding troughs; *11*, stalls; *12*, stanchions; *13*, hay bay; *14*, granary and storage level; *15*, drive floor; *16*, slots for feed. Plan (*right*): *1*, root cellar; *5, 6*, doors; *9*, tub for whey; *10*, pump; *a*, central aisle; *d*, ramp. Gurdon Evans, *The Dairyman's Manual* (Utica, 1851), 74, 71.

outbuildings.⁹⁵ Though ultimately many dairy farmers did integrate milking facilities into larger, multipurpose barns (see below), the milking barns were modern solutions to new problems. As with the selection of stock, innovation and tradition here merged.

The barns discussed to this point arranged functions horizontally: bay, floor, and stables were all on one level, and there was one primary entrance. Access to scaffolds above and to cellars below was from the barn floor and minor doors respectively. But another category of barns distributed functions both horizontally and vertically, in complex sets of interlocking volumes. The multiple levels were created either by raising a structure several stories above ground level, by building an earth ramp to the upper level, or by building the barn into a sloping site so that it could be entered from both the upper and lower parts of the slope. Of course, this type of organization had long been practiced in southeastern Pennsylvania, but it now attracted attention in a wider geographical area.⁹⁶

Within this basic organization the timber frames allowed for further subdivision of space. This is more clearly shown in visual examples such as one offered to the 1850 *Ohio Cultivator* by "K.K." of Belmont, Ohio. In this timber-frame structure the end was built into a slope, creating a basement story for cattle, horses, and roots. An entrance from the upper side of the slope led to the barn floor, flanked by haymows, both lengthwise. Between this level and the stable ceiling was another level. It was divided into three main parts corresponding to the barn-floor level: the central section had the barn floor for a ceiling and was a storage area; on either side the haymows extended to the floor. At the far end of this level was a carriage park entered from the side doors. This arrangement facilitated farm work in several ways. Perhaps most important, it permitted farm workers to "drive in with grain or hay, eight feet above the bottoms of the mow, and pitch the hay *down* instead of *up*. Two men can fill this barn as quick as three can one of the common kind." This was a crucial consideration in an era when haying time was the season of greatest urgency and scarcest help. Barns of this construction often had vertical chutes between levels, so that feed stored above could be deposited in mangers by gravity power, eliminating tedious hauling.⁹⁷

Central New York dairyman Gurdon Evans published a barn plan in his 1851 book titled *The Dairyman's Manual* (figure 1.3). It followed a multilevel pattern, raised from a level ground. The "lower floor" contained two inward-facing rows of chained cows, with central feed aisle (2). Though the rows paralleled the roof ridge, thus following the newer style, the doors (5) were still placed in the sides rather than in the gable ends. In addition to the usual manure gutter (4) and mangers (7) for hay and cut feed, Evans's arrangement had troughs (8) for "mess" of roots, whey, and so on. These troughs all connected to a single brace, so that they only needed to be pulled out into the feed aisle for loading, then "shoved back to the cow." At *c* a manure vault permitted waste removal. Adjoining the root cellar (1) was a pump. In the center of the feed aisle, at 9, a tub collected "whey, or any other slop intended for cows." A pipe from the cheese house conveyed whey to the tub. Above, hay was pitched down from the drive floor (15) into the bay (13). The middle level (14) below the drive floor was partitioned into two sections for granary and general storage. This barn coherently integrated much of the dairy work to make human labor efficient. On the stable level, feed, water, and manure cellar were all near the cows. From above, workers could pitch hay down into mows and from the mows down into the mangers (see 16). They would also use gravity to lower grain from 15 to 17, and thence to the cistern below at 9.[98]

The variety of barn forms indicates that, as in other aspects of dairy practice, farmers experimented with different solutions to meet their needs. Though no single dominant type had yet emerged, barns integrated cows and people into spaces that expressed a developing dairy culture. However, they also provided for grain storage and processing, root storage, space for some bad-weather farm work, and housing for other livestock. In other words, these barns were not focused entirely on dairying. Moreover, most farms still likely had numerous specialized outbuildings to accommodate a variety of other functions. Scattered around the farmstead, smokehouses, icehouses, pigsties, root cellars, corncribs, and other small buildings testified to diversification on the cheese-producing farm. In this regard they show that the label Dairy Zone did not convey the totality of central New York dairying. A yet larger scheme was essential to the way farm families conducted their dairying activity.

Diversification on the Cheese-Producing Farm

Paradoxically, the best way to understand the complete farming system of cheese dairying is to focus on its basic element, the individual farm. Only by understanding just what operating cheese dairying farms produced, and in what proportions, can we determine how the farming system functioned, how its components related to one another. To this end, the focus now shifts to a detailed study of all 475 of the cheese-producing farms in seven Oneida County towns (see frontispiece). The rural towns—actually townships rather than nucleated villages—of Annsville, Augusta, Boonville, Steuben, Trenton, Verona, and Westmoreland represent the varied geographical and ethnic makeup of the county. In the mid-nineteenth century they enjoyed varying degrees of access to transportation. All were formally established in the late eighteenth century. During settlement, speculators, prominent politicians, or missionaries owned large parcels of land, but despite their desire to lease rather than to sell, they eventually sold most of the county's lands on the market. Euro-American farming was thus established there according to a pattern typical for the United States during this era.

Situated in the northwestern section of the county, Annsville was settled in 1793 by people from Massachusetts and Vermont. Fish Creek and several other small streams provided good drainage. Perennial springs also promised abundant water supply. Fish Creek was aptly named, and early residents eagerly eyed an area in the town they named "the meadow," where the departed Oneidas had cultivated crops. In the southwestern corner of the county, people from Litchfield County, Connecticut, settled Augusta. The Stockbridge Indians originally occupied a reservation there, but the aggressive Yankees quickly pushed them out. Fertile soil and vigorous creeks, especially the Oriskany, attracted settlers. Boonville, a northern town, lay at the western edge of the Adirondacks. Its primary waterway was the Black River, which ran southward. A hilly, uneven surface and heavy evergreen forest signified that wood products would be important in the town's economy. Steuben, at higher elevations in the east central part of the county, possessed gravelly loam well adapted to grazing. Its population included not only Yankees but also a contingent from Wales who established a thriving community there. Trenton shared a boundary and physiographic characteristics with Steuben. Its original inhabitants included Dutch people, as reflected in the name sometimes used for the major town, Barneveld. Ultimately, however, Yankees dominated. The West Canada Creek formed its eastern border, and good roads connected it with Utica. Verona, in the west central section of the county, was bounded by Oneida Lake and Wood Creek. The Erie Canal ran through the town's northern edge. Finally, Westmoreland, near the county's center, contained part of the central valley and also bordered on the canal. Westmoreland had been first occupied by Judge James Dean, who lived with the Oneida Indians as a child, went to Dartmouth College, and became a missionary. He obtained a grant from the Oneidas for his services during the revolution. Besides its highly fertile soil, the town boasted iron ore deposits.[99]

Oneida County cheese dairying families amply exploited conditions there. Probably because cheese production demanded unusual skill and organization, cheesemaking farms tend to present the attributes of nineteenth-century dairying in magnified form. On the simple level of size, for instance, cheese-producing farms in Oneida County were large by New York State standards; their average total acreage in 1850 was 132, when the New York State average was 113. In 1850 cheese-producing farms averaged sixteen milch cows and forty-three tons of hay, when the state averages were only five and a half and twenty-two, respectively.[100]

Cheese dairying farms produced a characteristic mix of crops, dairy products, and livestock. In 1850 475 households in the seven towns were engaged in making cheese. They averaged forty-eight hundred pounds of cheese and five hundred pounds of butter. Cheesemaking farm families raised just a few bushels of small grains such as wheat, barley, rye, and buckwheat. But they grew higher-than-average amounts of corn (124 bushels, when the statewide average was 105) and potatoes (112 bushels, when the statewide average was 90). They grew about the same amount of oats as the average, 159 bushels. They usually possessed two or three horses, and occasionally a yoke of oxen. Besides milch cows, they averaged five other cattle, eight sheep, and six hogs. Of orchard products they reported an average of $22 worth, twice the statewide average. In the 1855 New York State manuscript census, nearly every cheese-producing farm listed an apple orchard; most had at least a few dozen trees, and many possessed one hundred or more. Some individuals also noted pears, quinces, and cherries. A reporter from the *Albany Cultivator* visited the county in 1849 and found that "not only apples and pears, but plums, peaches, and grapes flourish[ed]" along the Oriskany Creek in Augusta. The manuscript state census also reveals that most farm families reported selling some poultry and eggs. A few kept bees and harvested honey. Kitchen gardens supplied a substantial portion of the family diet; in some cases they were substantial enough to appear in the census manuscripts as "market gardens." In heavily forested areas such as Boonville the woodlands permitted farmers to make significant amounts of maple sugar.[101]

The pattern of crop and livestock production suggests that on cheese dairy farms most components contributed directly or indirectly to dairy production. Pasture and meadow made food for dairy cows. As we have seen, arable production too contributed to the dairy: carrots, corn, potatoes, and grains were fed to the cows. Oats fueled the horsepower that drew the plows. Similarly, pork was an item tied directly to cheese production, as whey from cheesemaking was usually fed to the hogs.

Yet at the same time, these components were not directed solely to cheesemaking; they also contributed to other ends. Flexibility was a hallmark of the cheese-producing farm. In its diversity the crop mix typified New York State agriculture. Pasture, meadow, and other crops also fed the other cattle, the sheep, and the swine. An abundance of hay or oats might well end up as fuel for city horses, for canal horses and mules, or for a neighboring farmer's stock. Barley might also find its way to local breweries. Except for wheat,

Agriculture in the Dairy Zone

a good many of the family's subsistence needs could be met; X. A. Willard remembered in 1879 that during this era farms produced most of the family's wants except for items such as tea, spices, and salt. A sampling of other items from Nicholas Gardiner's inventory suggests the range: "Nine sheep, flax brake, hay knife and fork, rake, scythe, harrow, two great wheels, one little wheel, quilts, woollen blankets, and linens." Corn and potatoes, of course, were popular items for human consumption; cheesemaking farms produced both in quantities well above subsistence levels. Indian meal was giving way to purchased wheat flour, to be sure, but it was still widely consumed. Beef too, of course, was raised for human consumption; on many cheese dairying farms one or two animals, sometimes including superannuated milch cows, were slaughtered each year. The fat was used for making soap. Cornelia Babcock, who with her husband, Peleg, made cheese in Oneida County, wrote in her diary in 1852, "Tried tallow, think we have a hundred lbs enough to last the year." Sheep provided wool for knitting family garments or for sale (raw or spun) to local mills; local papers during the 1840s and 1850s frequently advertised for wool or for knitters, and it was not unusual for cheesemaking families to report to the census that they had made quantities of knitted goods; Anthony Covery of Augusta reported "120 pairs of fringe mittens" in 1855.[102]

Thus the cheesemaking farms of these seven representative towns were at once specialized and diversified, paradoxically tied to a world market and at the same time immersed in an elaborate web of local exchanges. The subsistence base allowed cheesemakers to make sure that no matter what happened in the world market, they could still raise a multiplicity of items that they were not compelled to use in any single restricted way and that strengthened the local economy.[103] No clear line separated specialized production from the diversified subsistence base. On the farm, all activities circulated nutrients and energy endlessly. The farm family consumed or exchanged cheese, pork, beef, and field crops produced on the farm, and contributed their labor. Feed for livestock was grown on the farm; the animals provided labor, food, and manure. The last of these, of course, was returned to the land, both to tillage land and to grassland, especially meadows. Even in terms of energy it is quite likely that these farms, unlike most modern American farms, produced more calories than they expended. Energy inputs were modest by today's measures: human and animal muscle power, wind power, and wood for burning, all available on the farm. (By contrast, today's farms tend to run energy deficits, in the sense that large inputs from petroleum-based manufactured pesticides, fertilizers, and fuels far exceed the number of food calories produced.)[104]

The path taken by antebellum dairying suggests modifications to historians' portrayals of Anglo-American agriculture as invariably destructive to the environment. These arguments hold that Anglo-American cultural values—an emphasis upon spatially fixed private property, a view of nature as made up of commodities for market, a tendency to set humanity apart from nature—underpinned drastic ecological transformations. Wholesale forest clearing created conditions for flooding and even climatic change; the plow accelerated erosion and necessitated monocrop planting, which in turn invited pests and diseases; livestock destroyed indigenous herbage and served as the "linchpin [of] commercial agriculture." But nineteenth-century dairying seems to have reversed or at least slowed some of these processes. The plow and the axe were not absent, to be sure; but where agricultural practice was concerned, dairying, with its emphasis upon diverse pasture stands, recycling manures, and improving soil fertility through rotation, offered a basis for sustainability.[105]

For a short while, at least, it seemed as if central New York cheese dairy farming families had struck an elusive balance: between conservation and destructiveness, and between the independence that subsistence afforded and the prosperity that market participation promised.

✤ 2

Sharp Dealings
Cheesemakers and the Market

> Dairying, it is believed, offers great inducements for a judicious outlay of capital. It rests on no mere speculative basis—"up today, and down to-morrow," but is permanent in its character and prospects and sure in its rewards.—X. A. Willard, 1861

The meshing of "subsistence" and "commercial" production activity on cheesemaking farms is revealed most clearly in their relationship to the market. The nature of northern farmers' relationship with the market has been the subject of some original and provocative scholarship. Working since the 1970s and early 1980s, some historians have discovered considerable popular resistance to capitalism in rural America. They have explored values that they argue differed qualitatively from market values. In the system of trade by barter, seldom mediated by cash, they find that inheritance, family maintenance, or "use" determined the value of products. Similarly, at the household level family need, and not profit, was the goal of the farm enterprise. In communities citizens were bound together by the web of mutual debts and obligations this system engendered. Cooperation, not capitalist competition, ruled. The collective impact of these values was a local self-sufficiency that rendered inhabitants "independent," not of one another, but of distant markets. Adherents of this view maintain that the "market" was an intrusive, often unwelcome force.[1]

This approach calls into question some long-prevalent assumptions about the history of American agriculture and about how to study it. An illustrious background of scholarly work from the 1940s onward holds that from the beginning, American agriculturists were market-oriented—that is, they wanted to grow cash crops for sale. In this view even those on the frontier eagerly awaited transportation to make markets accessible. A more recent, more critical exposition of this viewpoint is William Cronon's work, which posits an immediate colonial connection between what he regards as capitalist agriculture and environmental change. Winifred Rothenberg has demonstrated that regional labor and commodity markets existed in the colonial era. Joyce Appleby has resuscitated the older interpretation, portraying Jeffersonian ideology as a buoyant agrarian philosophy based on the eager exploitation by small farmers of expanding European markets for grain. In Appleby's view the price system, far from occasioning dependence, was a means by which small farmers could acquire "independence" from their "social superiors."[2]

These two different points of view evaluate the implications of the market differently, but they share the notion that "self-sufficiency" and "market orientation" were fundamentally incompatible, whether because they were respectively "traditional" and "modern"; "precapitalist" and "capitalist"; "communal" and "individualistic"; based on "use value" and "exchange value"; or arising from values of "family" and "profit." The two interpretations also agree that eventually—the timing and means depending upon the region—a system based upon cash exchange came to dominate agricultural communities.

Yet often the evidence does not fit into such a neatly polarized pattern. Some scholars have challenged this dichotomous view, demonstrating for example that in colonial Massachusetts, trade in the market was essential to supply the shortfalls that nearly every family experienced. In upcountry Georgia in the nineteenth century, activity in the world cotton market complemented and even enhanced local self-sufficiency. Eighteenth-century New England systems of exchange have been characterized as "communal capitalism," balancing market and neighborhood exchange. In sum, participation in market exchange—even extensive participation—does not necessarily imply fully capitalist or commercial agriculture.[3]

The relationship of midcentury cheesemakers to the market captures a hybrid system at a point when the influence of the market in agriculture generally was intensifying but not yet fully ascendant. Central New York cheesemaking farms indisputably intended their cheese for the market; even assuming a high level of consumption, household members would eat little more than fifty pounds per year, a fraction of the Oneida County average production.[4] But they followed a production strategy that emphasized quality over quantity and made sure that they could rely on their farms for most subsistence needs as well. In everyday practice, market participation blended smoothly with subsistence exchange. In this respect they resembled the southern yeomen farmers who rarely produced the amount of cotton that would provide them with maximum profits, instead choosing to supplement market sales with a subsistence "hedge." In the South those with small farms tended to be marginal producers within the staple-based economy, and so the subsistence element was perhaps their primary concern. On northern dairying farms the inverse situation applied: the most self-sufficient farms also tended to be the most highly market-oriented.[5] This may explain in part why the transition to capitalism was so much smoother in the North than in the South. In the North this blend was underpinned by a corresponding set of values in which profit and subsistence happily coexisted: "competency" was the goal. In turn, the absence of a sharp contradiction between forms of exchange and the values they reflected paved the way for a relatively smooth transition to an agriculture more completely oriented to market production[6]—and ultimately, to wage labor and consumption.[7]

Three characteristics emerge from a study of how cheesemaking families marketed their cheese. First, the inclination to compete individualistically for high prices existed from the beginning. But cheese producers often chose cooperation over competition. This tendency is evidence not so much of resistance to capitalism as of adapting forms of capitalism to fit local circumstances. In this and in other ways "market" activity fit without disruption into the customary neighborhood system of noncash exchanges. A second aspect of this accommodation was that cheesemakers usually sold cheese but once a year, "by the dairy." The remaining months they lived within the neighborhood system based on barter and credit, cushioned by local security until they received their payment. Third, a dynamic element in the connections between farmers and market accounted for significant change over time. As merchants stepped up their demands for cash, and as middlemen competed with each other, the cheese business shifted from credit to cash terms. When cheese-selling middlemen became able to offer cash on delivery, farmers rarely offered resistance to the change, even though it put them into more direct competition with each other. There were several reasons for this. The new system required no drastic alterations to function with cash; farmers had all along gauged their trade, goods, and services in cash equivalents. A cash-on-delivery system also appeared to mitigate one important concern of farmers preoccupied with gaining a competency: risk. Thus paradoxically the concern for a competency propelled them further into reliance upon an ultimately unstable world market.

New York–made cheese was sent to consumers in seaboard cities, in the interior and on the western frontier, in the South, and overseas. At first U.S. markets accounted for the largest proportion of New York cheese sales. It is impossible to determine just which parts of the interior were the main destinations for northern cheese, but many contemporaries pointed to the South as an important market; some believed the southern market was as important as the export market. Dairying was little practiced in the South, partly for climatic reasons, and slave owners provided a steady market for the cheap supply of protein that cheese represented. Harvey and Alanson Baldwin traded cheese from Ohio farms to the Mississippi River cities of Memphis and Vicksburg for thirty years between about 1820 and 1850. Seeking to please southern consumers, cheese brokers advised makers to tint cheese destined for the South to a deep orange color, and to make the cheeses small. Thus the free-labor agricultural economy of the North, like many other northern industries, fueled the southern slave economy. Both were part of a market economy, yet they produced markedly different social systems.[8]

In writing about the southern market for cheese, some observers even went so far as to offer cheese as a remedy for sectional ills. They drew a political parallel to the notion of a benign, conflict-free Dairy Zone, seeing in interregional trade a mutual dependence that would "unite us closer in the bonds of fellowship and good feeling." An 1851 survey of the dairy industry concluded that cheese was potentially "one of the strongest links of our glorious Union." This author exhorted "our intelligent and patriotic farm wives" to "find herein an incentive to an improvement in the quality and flavor of the article, more potent, even, than the improved price, and twice as powerful with them as a political speech; for proverbially, these whole-hearted caterers to Southern palates are, to a man, in favor of Union."[9]

The Factor System and Brokerage Houses

Though some people sold cheese through the country store, the cheese trade was primarily organized and facilitated by middlemen or "factors."[10] Even small producers in the cheese districts could dispose of their products through these agents. In central New York these entrepreneurs appeared about 1815, when a pair of Yankees from Massachusetts turned up in Herkimer and Oneida Counties contracting for local cheese. Ephraim Perkins, a longtime dairyman from Trenton (Oneida County), recalled that these men, Robert Nesbit and Silvanus Ferris, had made contracts for various periods of time and also advised on the cheesemaking process. Their partnership brokered for much of the cheese made in central New York localities. They took it to New York City and there sold it themselves to grocers and produce dealers. The presence of buyers encouraged local production; in turn, more people ventured into the brokerage business. When the Erie Canal was completed in 1825, cheese production, and its associated marketing apparatus, escalated dramatically. The sight of aggressive factors "threading the dairy districts" had become a common one by the 1850s. After about 1835 "commission houses" specializing in cheese and butter also appeared. They helped to facilitate trade and in some cases to encourage production.[11] The *Farmer's Cabinet* reported in 1848 that "extensive houses, exclusively for cheese, are doing a large business" in major eastern cities and in the South. These companies sent out agents to the cheese districts, who shipped cheese to the headquarters, dividing the work of buying and selling. Some dealers even came from as far away as England; James McHenry came from Liverpool to buy American cheese in the 1850s. Eventually factors or agents operated in a wide region from New England west to Ohio's Western Reserve.[12]

Regardless of how the brokerage work was divided, the system operated according to several common characteristics. Factors (or their agents) usually visited the farm in springtime, before the cheesemaking season actually commenced. They would call on a farmer at his home and would contract for the coming season's "make"—that is, all the cheese produced—at an agreed price per pound or per hundredweight. There are some indications that both men and women participated in the negotiations over prices; one Ohio woman testified in 1846 that she sold her cheese herself, and another reminiscence referred to "a family council in the cheese room." The system was based, as Nesbit put it in a letter to Ferris, on "barter, trust, and credit." The farmer received only a small advance at the time of agreement, with the balance due him after his wares were sold, usually on January 1 of the following year. Sometimes the factor would furnish supplies, such as bandages or boxes. Individual farmers' testimonies show that within these general parameters, details varied according to individual circumstances. In 1847, for example, Rodney Wilcox of Herkimer County "marketed" his cheese "to be delivered in the fall at 7 cents in the boxes." In the same year Newbury Bronson, from Wyoming County, sold his cheese "at my dairy" to H. Burrell and Co., which provided boxes and agreed to take, on November 1, all of Bronson's cheese then at least twenty-five days old. This contract was made in August, when Bronson probably had ripening cheese on his shelves. Burrell agreed to pay 5½ cents per pound, with "no allowance for greenness" (shrinkage), and Bronson had to travel fifteen miles to deliver the cheeses.[13]

The papers of the Babcock family of Bridgewater (Oneida County) show more intricacies of the system. Peleg and Cornelia Babcock dealt with several different brokers in the 1850s. On February 10, 1852, they recorded in their diary, "Mr. Carey here to see the cheeses thinks them good." But Peleg Babcock did not close the deal immediately. On March 6 he went to the market town of Ilion, New York, and sold a batch of cheese for $44.21. On April 23 he went "to Mohawk and Ilion about selling cheese [but] did not sell." Evidently Babcock was looking for the best terms. Eventually he settled on Carey; on May 19 he "went to Ilion to sell my cheese have sold to Cary for

6½ cents per pound to be taken monthly." The very next day Peleg "carried off twenty cheeses brought $73 got $21 of what was back on last years lot." His record of monthly deliveries began on September 23: "Have carried off cheese carried 53;" and on the next day, "been off with cheese again carried 52 amounted to $522.31." About a month later, "have been to Ilion with cheese received $108." And on November 20, "sent off twenty boxes of cheese to Ilion the last I shall send this season amounted to $62.50 besides expenses." In this case payment was made in several installments at the end of the season rather than at the new year.[14]

Once the cheese was delivered in the fall, the factor's work of selling began. In his job the factor encountered numerous uncertainties. Since he was dealing in a product variable in quality and perishable, the dealer was always preoccupied with the condition of his wares. Even if the cheese started out in good condition, unreliable transportation, inadequate storage facilities, and unpredictable weather swings could easily turn it into a reeking mess. If the quality was poor in the first place, the dealer had an even worse time of it. One Ohio writer moaned that "when sent to market, the purchaser will have to pay *funeral expenses* on [it], in order to get rid of it. [The cheese] will have to be carried beyond the city limits and be buried, or carted to the river . . . and consigned to polliwogs and alligators." Indeed, the private correspondence of dealers is full of references to cheese ruined in all sorts of ways. In 1821 Robert Nesbit was trying to sell Herkimer County cheese in New York City; a large lot of it had literally disintegrated, and he (not surprisingly) had a difficult time peddling it. In 1828 he reported to his partner Ferris that the dairy of one "P.E." looked "as if it had been run over by a cart wheel." The casks belonging to "S.W.F." were in pieces and the cheeses broken. And "J.M. and S.M." had not cured their cheese long enough, so it was "by no means fit for market." Nesbit was particularly angry about this: "I do think that they have not done right to put up so much[,] for my purchase was on terms not to take any late made cheese until it was cured. . . . I think sellers ought to have some conscience."[15]

The difficulties dealers had with marketing cheese were reflected in a letter entitled "To the Dairywomen of Our Country" that a dozen agents circulated in 1838. In it they referred to "the existence of general and just complaints in regard to the quality and condition of both butter and cheese . . . together with the packages." Following detailed suggestions on technique and storage, the petitioners concluded that following their recommendations would "secure to [dairywomen] the advantages of a fair price and good reputation for their labor and pains." Of course, dealers tended to emphasize cheesemakers' culpability in accounting for poor quality, but in reality poor storage and transport probably ruined as many cheeses as dairywomen did. In any case, dealers had little hope of effecting a thoroughgoing uniformity in cheesemaking practice.[16]

To add to the uncertainty of the condition of cheese, activity in the market fluctuated unpredictably. Sometimes business was "dull" and cheese a "drug on the market." Nesbit would write to Ferris with discouraging news from around the eastern cities—he heard from a colleague that the Philadelphia markets were slow, and that one broker had sold no cheese at all. At one point late in the year Nesbit estimated that there were eight hundred thousand pounds of unsold cheese in New York City alone. He anxiously reported, "I think it is about equal to anything to see what work the cheese mongers are making[;] they seem distracted to sell." Though over the antebellum period prices were steady (figure 2.1), the day-to-day trade still varied considerably. M. C. Burrell wrote in 1852 that "there [are] some parties who will [sell] at all hazards. . . . Some of our neighbors [traders] have been most awfully frightened and have pushed their cheese in all sorts of ways." At other times, especially when new cheese came on the market, trade was brisk and dealers rushed to take advantage of the "advance" of business. Then they enjoyed making the grocers squirm by demanding high prices: "They hate to come to us to get their supply." In 1857 Anson B. Ives, agent for a Philadelphia produce firm, wrote that "this trade is very exciting and keeps a man on his toes every minute. Some days it pays a big profit other days not so much."[17]

Under these circumstances cheese middlemen engaged in constant posturing that sometimes reached outright deception. Agents went to the farmers with long faces, criticizing the quality of the cheese and complaining of low prices; one unverified but popular story in Herkimer County local lore portrays Nesbit and Ferris operating in a secret partnership, Nesbit making an initial circuit with bad tidings intended to set up the farmers for Ferris, who followed up with tales of more sanguine prospects and offering very slightly higher prices. Whether or not this story has been exaggerated, there are other clues in dealers' private correspondence that indicate the selective way in which they shared information with their suppliers. When Nesbit negotiated in 1820 with an English agent who had "offered to buy . . . 400 casks," he cautioned Ferris that "it will not do to tell this story to the country folks." Dealers kept confidential their evaluations of individual dairies. "The GS dairy is a fair dairy, nothing nice in it at all," wrote Anson Ives. "Privately, there was never any money made on it."[18]

In the city market, where dealers and buyers congregated in close proximity, dealers constantly competed with each other. They traded in rumor and suggestion to gain advantage. The Nesbit-Ferris correspondence shows how

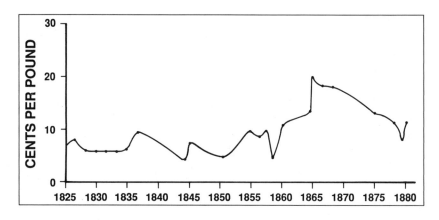

Figure 2.1. Cheese prices, 1825–80. Based on X. A. Willard, "Prices of 54 Years," *Country Gentleman* 44 (Apr. 10, 1879): 235–36.

Nesbit was always talking with other dealers, hearing about who got what price from what firm, and attempting, often unsuccessfully, to procure the same prices. This psychological competition seems to have been a constant accompaniment to the business; in 1860 P. W. Stebbins wrote almost frantically to his agent in the dairy district, "Snell has not offered [farmer] B. Timmerman 11 cents I know he has not," and warned him, "Do not let Crist get the start of you on Fred's lot." [19]

As well as competing among themselves, cheese brokers jockeyed with wholesale buyers. When Nesbit negotiated with the English agent in 1820, he used the episode as an excuse to say "within [the] hearing of grocers that we had a call for 400 casks of cheese to ship to Great Britain." (He never did make the sale.) And though he might complain to farmers about their cheese, he revealed a different evaluation when enticed with the prospect of selling cheese to England: "It does us good to think that we have it in our power to send to the mother country the cheese of golden hue, that in a short time we hope will equal, if not surpass the finest Gloucester itself." [20]

Not surprisingly, relationships between factors and producers back "in the country" were often adversarial and mutually manipulative. For the farmer, much anxiety must have attended the once-yearly negotiations by which his family's entire season's output of cheese—often not even yet existing—was sold. In exchange for the convenience of a guaranteed outlet for their entire lot, farmers risked total loss in the event that the cheese went unsold—for example, if it spoiled in the factor's care or if a brokerage firm failed. Friction also arose because the broker, not the farmer, received the benefits from a price rise after contracts were signed.

Pricing and "Profit"

Given the chancy nature of the brokerage system, and the fact that these farmers were making cheese expressly for the market, we might expect to find individual farmers negotiating aggressively for the highest possible price. There is evidence that some farmers managed to secure a price 1 or 2 cents per pound above the usual price. Peleg Babcock is an example. He sold his dairy's output to a broker in Ilion one year for 6½ cents per pound. The next year he successfully negotiated a contract for "7 cents per pound and one half profits over and above that." Excitement over prices is also very apparent in the dairy of James Kilham of Turin, Lewis County: "Butter sold the highest... that I ever knew it from 25 cents to 26 cents a pound by the dairy the first of December it fell in market after that it fell and you could not sell it over 16 cents by the dairy in Turin the same month cheese sold from 6 [?] to 9 at Boonville." And reminiscences of the Ohio frontier noted that cash for cheese was "a greatly prized article in those days."[21]

Yet to characterize cheesemakers as devoted wholly to market competition for the best price would be erroneous, for evidently a great many did not behave as autonomous individuals in the marketplace, nor did they regard the market with unalloyed enthusiasm. Their production strategies attest to this mentality: between 1840 and 1860 few households substantially increased output, and in fact many curtailed production. Many deeply suspected the values the factor seemed to represent. In this regard they behaved in much the same way as the precapitalist farmers and traders who did not base prices on abstract market conditions but instead took into account individuals' particular circumstances.[22]

One New Yorker's recollections reveal this view. He remembered that in his boyhood home in Massachusetts, factors would travel twenty miles to see a dairy of five cows, "with a zeal known only to speculators." The term *speculator* was often applied to cheesemongers, always with pejorative connotations of grasping cupidity and ill-gotten profit. Partly this was age-old resentment of the middleman, but it also indicated ambivalence about the developing market system and about the values it seemed to express. Eighteenth-century Massachusetts farmers often regarded capitalist business methods as simple greed. The same sentiments appear to have been operating in mid-nineteenth-century New York.[23]

Many farmers, even large producers, regarded "speculative" activity as fundamentally different from their own ideals of economic and communal life. They harbored a conservative aversion to risk. This preference for stability was part of a wider concept popular at the time: that of a "competency." The term derived from English usage, which Webster's *American Dictionary* in its 1852 edition essentially replicated: "Sufficiency; . . . property or means of subsistence sufficient to furnish the necessaries and conveniences of life, without superfluity." The term obviously admitted a good deal of elasticity; one family's competency might conceivably be another's poverty. Nonetheless, to seek a competency clearly meant to accept limits to accumulation, and to stress necessities, however defined, over superfluities. This notion was widely held, even among those customarily labeled agricultural reformers; the New York State Agricultural Society in 1858 refused a premium to an otherwise highly successful farmer because he told the prize committee that he farmed only for money. In America a competency was also associated with moral virtue; the *Oxford English Dictionary* quotes a nineteenth-century American geography text as attributing a particular region's "general competency" to its people's "contempt of tyrannical fashions."[24]

Indeed, the rhetoric of the Dairy Zone reflected these ideas. Moral worth —as expressed in immunity from corruption—was linked to independence, and to the belief that competency mitigated the acquisitive drive for profits. "If not the richest in dollars, we think the [dairy] district . . . is to become the richest in moral worth, in republican virtue—in the treasures which improve society and render man happy."[25]

Significantly, some explicitly declared dairying, of all branches of agriculture, to provide the ideal competency.[26] Cheese prices fluctuated relatively little in the antebellum period, and so the very choice of cheese dairying was indeed a conservative one. The agricultural reformer Henry Colman characterized the cheesemaking farmer of Cheshire County, Massachusetts, as "substantial and independent" because he had a competency and was not afraid of the "caprices of trade and speculation." In 1838 an article in the *Albany Cultivator* entitled "Cattle Husbandry" argued that the rise of dairying in the hilly districts west of the Hudson and south of the Mohawk was a harbinger not merely of economic but of moral improvement. Under dairying, the author hoped, the population would enjoy "substantial comfort . . . equality, and social enjoyment" and not succumb to "ostentation and extravagance." Even dairymen portrayed their work as an agricultural choice that possessed the virtue of being "regular" and "sure," even if it did not pay spectacularly. X. A. Willard wrote in the 1861 annual report of the New York State Agricultural Society that dairying "offers great inducements for a judicious outlay of capital. It rests on no mere speculative basis—'up today, and down to-morrow,' but is permanent in its character and prospects and sure in its rewards." Because grain-based farming had been associated with ecological and economic crises, Willard's characterization probably possessed considerable credibility for an audience made up of people who well remembered the problems associated with wheat culture.[27]

Certainly nineteenth-century advocates of dairying praised its "profits," and as the following chapters will show, dairying families rigorously rationalized their operations so as to exploit lands, cows, and human labor efficiently. But their idea of profit was nonetheless consistent with the notion of a competency.[28] By *profit* they meant something different from the modern notion of a rigorously calculated surplus of revenue received over that paid out. So many items lacked identifiable prices that a strict notion of profit is impossible.[29] Published accounts of profit and loss for cheesemaking farms occasionally appeared in agricultural journals and in transactions of agricultural societies. They differed substantially from one another in how expenses and income were calculated, partly because quite a few of them were compiled not on the farmer's initiative but at the request of the agricultural societies or journals. (This in itself is revealing in that it shows that even farmers who operated on a comparatively large scale were not thinking about their farms in terms of strict income and expenditure.) These accounts show that many farmers had but a vague idea what their expenses were. Some recorded mortgage interest as expenditures, while others did not. Some calculated the cost of feed, tending, and so on against each cow's production. Some included manures as credits, but most did not. Another way of measuring profits was to compare the value of stock, feed, and crops at the beginning of the season with their value (enhanced) at the end. Most knew far better how much they produced and what they got for it than how much it cost them to produce it. The farmer's idea of economy was conceptualized as a series of relationships with other people rather than as a process with the farm at the center where money came in and went out. Thus farmers entertained a flexible notion of profit based only partly upon cash flow.[30]

The notion of competency accommodated nonmaterial values in another way. E. H. Arr remembered in *New England Bygones* that few dairying families had ever made cheese on Sundays: "A woman minding her dairy or a farmer storing his hay made a scandal." Yorkers maintained this tradition in the first half of the century. In this way too dairying families set limits on profit.[31]

To a great extent, then, Yorker dairymen's outlook was quite conservative. In this regard they shared important values with the agricultural reformers whom Henry David Thoreau subtly lampooned in *Walden*. The Yankee iconoclast satirized the "improvers'" humorless, soulless obsession with market crops and the intensified work regimens that came with commercialization. From their agricultural practices to their market activity, their goal was to maintain order and stability.[32] Yet Yorker dairymen's rhetoric and behavior were more ambiguous; they repudiated, at least rhetorically, social hierarchy and deference and did not fully embrace the acquisitive mentality Thoreau so caustically condemned.

In some ways the brokerage system seemed inimical to a mentality in which security or low risk assumed priority, while in other ways it paradoxically suited this worldview, because cheesemakers did not depend on continuous cash income and functioned reasonably well in a credit-and-barter framework. In such a system motivation to compete and to accumulate via increased production was surely not absent, but it was tempered.

Evidence that even substantial producers did not always compete aggressively for the highest prices comes from cheesemakers' critics. Observers from all of the major cheese districts complained that cheesemaking families lacked ambition; too often they failed to strive with one another for the best price. Moses Eames of Jefferson County, New York, voiced this opinion to Benjamin Johnson, secretary of the New York State Agricultural Society. He decried the lack of uniformity in the quality of cheese and blamed this on the attitudes of both buyer and seller. The first question a cheesemaker asked the dealer, Eames charged, was "What did you pay my neighbor?" He thought that buyers too frequently offered the same price to every cheesemaker in a given locality—for the "good" dairies and the poor. Then, Eames continued, buyers were reduced to pleading with the cheesemakers that because they had been promised good prices, they should take care to produce good cheese. In another farmer's recollection, at the annual negotiations the dealer would offer "2½ or 3 cents for skimmed milk cheese, and 2½ or 3 cents for that which was not skimmed so much." An Ohio writer reporting from Ashtabula County noted in 1851 that the practice extended to the "cheesedom" of the Western Reserve. He complained that there was "no moral rule" stipulating that the buyer ought to offer a uniform price, but he concluded with regret that "such is the relation of the parties . . . that at home the worthless article brings the same as the valuable." He concluded that the market was not based on "inherent merits" because farmers refused to repudiate their "stolid ignorance," and suggested a secret bonus on delivery for better cheeses.[33]

These criticisms, approached from the viewpoint of the cheese producer, suggest that even though each family bargained individually, a collective community pressure was exerted upon the factor. He knew well enough that through busy local lines of communication, everyone would quickly learn the prevailing price offered. Ephraim Perkins noted a "community of exertion" and a constant circulation of information among Mohawk Valley cheesemakers. In confronting the market, personified by the factor, cheese-producing families thus sometimes presented a united front; they tempered market orientation with a practice that minimized distinctions among families. Many participants in this mutually protective system had a superior product and could have realized more by opting out. They instead evidently preferred security and rough neighborhood parity; that any premium over the usual price must be awarded secretly tells much about the values involved. In this instance the "capitalist transformation" called forth cooperation, further proof that this process cannot be neatly dissected, nor can a competitive ethos be assumed even for large-scale market producers.

When farmers did seek a higher price, their efforts sometimes took a collective form. In 1845 sixty dairymen in thirteen central New York towns united in a "vigorous effort" to produce uniform cheese for the foreign market. They all agreed to follow a set of prescribed guidelines on temperature, cleanliness, and curing procedures. Rather than competing as isolated individuals, they cooperated. Their use of the term *emulation* bespeaks this idea, conveying the sense of imitating a model for collective improvement rather than competing for individual advantage.[34]

Neighborhood Networks of Exchange

The brokerage system accommodated other preexisting customs. For example, farmers could wait for payment until the new year because they did not need large amounts of ready cash in order to carry on their everyday lives. They and their neighbors constantly exchanged debits and credits in the form of goods, services, and labor, sometimes extending accounts over years before "settling up." A value was assigned to every item, but this did not mean that cash changed hands. Local people had accounts with one another and with local stores, which also often accepted goods in payment of debts. Even local manufactories bartered extensively. The 1832 McLane Report on American manufactures shows that several Oneida County textile factories exchanged "considerable . . . for agricultural productions," and then paid their hands in farm goods.[35] In the central New York agricultural economy, dairy products tended to be sent to distant markets while items such as oats, wool, and barley were exchanged locally.[36]

Peleg Babcock's diary shows more concretely how market activity and local exchange networks intersected. In 1852 he contracted for his cheese, then carried on a variety of daily activities and exchanges during the year. For example, on July 14 he "drove 14 young cattle to Olivers to pasture," fetching them back on August 30. In return "for keeping," Babcock "gave him two." At the same time, he "traded the old Durham cow for two two year old heifers" and was "to give $2 to boot." Babcock recorded borrowing small amounts of cash from friends as well as borrowing substantial amounts from relatives—at one point, $250 to buy cows, "if I can find them cheap." (Undoubtedly the $250 borrowed was in the form of a note.) These entries show Babcock trading, and incurring debts, but with very little exchange of cash. In late September the receipts from his cheese contract began to come in, and the very next day he began to settle accounts: "$263 to Uncle Joshua, $75 to Willard, $17.50 store bill—duWolf; $105.83 to Patrick M." In December, after selling eleven hogs for $280, Babcock "settled with Patrick owed him $158 paid him in book account and 320 pounds butter and cash settled up to Dec first." He continued to settle with creditors; on the last day of the year, "paid Katharine $3.50 she let me have a year ago last summer"; on February 3, 1853, "settled with Uncle Peleg Brown."[37]

Babcock mixed trade in goods, services, credit, labor, and cash; all were equivalent items in the farm operation. Babcock was producing large amounts of cheese—thousands of pounds per year—using hired labor.[38] Yet even on this highly market-oriented farm, the customary system of exchange—known as "changing works"—and the income from the market meshed together without obvious conflict or disruption. Part of the reason was that the factor system, though it facilitated long-distance trade, possessed many qualities associated with local trade. The factor system preserved the face-to-face interaction and long-term, flexible payment schemes characteristic of local trade, rather than exerting the more stringent conditions usually associated with long-distance trade. Thus the factor system did not destroy earlier exchange customs; the "market" did not necessarily disrupt the old ways. Rather, both continued together.

The factor system could be manipulated in other ways to fit more closely with customary, nonmarket social ties, such as kinship or geographical proximity. Several examples from Oneida and Herkimer Counties are suggestive. One of the most successful cheese brokerage firms in New York State was operated by the Burrell family. This was a pioneer Herkimer County family, well established locally, with many kin and commercial ties. Henry Burrell commenced in the brokerage business in the mid-1820s and soon built up a large clientele. Local historians rightly attribute his success to his honesty and integrity. Yet it is also significant that the Burrells offered a local alternative to the despised "Yankee" dealers; at least some of Burrell's success may be

Cheesemakers and the Market

attributed to his local origins and contacts. Other local firms, such as that of J. H. Ives, shared the trade with Burrell.[39]

Still more concrete evidence of how kinship worked in the brokerage business is revealed in Charles S. Brown's 1859 letter to his "Dear cousin Cornelia" Babcock. Writing from New York City, he reported, "Your cheese[s] have been received as advised. The market at present is quiet but firm for good and well cured cheese. We are getting more green late cheese than we expected." The contrast with Nesbit is notable, and perhaps an indication of how kinship could mediate the confrontation with the market.[40]

The Shift to Cash Terms

There thus were powerful forces binding the brokerage system, and hence the market, to traditional networks of kin, neighborhood, and exchange. Yet forces for change were at work that would significantly alter the relationship between cheesemaker and broker, and in turn between producer and the market. By the late 1850s cheese factors, increasingly in competition with one another, had introduced a new way of doing business that departed significantly from the old.[41] The correspondence of the Ives firm documents it:

> October 1857, Sir: We this day sent you by express at Little Falls five hundred and eighteen dollars ($518) which amount you will please buy cheese with. You know what article of cheese we want and you will please be careful and buy only such. You say Burrells and Company are buying from 7 to 7½ cents payable 1 January if they can buy at those prices on time we think the ready money should do full well if not better.[42]

The account book of Samuel Hinckley, a Herkimer County storekeeper, also shows this shift. Hinckley rented his farm out on half-shares to a tenant who made cheese. From 1844 until 1859 Hinckley kept meticulous accounts of the farm's production and of his transactions with tenants and brokers. From 1844 until 1858 he sold the cheese by the dairy, at a single price, on contract for payment January 1. In 1846, for example, he "kept 17 cows made 6208 lbs cheese Sold at $6.50 [per hundredweight]." Hinckley dealt with at least eight different brokerage firms during this period, so that nearly every year he switched brokers. But the same basic structure applied with them all. In 1854 the books show a change. Payment was still made in January, but the price was no longer uniform throughout the season. Instead, thirteen separate lots, apparently delivered between May and December, were sold for prices ranging from $9 to $12 per hundredweight. This continued until 1858, when the firm of Betticher and Coon began not only to buy more often but also to pay for cheese at shorter intervals; Hinckley received cash on June 23 for a lot delivered May 21, and at more or less monthly intervals thereafter. The same was done in 1859, with another company. By 1860 X. A. Willard noted that cheese dealers preferred, and makers were coming to prefer, cash on delivery, the transaction taking place at the railroad or canal depot, where dealer and producer met.[43]

What were the reasons for this shift? Increased production does not account for the change, because the amount of New York State cheese produced actually dropped slightly between 1850 and 1860. (Indeed, Oneida County production fluctuated sharply during the 1850s, partly owing to drought. Production there dropped from 5.2 million pounds in 1850 down to 3.3 million in 1855, only to rise again by 1860.) Rather, an explanation lies in developing conditions in the larger economic system. Transportation innovations made possible the wider distribution of goods from expanding industrial enterprises. Rail links were established among Rome, Utica, and Syracuse in the 1830s and 1840s, and the Utica and Black River Railroad opened in 1854, linking Utica with communities to the east, west, and north. The Utica and Schenectady Railroad, which ran through Oneida County, achieved freedom from legal restrictions on freight hauling in 1851. Boonville was linked by canal with the rest of the county in 1855, and almost immediately the local papers were advertising recent shipments of fashions from New York City. By 1861 all of Jefferson County's cheese was shipped out by rail. Thus the way was opened for the receipt and shipment of ever more goods.[44]

At least partly in response to this widening circle of trade, merchants' ads in Oneida County papers began to stress the advantages of cash transactions. Merchants involved in long-distance trade had to operate differently from those who were connected mostly to local institutions and people. Oneida County merchants pressed for tighter economic discipline, urged quicker settlement of debts (preferably by cash), and exploited the weaknesses of the old system. They used various pitches to persuade customers. They argued, for instance, that the old credit-and-barter system meant high prices, because of the cost of credit. S. Phelps of Rome counseled "farmers from adjoining towns, who have paid the high prices of credit and barter merchants long enough," to "call at our cheap cash store." Some merchants claimed to offer "greatly reduced prices for cash." Others made a point of having bought their stock "new purchased entirely for cash"—that is, without the extra costs added by numerous intermediaries.[45]

These arguments quite obviously catered mainly to merchants' own self-interest, but their appeal to dairy farming families should not be underestimated. Cheese dairying families were generally well-off and potentially able to take advantage of low cash prices as they gained access to ready cash under the new system of brokerage. Furthermore, though the old system's merits were considerable, its longstanding obligations, monetary and otherwise, could grow onerous. Rome merchants Parker and Mudge implicated these weaknesses when they hectored customers with a pitch for cash as the "most convenient, most proper and altogether the most desirable form of exchange." Credit, on the other hand, was "the parent of many evils, viz., high prices, poor goods[,] *duns, executions,* and bankruptcies." And John Thayer, in 1845 owner of the "only cash dry goods store in Rome," bid to exploit conflict over unpaid debts by appealing to those who did not care to be "overcharged equivalent to the delinquencies of those who NEVER PAY." As an added advantage, cash prices were often fixed. "Do you want to buy without jockeying or bantering?" asked one vendor. In the Connecticut Valley the old system in part collapsed under its own inconsistencies, and events in the New York State dairy districts suggest that a similar process occurred throughout the North.[46]

Two more developments peculiar to the industry affected cheesemakers. Helped by a few aggressive brokers, especially the enterprising Harry Burrell, the export market for cheese expanded dramatically. In 1840 about a million pounds were exported. The boom that occurred in the following decades owed its vigor primarily to an increasing appetite for American cheese in England. After duties were reduced in 1847 under Robert Peel, the flow of American dairy products increased. Export volumes then fluctuated between 1850 and 1858, from four million to ten million pounds, but in 1859 an astounding fifteen million pounds of cheese were sent to British ports, and by 1860 the figure was twenty-three million pounds, an increasing proportion of a more or less steady domestic production (ninety million pounds). This intensification of export activity probably influenced a substantial rise in prices starting in 1858. After a sharp drop in the mid-1850s, prices recovered again, rising from 6 cents per pound in 1858 to 11 cents in 1860.[47]

In 1859 the weaknesses of the existing cheese brokerage system were disastrously exposed as the large firm of Samuel Perry failed. Ambitious to corner the entire cheese market, Perry offered higher prices than his competitors and attempted to ship the cheeses earlier, during the summer months. Huge lots of cheeses spoiled en route, and British agents refused to accept delivery. As a result, Perry could not meet his obligations, leaving hundreds of cheese producers without any reimbursement for their year's work.[48]

All of these developments stimulated the shift to a cash-on-delivery system that in turn allowed still more changes. Since the transactions involved actual cheese rather than a product to be manufactured in the future, it was probably easier to justify differential prices.⁴⁹ Farmers had less opportunity to communicate with one another on the spot, and companies reminded their buyers to seek only "prime" dairies. Evidence of differential prices appears in Hinckley's account book; one batch fetched only 4 cents per pound. The factors' growing ability to insist upon their standards of quality meant that farmers were more obviously in competition with one another. It may also have meant a heightened awareness of the market and of current prices, the result of greater frequency of contact instead of the old once-and-done system. But the farmer was assured of getting his money—not a trivial consideration. It still seemed possible for cheesemakers to partake simultaneously of the cash economy and of a competency.

If testimonials from "cheesedom" are to be believed, this competency was handsome indeed. Evidence from the census of agriculture affords perhaps the clearest and most comprehensive documentation for the economic success of home cheesemaking. In farm value, cheesemaking farms exceeded the average for both the state and the county by a substantial margin. In 1850, for example, cheesemaking farms (in the seven Oneida County towns, 475 households) averaged $3,883 in value, as against a statewide average of $3,250. By 1860 this gap was more pronounced: a sample of 393 Oneida County cheesemaking farms averaged $5,827, as against a statewide average of $4,466 and a county average of $3,081. Data on average farm size reinforce this picture. Not only were acreages on cheesemaking farms higher than the state average, but cheesemaking farms actually increased in size from 1850 to 1860, to as much as 245 acres (in Boonville). By contrast, the New York State average dropped from 113 acres in 1850 to 106 acres in 1860.

Other evidence also points to prosperity in the dairy districts. The Babcock boys, for example, were sent to private academies rather than attending free public schools. The family also had enough invested in the farmstead to insure it; upon Peleg Babcock's death in 1857, his wife Cornelia took out an insurance policy with the Agricultural Mutual Insurance Company in Watertown. Her dwelling and wood house were valued at $400; "barn No 1" at $300; "produce in barn" at $200; and the cheese house at $100.⁵⁰

Observers too saw a marked improvement. In 1859, for instance, a traveler wrote in the *Ohio Cultivator* that the dairying regions of central New York had been transformed in the thirty years of his acquaintance with the region. A generation ago, he recalled, the local farmers' houses went unpainted, and their families did without even crude wagons. But now that dairying has taken firm hold there, "they ride in spring carriages and fine buggies." A correspondent of the *New York Express,* on a tour of the central New York dairy region in 1845, thought that the dairy farms were "among the finest in the state" and that "very poor men [were] scarce" there. John Gould remembered later that "dairy pioneers had a great advantage in prosperity over others who had no dairies, and it is said that the dairy communities developed three times as fast as did others, in the way of frame houses, barns, school houses, and . . . churches." Thus cheesemaking allowed New York State farm families to involve themselves more and more in the consumer economy, even as they preserved their competency. Again, this ability simultaneously to participate in the market and to maintain a measure of family independence helps to account for the smoothness of the "transition to capitalism" in the rural North.[51]

A correspondent of the New York State Agricultural Society in 1849 summed up the reasons for dairy farmers' "eminent success": "They have doubled their capital, by the aid of their sons and daughters. They have now money to lend, instead of being under the whip and influence of the banks; a case not unfrequent in Oneida County in former times."[52] The reporter had pinpointed an element crucial in both aspirations to and achievement of competency: the farm family. We now turn to the distinctive family structure that undergirded cheesemakers' prosperity.

3

The Social Organization of Cheesemaking Households

> Children . . . [should never] hang around those who are engaged in a dairy . . . to the annoyance of all.—*New England Farmer*, 1829

Few farm couples who achieved a "competency" through cheese dairying did so without the help of sons and daughters. But merely to note the presence of children in cheesemaking households would present too simple a picture. Home cheese production was sustained by an intricate social organization that possessed a distinctive age structure and, in many instances, a complex mix of family and hired labor. This social structure itself made up an integral part of the farming system. The composition of cheesemaking households offers insights not only into the nature of the work but also into dairying families' motivations and strategies. Moreover, the position of hired help reveals again the hybrid character of cheese dairying, mixing customary forms of trading work with labor hired for wages.

A production profile of the cheesemaking farms in seven Oneida County towns in 1850 has already been established; to understand the household economy that made that production possible, we must link the agriculture census to the population census. This was possible in 348 cases (73 percent of the total). The results show that cheesemaking households were demographically quite distinctive.

Perhaps their most striking characteristic is their tendency to represent a particular stage in the family life cycle. If we compare the age structure for cheesemaking households in 1850 with the same information for all Oneida County households outside of Utica in 1855, a striking contrast becomes evident between the two groups. A graph for the rural towns in the county would begin at a wide base and taper upward fairly regularly. This profile is what might be expected for the mid-nineteenth century, when in general the population was young. But for cheesemaking households in the seven towns, the profile would bulge in the age groups 10–19 and 20–29, while the group at the

The Social Organization of Cheesemaking Households

bottom, the 0–9 age group, would be smaller than either of the two groups immediately above it. To put it another way, fully 40 percent of cheesemaking households had no children under 12; another 25 percent had only one child under 12; and still another 20 percent had only two. Thus 86 percent of cheesemaking households in Oneida County had two or fewer children under 12 living in them.[1]

The absence of young children in cheesemaking households was due to several circumstances. For one thing, cheese dairying on a substantial scale required capital to purchase livestock and equipment; families at an advanced stage in the life cycle were more likely than young families to have accumulated the means necessary to take up cheese dairying. But perhaps more important, as we shall see in more detail later, cheesemaking was such demanding work that small children were a decided liability. The nineteenth-century dairywoman was often advised never to allow children to "hang around those who are engaged in a dairy . . . to the annoyance of all." Conversely, older children could help with the cheesemaking.[2]

Thus cheesemaking was a life-cycle phenomenon, taken up as a family reached the point when its children were grown but not gone. But the figures present a still more complex story. The overall sex ratio (106:100) was skewed to overrepresent males. Yet cheesemaking households invariably had both men and women in them. This indicates how crucial was a shared gender division of labor in the work. Cheesemaking households can be classified into four main groups based upon the mix of people who would be doing the work. (For these purposes, I have counted as workers those aged 10 and older.)

Comparatively few households listed only a husband and wife. In the town of Augusta, for example, lived Sullivan Bridge (30), his wife Lucy (29), and their 3-year-old daughter; from twenty cows they made six thousand pounds of cheese. But this was unusual; most households fell into one of three other categories.

Those I have classified as using only "family labor" had parents and children living in the household.[3] John and Mary Maclusky of Boonville (50 and 45, respectively) milked thirty-one cows and made six thousand pounds of cheese with the help of sons aged 11, 14, and 25 and daughters aged 17 and 20. They also had younger children. Households of this type made up a little less than half of the total (45 percent).

About 15 percent used "hired labor only." Besides the husband and wife, the only other household members between the ages of 10 and 29 were people listed as "laborers" and bearing different surnames from that of the head. In the town of Boonville, William Reymond (37) and Phebe (47) had three young children; they hired Mary Waterman (18), Thomas Barns (18), and Catharine

Christian (19). Barns and Christian had both been born in Ireland. An elderly couple (81 and 75), probably Phebe's parents, also lived with them. From twenty-seven cows this household turned out six thousand pounds of cheese and one thousand pounds of butter.

Finally, households using what I term "mixed labor" had both hired labor and children aged 10 to 29 living in the household. These made up about 35 percent of the total. In Verona, Rufus Eldred (53), his wife Maria (50), and son Moses (14) employed Martin Menope (20), from Ireland, and Phebe Bergen (14), a native New Yorker. They made over nine thousand pounds.[4]

All four household types were represented among both small and large cheese producers. Among producers making less than five hundred pounds per year (a little under half the total), the proportion of types was the same as that for the entire sample—that is, 35 percent of these households had mixed labor, 15 percent had hired labor only, and so on. But most households (about 55 percent) made a thousand or more pounds per year. Among these households those with hired labor tended to produce larger quantities (more than five thousand pounds per year), as might be expected; this was true for both mixed-labor households and those with hired labor only.[5]

In the age group 20 to 29 the proportions of hired and family labor were significantly different for men and women. Of all of the young men aged 20 to 29 in cheesemaking households, only about half (52 percent) were sons of the household head, 6 percent were heads, and the rest (42 percent) were hired hands. By contrast, 61 percent of the young women were daughters, 15 percent were wives, and only 24 percent were hired workers. Almost never were daughters and hired women present in the same household. In households with family labor only, the numbers of sons and daughters were roughly equal. But in the households with mixed labor or with only hired help, hired men aged 20 to 29 made up a greater proportion of all workers (excluding heads and wives) than any other group. Thus among families that employed nonfamily labor, hired men dominated.

The household structure of cheesemaking suggests how social considerations may have operated in the decision to pursue cheesemaking. Grown sons enjoyed a considerable latitude of life choices, including, of course, migrating off the farm. Daughters, on the other hand, found themselves in a different position. For them, life in the cheese dairying household was filled with ambiguity. On the one hand, the valuable skills they learned there allowed daughters to sustain themselves while remaining in a family setting, and to avoid emigration or off-the-farm employment in the few, mostly unattractive, occupations open to women. They might even be enabled to postpone marriage, should they so desire. At the same time, though, their status as daughters in the patriarchal household also meant subordination, in work and in everyday life.

Workers brought in from outside faced conditions and expectations different from those faced by members of the family. Because all home cheesemaking operations involved family workers, even if only husband and wife, the intricacies of the connections between family and work will be treated separately. But before detailing the actual work, it is worth considering the place of hired help in the household structure. Relations between farmers and their hired help were in a transitional phase: ways based upon ties that depended on kinship, neighboring, reciprocity, taskwork, and egalitarianism coexisted with relationships characteristic of later agricultural capitalism—more impersonal, based upon wages more than upon neighborhood reciprocity, undertaken for specified periods of time more than for discrete tasks, and accompanied by deep cultural and social differences. As with the broader interaction between market and rural society, these two sets of characteristics often overlapped, creating ambiguities of expectation and status.

Sometimes workers were hired in family groups, reflecting the importance of the family unit to accomplishing the work. In 1858, for example, Norman Gowdy of Lewis County hired a couple at $475 for the year to live in a tenant house and make cheese on a thirty-cow dairy farm. Peleg Babcock also engaged a family to make cheese on his farm. On April 1, 1853, Babcock "hired Mrs. Cole to make cheese am to pay $2.50 a week am to give Mr. Cole and boys $28 a month for 8 months to work."[6]

But more typically, male and female hired laborers in the central New York dairy districts were hired individually, came from different social backgrounds, and were engaged in different patterns of work. Women dairy workers were, by some accounts, difficult to obtain; Lavinia Mary Johnson remembered that when the Deans sought a woman to make butter and cheese in 1814, they found it "difficult to get good women's help." Indeed, the census shows that female hired workers were comparatively rare in cheesemaking households. The relative scarcity of hired women workers underscores that familial reasons had been an important factor in central New Yorkers' decisions to take up dairying; single farm women used their skills in their position as daughters rather than seeking paid work on other farms.[7]

Only when it lacked enough female family members to produce cheese did a household absorb hired women workers. But then those workers often occupied a critical position. Perhaps partly because they were indispensable, women hired to make cheese worked in a situation that more closely resembled the traditional position of "helps" than it did that of wage labor. In the custom of hiring women or girls as "helps," most often the person who "helped" was a relative or neighbor, hired—usually by the mistress of the house—to work with her at specific tasks. "Helps" came into a household at busy times or when special circumstances, such as the birth of a child, made it difficult for the mistress to do all the work herself. "Helps" usually were hired to work at tasks that directly or indirectly contributed to farm production. They lived in the household and ate with the family, sharing family status. The custom of "helping" was often reciprocal, part of the wider exchange of labor among families. A young woman who "helped" could avail herself of a number of protections in case of conflict or dissatisfaction; her family could intervene on her behalf, and she had the ultimate resort of returning home if things got too bad.[8]

Among some dairying women, "helping" and large-scale cheesemaking had gone hand in hand for generations. In some New England villages women had pooled their cows' milk and alternated in the work of turning it into cheese. Each person received compensation in proportion to the amount of milk she had contributed; the exchange of work also became the occasion for exchanging technical information and for reinforcing social ties.[9] Patterns of hiring and work in New York State dairy households also show how cheesemaking was consistent in many ways with "helping." In Oneida County, for example, 78 percent of the women working in cheesemaking households were native-born, probably from local families. Andrew Hurlburt hired local people known to the family; the Hurlburt women made a good deal of cheese, so it is not surprising to read in Andrew's diary that they "churned and did chores in forenoon and in afternoon did chores and went and got Sophie Palmer to help mother." Hurlburt "drawed hay" from the Palmers' a while later, in the same work exchange of which Sophie was a part. Sophie Palmer was not the only local woman "helping" at the Hurlburts'; in October 1865 Andrew noted that "Adaline Avery came here to work," and a few weeks later he "carried Adaline Avery home." Rosetta Hammond Bushnell of Chenango County worked for relatives named Bushnell in 1858. She churned, milked, and "crushed cheese" (probably broke and cut curd) for a dollar a day. She also baked and did other kitchen work.[10]

In other ways, though, the situation of women workers in cheesemaking households diverged from customary patterns of "helping." Though cheesemaking was a much amplified, elaborated form of taskwork, dairywomen were not employed to meet a temporary exigency; the extent of the cheesemaking season was well defined, in contrast to the more sporadic and limited tasks of "helps." Most of the manuscript and published accounts of female hired help in cheese dairies noted that women hired as dairywomen worked an eight- or nine-month season, usually from March, when the cows began to calve, to October, when the last cheese was stowed in the cheese room to cure. In central New York these general terms prevailed. Chauncy Beckwith of Herkimer County hired a "girl" for thirty-five weeks in 1847; Abraham Hall of Floyd, Oneida County, hired his "girl" for thirty-two weeks at a little more than a dollar a week. Ann Diggems was dairymaid to William Thompson in Herkimer County from 1853 through 1856, each year for eight or nine months. Each of these years she was hired at the season's beginning for a specific time period and a specific weekly wage—from $1.50 her first year to $3.12 her last. She received board and monthly transportation to Little Falls and occasional necessities such as money for shoe repairs.[11]

Another difference from customary associations of "helping" was that though women cheesemakers might work under the mistress of the house, most records that survive show the actual hiring being done by men. Andrew Hurlburt traveled himself from Ava, where he resided, to the towns of Lee and West Branch "after a hired girl." It took him several trips to find one, and when he did, he noted her name in a way that implied he was unacquainted with her. Where neighbors had "come" to the Hurlburt homestead to "help mother," the new hired girl, Barbara Saintiff, evidently came to work for Andrew. The correspondence of a Herkimer County cheese buyer, J. H. Ives, also shows men hiring women cheesemakers. Ives frequently received inquiries about dairymaids. In 1861, for instance, a cattle broker named J. F. Knox wrote Ives asking for a recommendation, as he had decided to turn his 135-acre farm into a cheese dairy. He turned down Ives's first suggestion, a young woman with a small child, and accepted Ives's second choice, a single woman. Her duties were to milk twenty-five cows, to make all the cheese, and to do the housework, for $2 per week with board.[12]

Hurlburt traveled away from his home town of Ava to find a dairymaid; Ives's contacts obtained their dairymaids via correspondence. Either way, a more formal mechanism substituted for the informal summons from a neighbor. Young women hired in this manner may have had fewer opportunities to assess their potential situation than did "helps" who not only knew the family that would employ them but had relatives nearby to provide protection and oversight. Also, the formality of these contacts probably rendered the system less flexible than the old ways; the young mother seeking work as a dairymaid had no opportunity to plead her own case personally. Many of these contacts acquired a still more definite business nature when novices, without themselves knowing much about the work, hired experienced cheesemakers. The relation between the person hiring and the person hired was more one of employer and employee, because the major mediator was money.

In situations like these the reciprocity that had characterized the work of "helps" could erode, with varying results. For skilled dairywomen, especially those hired by employers ignorant of cheesemaking, competition in a wage-labor market may have opened up opportunities superior to those that could be had in the old ways. Ann Diggem's experience suggests that she was able to negotiate better wages each season by virtue of her success in the last. Others may have sought mobility via marriage. X. A. Willard complained of a shortage of good dairywomen and of the high wages necessary to secure them. He believed that once they learned cheesemaking skills, they left hired help to become farmers' wives.[13]

The erosion of mutual self-interest and reciprocity in the face of the cash nexus also could bring a freedom from customary obligations that might impede the dairymaid's search for better compensation. An example suggests how this might happen. In 1861 John Knox of Montgomery County wrote to Herkimer County resident James Ives that he had encountered a problem in his search for a cheesemaker: "John could get us a girl for $2 per week but [she] would do nothing but make the cheese and milk and take care of them we can not afford to pay that for milking and making cheese of 15 cows if she will help wash and cook and do all kind of house work as it comes in course I will give her the $2." While the outcome is unfortunately not recorded, this intriguing bit of correspondence suggests that the young woman in question was taking advantage of her skill to try to set limits on her duties, gaining exemption from the assumed obligations of female "helps."[14]

At the same time, the atrophy of customary protections could deprive those who were less well positioned. In larger-scale dairy establishments relationships of hired help with the farmer could be increasingly impersonal; the *American Agriculturist* wrote in 1854 that in these operations poorly paid maids were under considerable pressure to milk large numbers of cows and to produce huge quantities of cheese with less regard to quality. These wage earners were simply factors in the farmer's accounting system. Especially where the mitigating factors of kinship, proximity, and shared work no longer obtained, the wage nexus predominated. In the case of women's employment this happened mainly on very large farms, when the worker was doing less skilled work, or when the hired worker was an immigrant. The trend was summarized in an 1854 piece in which the *Ohio Cultivator* mourned the demise of "The Hired Girl on the Farm." The "ancient spirit of the profession is lost," the author complained, and a wide gulf now separated hired girl and employer, particularly if the hired girl were foreign-born.[15]

The Babcock family papers show the influence of ethnic difference in creating tension between farmer and laborer. Peleg and Cornelia Babcock operated a large farm and had a number of people working for them, several of

The Social Organization of Cheesemaking Households

whom were Irish—in 1852–53, a young man and two young women. Cornelia was evidently responsible for dealing with the women. She did so with obvious distaste. "Paid Catherine $11 for making cheese," Cornelia wrote in her diary in 1852. She added wearily, "Give me anything but an Irishman to settle an account with." A few weeks later, when her other Irish hired woman, Bridget, had gone to Utica for the day, Cornelia rejoiced: "Have enjoyed myself extremely well doing my work alone. Wish I could be alone with my family all the time." And in an undated letter she expressed further her mistrust of hired wage laborers, with a reference to a relative who was "improving her farm by hiring help which is quite an uphill business in these times."[16]

Cornelia Babcock's remarks, together with information from other family accounts and papers, help to establish how immigrant wage labor affected the internal workings of one household economy. Cornelia evidently found the managerial burdens of dairying not to her liking; she did not express distaste for her work, but she disliked dealing with hired help. Cultural differences figured prominently in the tension between Cornelia and her Irish employees; so did the bargaining process involving wages. Peleg Babcock included "house rent" in his agreement with his employees, so we can infer that the immigrant hired hands were segregated from the family; this is another concrete sign of a clear divergence in status and interest. Finally, the year-round work force of the farm consisted of immigrants, while neighbors and relatives came and went as "helps." It is not too speculative to imagine that cousins Laura or Emily entered the household workings with a status superior to that of the immigrant Irish working for wages, and that this hierarchy must have been plain to all.[17]

For male farm hands, in contrast to the experience of women, the conditions of employment typically partook more of the newer system of wage labor than that of "changing works." In Oneida County the most notable indicator of male farm hands' status within the household is their ethnic background. In the dairying country among men as with women, local labor exchange—both paid and in-kind—coexisted with wage labor, but the proportion of the male labor force from outside the local customary system was twice as high as among women: over 40 percent of hired men in the seven towns in 1850 were immigrants, from Ireland, Scotland, England, Wales, Germany, France, and Switzerland. Most probably arrived with few financial resources and lacked education. All would come from cultural backgrounds quite different from those of native-born Americans; some were Catholic, and most had a language other than English as their native tongue. The Oneida County Welsh in particular clung to their language in an attempt to preserve their identity from Anglo influence. Indeed, the Welsh-speaking community persisted fairly vigorously; by the later nineteenth century, Welsh voluntary societies and social clubs existed in Utica and in the towns of Remsen and Steuben. Irish, Scots, and Welsh agricultural immigrants would likely have come from preindustrial agricultures; farming in these culturally and economically marginal areas of British agriculture was still small-scale, often barely at a subsistence level, and under enormous pressure from population and encroaching capitalism that would substitute sheep for people. Many came too from urban areas where wage labor was predominant. Their poverty made them vulnerable to impersonal, wage-based, low-status employment.

Again, the Babcock papers show how cultural differences contributed to tensions in the household economy. Just as Cornelia Babcock and her Irish help conflicted, so Peleg and his hired men experienced friction. Evidence for this is more oblique but still suggestive. Frequently Peleg would note in his diary that Patrick had "lost the day" and had not gotten any work done. Peleg rarely detailed the reasons for these lapses, but they evidently bothered him, for he mentioned them as a special point. He also seemed to have placed a rather low value on Patrick's work, if his winter wages of $9 per month in 1852 are any indication; Babcock paid Mrs. Cole more to make cheese. Here we can see the systematic farmer becoming frustrated with what he regarded as laziness on the part of his hired hand, himself perhaps acting according to preindustrial mores of agricultural work.[18]

Peleg Babcock's troubles with his hired help point to some disadvantages for the farmer of the wage system. Farmers took the opportunity to substitute a purely wage connection for the old multiple obligations, but not without ambivalence; they at once regarded their wage laborers as an item in accounts and also expressed doubt and regret about whether the transition was altogether beneficial and positive. Under the "changing works" system, of which local hired help was an extension, farmers probably got a good deal of "free" help; neighbors and neighbors' children knew they were accountable and would themselves eventually need help in return. Immigrant workers' critics were inclined to contrast the departed "Yankee" hired man, adept at any number of skills, with his allegedly incompetent immigrant counterpart. Certainly cultural differences, and individual variations in skills, figured in these perceptions, but it is also likely that the perceived gulf between immigrant and Yankee arose not so much because the old-time hired man had been a Yankee as because he had been a neighbor, perhaps even a relative.

Once money had become the medium and hired hands were no longer bound into the web of mutual obligation, employers not infrequently received just as much work as their employees thought was appropriate. In this regard traditional standards of neighborly "mutuality" were more exacting because potentially infinite. The *American Agriculturist* disapprovingly contrasted the dairy establishment run by its mistress with one staffed by "a certain number of lads and maids, [who] have to milk a certain number of cows, as long as they bring in a fair quantity of milk, few questions are asked." The *Ohio Cultivator*'s statement that the "spirit of the profession" was lost also associated the old ways with adherence to accepted standards of work and conduct.[19]

The relationships between employers and workers are yet another indicator that cheesemaking households stood at a transition point between old and new. Indeed, within most households family members worked closely with hired hands. The status of these hired workers ranged from essentially full membership in the family to a position mainly mediated by the cash nexus. Yet all household members worked together to produce the myriad output of a typical cheese dairying establishment. Cheese, of course, predominated, but the subsistence base too was substantial. The combined demands that these varied tasks asserted pushed all the laborers in the system toward a highly coordinated, disciplined set of work patterns.

4

"Intense Interest and Anxiety"
The Women's Work of Household Cheesemaking

> Thought and care are essential in all the various operations. Intense interest and anxiety are necessary *to do all these things well.*
> —*Country Gentleman*, 1859

All the elements of the farming system came together in work. People, organized in accordance with cultural patterns of family and gender, labored on the land and in dairying spaces. In the cheesemaking household, farmers, farm wives, daughters, sons, and hired hands worked together to turn prodigious quantities of raw material into useful or salable items. Sex, rather than status or age, was the principal basis of work organization. However, the work demanded a significant sharing of tasks between men and women, a reminder that the reality of everyday work in America has rarely reflected the ideology of "separate spheres." Moreover, this sharing extended across the hazy line between "subsistence" and "cash" production.

Scholars have frequently associated women with subsistence production and men with cash-crop production.[1] But for nineteenth-century dairying, at least, across the mid-Atlantic region, this association does not hold. Because cash and subsistence uses were so often interchangeable and because dairying was traditionally women's work, it is impossible to draw a line between market and subsistence production that parallels the gender line.[2] Indeed, if anything, the direction of change in cheese dairying was to invert customary associations: during the wheat-farming era men had been responsible for the main cash crop, but under the dairying economy women assumed that responsibility.

This shift in women's work, toward increased involvement in market production, partly represented a shift to dairying from textile production. The era of settlement in Oneida County was also the "age of homespun" there, as huge numbers of sheep supplied wool to keep the spinning wheels turning.

Beginning early in the nineteenth century, textile mills were erected in Oneida County. Migrant textile manufacturers from Rhode Island found the county's waterpower attractive, and by 1827 factories in New Hartford, Whitestown, Paris, Waterville, Kirkland, Rome, and Verona employed over three hundred people and turned out in excess of two million yards per year. Most produced cotton goods, and as in other regions, woolen manufactures lost ground; by 1855 the number of sheep in the county had dwindled fast (to 50,000, from 195,000 in 1845). Wool played a correspondingly smaller role in the farm economy, likely limited to filling the need for warm winter garb and possibly serving as a minor source of cash via the sale of raw wool to woolen factories.[3]

The disappearance of home textile manufacture, of course, had major implications for farm work and especially for women. The fabled "age of homespun" had been accurately named, as its historians well understood that prodigious effort in spinning that had kept families supplied with warm winter clothing and blankets. With industrialization, some of women's time was freed for other pursuits. It was no coincidence that the rise of dairying in central New York occurred at the same time as textile manufacture industrialized.

The centralization of textile manufacture thus combined with the pressure of competition from western wheat to make dairying an attractive alternative. But in moving to dairying, families did not just make economic choices; they also responded to social and cultural circumstance, because they could easily adapt the existing family structure to dairy production. In many respects the shift to dairying was born of conservatism: families could stay put, instead of migrating westward to start new farms or leaving agriculture altogether. Daughters especially could remain in the household and need not be displaced.[4] Local observers made this connection directly; the *Rome Sentinel* reflected in 1851 that "the wheel and the loom were relinquished, but the milk pail, cheese tub and the butter ladle took their place."[5] The shift to commercial cheese production allowed women to preserve, possibly to enhance, their importance to the farm enterprise, but at the same time it kept them under the patriarchal yoke—a situation that would eventually give rise to tensions within the family.

Work on the Cheese Dairying Farm

The elements of subsistence, and of tasks indirectly related to dairying, presented intricate considerations of seasonality and labor allocation. Men usually planted, tended, and harvested the grain crops. Other field crops, especially potatoes and carrots, occupied a seasonal niche in the cheesemaking economy, since they could be harvested when cheesemaking slowed down in the fall. Men undertook the strenuous and time-consuming labor of breaking up grasslands when rotation schemes called for it.[6] Men cut and hauled forest products—bark, lumber, and cordwood—in the winter; Andrew Hurlburt, a young farmer in Ava (Oneida County), spent many of his winter days in 1865 stripping bark and hauling it to the local tannery. Care of livestock was also regarded as mainly men's work.[7]

Haying was primarily the province of men, though women probably "helped" frequently. Because hay was such an important crop, and because speed was essential to ensure high quality, harvesting the meadows called forth especially frenetic activity. Cutting, stacking, and storing were done by hand. Time was precious. All available hands were pressed into duty during these frenzied days, mowing with scythes, pitching the cut hay into wagons, turning the hay, and unloading in the barn or in some cases stacking in the yard or field. By the middle decades of the century the revolving horse rake was coming into popularity, reducing work in one facet of the harvest. This device eliminated the need to stop and shake the hay into windrows. Ephraim Perkins, an Oneida County cheesemaker, claimed to have invented a revolving rake around 1811.[8] Some observers suggested that women's participation was curtailed; the *Genesee Farmer* remarked in 1842 that in Herkimer County many farmers possessed these new implements, but those opposed to "improvement" still "let" their wives and daughters do the raking.[9]

In Oneida County, as elsewhere, most farmers made hay more than once. Those lucky enough to possess "moist" lands planted redtop and cut it early in the spring. They left the newly cut hay to dry, spread a foot or two thick, over fence rails or in sheds. They fed this early crop to cows just coming into milk. Farmers usually took a second cutting of redtop before they moved on to the main crop of timothy and clover hay.[10]

The gender division of labor allowed a considerable degree of flexibility. Women often drove the cows to pasture, for example, and they often raised swine.[11] Overlap was even more pronounced in the orchard, garden, and sugarbush, where men and women worked together. As we shall see, the work of dairying required constant cooperation.

On the other hand, food processing and preservation, clothing manufacture, cleaning, cooking, child care, and tasks such as making soap and candles fell mostly into women's province. An 1820 probate inventory from Oneida County is a reminder that textile manufacture did not disappear altogether: among the decedent's belongings were a flax brake, two "great wheels" (for spinning woolen yarn) and one "little wheel" (for spinning linen), and various quilts, woolen blankets, and linens. Even as late as 1855 cheese producers appearing in the Oneida County manuscript census occasionally listed items such as "120 pairs fringe mittens," "170 yards flannel," "53 yards rag carpet," all likely made by women. But poultry and eggs, another common item in women's farm production, were often absent from cheesemakers' returns and were usually listed in small amounts when they did appear.

The evidence indicates that dairying women had little time for extra work, but that when they expanded their domestic manufacturing outside of the cheese room, they followed a strategy consistent with the farm's overall crop production mix, stressing items with interchangeable uses, among which was family subsistence. Seasonal considerations also played a role; women could knit, spin, and weave in winter, when the demands of cheesemaking slackened. Seasonality may also explain why cheesemaking women kept few hens, since that work would have conflicted with cheesemaking.[12]

Farm family members' participation in local and state fairs reinforced this pattern. Men from dairying farms competed frequently in the 1840s and 1850s for premiums in the livestock, dairy, and crop categories, while women from the dairy districts were far less visible. If dairying women appeared at all in the premium lists, they tended to win prizes for items such as woolen stockings, feed bags, and quilts. Their town counterparts were more prominent at the fairs during the 1840s and 1850s, garnering prizes for wax flowers, embroidered flowers, and bead purses. That rural women tended to seek premiums for practical, useful articles rather than for decorative ones reflects the ethos of "competence" as well as time limitations.

The relative absence of cheese dairying women from premium lists for dairy products is, of course, not a sign that they were uninvolved in that endeavor; rather, it is a reminder that their efforts were seldom formally recognized within the fair structure, which assumed that livestock and dairying were the provinces of men. For example, fair organizers in the 1850s separated cheese and butter from domestic manufactures. But occasionally the agricultural societies would recognize women's contribution, as when the New York State Agricultural Society commended both Mr. and Mrs. William Ottley for their dairy farm operation in 1846.[13]

Of all the activities on the cheese dairying farm, cheesemaking claimed the most sustained and devoted attention. Before the cheesemaking season actually commenced, certain preliminary tasks had to be completed or at least started. As we have already seen, the milk was the end result of a complex series of tasks that encompassed growing, harvesting, storing, and feeding; ensuring that cows calved at the desired time, usually around March; and pasturing the animals about May 1. As in other endeavors, men's and women's responsibilities and work often overlapped.

Women cheesemakers had a direct interest in ensuring that the raw material they worked with came to them in good condition. There is some evidence that women acted to pursue this interest. For example, long-accumulated experience had taught cheesemakers that the cow's diet could affect the quality of the milk and hence the quality of the finished cheese. Certain weeds, such as wild garlic, imparted an undesirable flavor to milk; so did turnips. The folklore of cheese dairying held that rich pastures resulted in less flavorful cheese than pastures of middling quality. A historian of Chenango County recalled that in the early days of the county's settlement the pastures had been overrun with wild "leeks," so "annoying to the dairy-women." After long, tedious work, lasting all the way into the 1860s, these harmful plants were weeded out of local pastures. Sometimes women participated in discussions of feed in the agricultural journals, and undoubtedly also during everyday work.[14] The Babcock family papers show that Cornelia Babcock carried on buying timothy and clover seed after her husband died in 1857. Women also ventured to make judgments on feed expenses as against the potential income from dairy products. An anonymous Oneida County woman kept a diary in 1868; at one point she complained that "according to the way the hay has been guessed off to sell[,] our two [milch] cows will consume 800 [illegible] from the second week in February. . . . I think they ought to be beef at that rate."[15]

Another way in which women were involved with the care of livestock was that often they selected and raised the few calves intended to continue the herd. In a farm operation dedicated to processing milk, calves were either killed very young or taken from their mothers after a month or two and fed by hand. Maria Whitford of Almond, New York, wrote in her diary in the spring of 1859, "Our cows are all calved now, we are raising them and give the oldest sour milk." Most discussion in the farm journals about raising calves was dominated by men, but women sometimes participated.[16]

In the weeks before cheesemaking commenced in earnest, men and women busied themselves around the cheese house, making repairs to ward off pests, scrubbing, and readying the boxes, bandages, and other equipment. Peleg Babcock repaired equipment and saw to it that his cheese house was vermin-proof. Andrew Hurlburt spent a good number of his winter days in 1865 making cheese boxes and a log cheese press. Together, men and women rigorously cleaned all dairying spaces and implements; farm writers recommended "hot water, soap and sand, and hard work," and diary entries of women cheesemakers like Maria Whitford indicate that these prescriptions were taken seriously: "April 30, 1858: Mopped the cheese room floor. . . . Sam'l helped me put things back in the cheese room."[17]

Women generally prepared in advance a critical element in producing cheese: the rennet. Derived from a substance—now identified as the enzyme rennin—present in the stomach of a young calf, this liquor made milk curdle. Making rennet was a nasty job, because it involved killing (or "deaconing") the calf and removing its fourth stomach ("maw"). The dairywoman then cured the stomach in a special solution. Every dairywoman had her favorite recipe for rennet; a "New York Dairy Woman" claiming thirty years' experience favored this one: "To six gallons of water add salt enough to make a brine sufficient to bear up an egg; scald it and skim it. Then add 12 rennets (having emptied, rinsed, and salted them first), six lemons, and an ounce each of cinnamon and cloves."[18] This mixture was usually made a year before use.

Dairy farming, of course, meant incessant early-morning and evening milking. The art of coaxing milk from the cow was on some levels fairly straightforward; but this work took place in a context marked by changing cultural perceptions of the work, and apparently also by changing social allocations. In this sense milking the cows expressed widely held assumptions and cultural values; it is an avenue to historical analysis.

The most noticeable aspect of the social context of milking is its strong early association with women, in image and in fact. The work itself was endowed with gender qualities, and ideas of the dairymaid evoked vivid imagery. This aspect of the "gendering" of labor derived from European, especially British, custom, where a well-developed image of the dairymaid existed in pictorial and popular literary culture. In popular plays and songs, and even in high-style painting, the dairymaid was young, invariably robust, often physically large, with rosy cheeks. She sometimes also represented fecundity and its accompaniment, lactation. She seemed to personify the virtues of simplicity, sometimes of independence, and of harmonious union with nature. In England the relationship of the dairymaid to her social superiors also played an important role in the imagery.[19]

In America this image persisted, but in considerably diluted form. For one thing, the class element of the English version was largely lacking; the American milkmaid usually appeared as the farmer's daughter. Neither did American songsters, popular fiction writers, or artists display the same affinity as the English for the dairymaid theme. But the milkmaid appeared frequently in the American farm press, with the same rosy cheeks, sturdy constitution, and virtue transmuted into filial devotion. Her status as daughter, of course, diminished her association with fertility. These word pictures often contrasted American milkmaids invidiously with sallow, feeble, idle city girls, praising daughters who "go forth morning and evening, with flushed and rosy cheeks through the pastures and green meadows." Yet despite the implication of healthful, vigorous exercise, in most of these depictions milking hardly seemed like work at all. The dairymaid, "with soft and willing hands," was credited with "almost charming the milk into the neat milk pails." The cow submitted without protest, effortlessly yielding "her lacteal treasures."[20]

In addition to perpetuating the Americanized image of the milkmaid, the agricultural press applied conventional gender associations of the time to justify assigning the milking to women. The *Genesee Farmer* of 1840 expounded on the suitability of women for the work:

> Men are seldom neat enough in their habits to be trusted with milking. They have not patience to wash their hands or to wash the udder before milking. They are not gentle, and often abuse the animal by their kicks and thumps. They are in a hurry in the morning to get through a business which they dislike; and they come home tired at night; the cows are unnecessarily milked at an unseasonable hour; and the business is very often badly performed. Women, on the other hand, are more patient; more gentle; more neat.[21]

In the *Genesee Farmer* in 1852, in an article entitled "Milch Cows, and Calves," the author asserted that

> a cow will give a dairy-maid more milk than she will to a man. We think a man, during the milking season, has seldom any business near a cow, his great, rough, hard hands, and still harder heart, rendering him unfit for a good milker; while a gentle, rosy dairymaid, with her kind words, soft hands, and "So, so, my good bossy," seated on a three legged stool, will fetch out the milk till the froth runs over the pail.[22]

The dairy writer Charles Flint thought that milking "should be entrusted to women. They are more gentle and winning than men."[23]

This "gendering" of the work in cultural perception and imagination both reflected and reinforced a definite social allocation of labor. Historians of colonial and early national agriculture have found that in rural areas women almost always did the milking (figure 4.1). Into the 1820s and even the 1830s most references in the farm journals concerned women milkers. The 1823 *New England Farmer,* for example, noted that women were "anxious" to make butter and that they milked their own cows for this purpose.[24] A correspondent in 1828 wrote that his wife milked four cows "expeditiously" and made butter and cheese. This case is notable because the couple hired a man by "day's work" but assigned him to the field work.[25]

Perhaps not coincidentally, in the early dairy era the job of milking was informal and unsystematic, in the sense that it was usually performed in an open yard, and the cow was seldom tied up. One of a set of milking "Rules" published in 1848 suggested this: "If she runs about, have patience, talk kindly to her, and tie her up as a last resort till she is not afraid."[26] The milker carried a little stool around to the different cows. This informality, though it surely

The Women's Work of Household Cheesemaking

Figure 4.1. Early in the nineteenth century, women milked cows in the barnyard. *The Progress of the Dairy* (New York, 1819), 4, from the collections of the Henry Ford Museum and Greenfield Village, Dearborn, Mich.

partook of no romantic imagery, reinforced women's cultural association with the work because it depicted milking as in association with nature, and as a task not strictly regulated by time or procedural constraints—qualities culturally associated with women's work.

In the mid-Atlantic region, especially Pennsylvania, women apparently continued to do most of the milking as the nineteenth century went on. But in New York and New England, as dairying grew more rationalized, men more and more frequently joined the women at milking time, and in many cases women ceased milking altogether. In the well-to-do "Yorker" household of Judge Dean in Westmoreland (Oneida County), for example, the men and boys had already assumed the task of milking by the early nineteenth century. "Thirty years ago," lamented the *Genesee Farmer* in 1840, "it would have been almost as difficult to find a man milking as a woman mowing.... In this respect matters are greatly changed; and any hope, for aught we see, of getting back to the old practice, would be in vain.... '[The] rosy milk-maid' is an animal not known in modern natural history."[27] Yankees in particular came to have the reputation of "sparing" their women from milking. "Mothers here," reported a resident of the New England–settled Western Reserve, "have taught the men to be careful of their wives and sisters." Though doubtless some of these accounts exaggerated the degree of change, the evidence suggests that in many households women were consciously moving away from regular milking. One woman writing in the agricultural press, for example, declared that she believed in "knowing how to milk," in case she ever needed to. Or the "mistress of the dairy" would be prescribed a supervisory rather than active role in milking: "See that the milkers' hands are clean, and also the cow's udders." By the 1830s advice in the agricultural journals prescribed that "the milker, whether a man or a woman, ought to be mild in manners, and good tempered."[28]

Though a more thoroughgoing revision of cultural ideas relating milking and gender would not occur until later, there are some signs that as men entered the occupation, the qualities supposed to facilitate the work changed to reflect conventional "male" gender traits. Dairy cows, of course, invariably impose routine on their human keepers to a significant extent, but perceptions of their physiology has varied considerably, in conjunction with cultural ideas about human gender. By the 1860s the connections between female nurturance, lactation, and the milking of cows were replaced by suggestions on milking technique and procedure that increasingly showed a concern for discipline and system. This shift paralleled the rise of the notion of the cow as a machine: if cows were machines, then according to prevalent cultural canons they were properly men's province. Now the job was regarded as a "regular task, . . . not . . . an odd job." Dispatch was important: "Don't go through with some long yarn, and be ten or twenty minutes."[29] To get the most milk, workers ought to "milk fast, and milk clean"—that is, make sure they obtained the rich "strippings" that came at the very end. Strict regularity, silence, and concentration were also highly valued: "Never let two persons, while milking, talk to one another."[30] "Think of nothing else while milking."[31]

However obtained, raw milk was the basis for cheesemaking. Wives and daughters made the overwhelming proportion of home-produced cheese. Indeed, the *Rural New Yorker* joked that a man who had inquired in the journal about cheese dairying must be contemplating matrimony. A typical description comes from an anonymous traveler in 1844, who visited a farm where the forty-cow herd was tended by the farmer and his three sons, and the "two charming daughters" had "under the direction of their kind mother" made "a show of cheese . . . as I had never before seen." Women of the "cheese-making sisterhood" learned from one another, sharing recipes, remedies, and advice on technique. References to the "never-ceasing duties of the dairy-woman" abounded. On larger farms the farm wife performed some of the labor but also supervised hired help in the production process; Cornelia Babcock filled a role of this type.[32]

Men were not entirely excluded from cheesemaking, however; in some cases the men and women shared the work, especially on the largest farms. In 1863 the *Black River Herald,* for example, recognized Boonville native Timothy Jackson as "a veteran cheese-maker." "He informs us that for thirty years [c. 1830–60] he never missed making a cheese every day during the season, and was never absent from a milking."[33] At a Utica trial in 1873, Franklin Wellington testified that in 1849 and 1850 he had resided on the large Madison County dairy farm of F. B. Hoppin, where "I made cheese . . . in his cheese dairy; I was the chief maker." This was in a dairy of seventy-five cows.[34] Moses Eames of Herkimer County included a note with his 1851 plan for a cheese house, saying that he had "seen to the manufacture mostly and for some years made [the cheese] myself."[35]

The presence of men in American cheese dairies was a notable departure from English tradition, in which male cheesemakers were virtually unknown. English cheese dairying families tended to occupy small farms and to command modest resources. They produced for the market but did so within the framework of a rich traditional work culture in which the sexual division of labor was a prominent and cherished element. Cheese dairying families passed on skills from generation to generation, intermarried among themselves, and subscribed to a set of beliefs that rendered women's participation crucial to the survival of their entire way of life. Cheesemaking women aggressively protected their turf, the cheese room. A crucial element in this social system was that cheesemaking was unequivocally "gendered" work—that is, women were supposed to possess the inherent instinct for it.[36]

In the United States, by contrast, cheese dairying lacked a continuous tradition. No regional identity bound American cheesemakers together; little generational continuity existed either, for the people engaged in it were usually making cheese on a scale unprecedented in their family backgrounds. This temporal and social discontinuity helps to explain why cheesemaking was not so extremely gendered in the United States as in England. Though cheese dairying complemented the customary family economy, its origin as a commodity-producing business permitted the work to be associated with men as well as with women.

Even today, as a recent writer in the *Scientific American* observed, cheesemaking is "more of an art than a technology." This adage was still more true of nineteenth-century cheesemaking. New Yorkers made hard cheese, intended for shipping a long way to market. In the nineteenth century, of course, they began with raw (i.e. unpasteurized) milk teeming with microorganisms from plants, soil, and animals. As soon as milk was drawn from the cow, bacteria and fungi in the milk began to grow, altering its chemical makeup—most importantly, by making the lactic acid so crucial to good flavor. Dairymaids first strained the milk into a container and then adjusted its temperature, sometimes by adding heated evening milk, sometimes by applying heat directly.[37] Though cheesemakers were ignorant of microorganisms, they knew they got better-flavored cheese if they started with milk warm from the cow—that is, at a temperature between 80 and 90 degrees Fahrenheit. Most dairywomen relied upon their sense of touch in determining the correct temperature.

Many cheesemakers, lacking enough cows to make cheese from a single milking, made "two-meal" cheese, combining the morning's milk with that of the previous evening. In this case it was important to cool the evening's milk, for if the fermentation process went too far, the cheese would spoil. Many had dairy houses cooled by spring water or ice. Others set pails of cold water inside the containers holding the evening milk. A "two-meal" cheese also required dealing with the cream that would inevitably rise overnight. Some people skimmed it for butter, but this impoverished the cheese; others tried to work it back into the milk by heating. To get around this problem, some made a curd in the evening and mixed it with the next morning's curd.[38]

Once the milk was at the right temperature, the dairymaid added rennet, covered the milk, and left it to curdle, which usually took about forty minutes or an hour. The amount of rennet added depended upon how the rennet had been prepared and stored, and on the quantity, richness, and temperature of the milk. These in turn depended to some extent upon the season, because the butterfat content of the milk varied with the cow's diet and with the stage of her lactation.[39] After a time the dairywoman would check to see if the curd was "set," a determination requiring much nicety of judgment; the curd should cleave smoothly when she ran her finger through it. Then she would cut the curd into small pieces in order to expel the whey. The curd would sink, leaving the whey floating above it. Then the curd was "cooked," or "scalded." Again, the time and temperature for cooking varied substantially with the individual maker and with day-to-day weather conditions. Cheesemakers cooked the curd at anywhere from 100 to 140 degrees Fahrenheit for as little as half an hour or as long as three hours; most raised the temperature slowly. When the cooking was done, the curd was supposed to "squeak between the teeth." Then the cheesemaker carefully drained off the whey, broke up the curd, and salted it. Salting, in ratios ranging from 1:25 to 1:50, was important both for flavor and for safety; a poorly cured cheese could poison its unfortunate eater.[40]

The process then moved to pressing, to expel more whey and to make the cheese firm and of a desired shape. Pressing took about twenty-four to forty-eight hours, during which time workers removed the cheese periodically, then returned it to the press and gradually increased the pressure. Then they finally removed the cheese from the press, trimmed it, and oiled it with whey butter or grease to keep flies out and to promote the formation of a rind. Then the curing began. In this phase the cheeses ripened and developed their characteristic flavor and texture, as enzymes and bacteria transformed the protein, fat, and carbohydrates. While curing, the cheese was frequently turned—men often did this heavy work—washed, and rubbed.[41]

Cheesemakers knew from long experience the results of these procedures. They nevertheless confronted a host of problems attributable to the complexity of the process. Even the most careful cheesemaker could not always prevent unwanted organisms or an incorrect guess from spoiling the cheese. Most of the problems did not become evident until after the cheese, innocent-appearing, had sat curing for a while. Some cheese would "heave" or puff up, resulting in a "blower." This, according to many, was a sign that too much rennet had been used. Cheeses would occasionally crack; people thought this happened because the cheese had been exposed to air currents. Insect pests laid their eggs on the cheese and in cracks in the rind. Numerous recipes appeared in the farm press for deterring mites and "cheese-flies," the most popular being a concoction containing a heavy dose of cayenne pepper. Sometimes cheesemakers found their cheeses leaking. A correspondent of the *Country Gentleman* thought he knew why: "The process of *pressing* is more important than many propose. *If you leave whey in your cheese, you may be sure it will find its way out,* and if in warm weather, you will have a worthless, stinking cheese." At the other extreme, if made from milk that was warmed too much or skimmed too much, the cheese would turn out hard and dry. Too little salt and heat could result in poor flavor. And there were cases when cheese turned out to be poisonous, making people sick and sometimes even killing them. Many theories were offered when this happened, and it is difficult to know in retrospect—indeed, as it is today—what made cheese harmful to consumers, but it is possible that the curing process went awry. Harmful bacteria (pathogens) are normally destroyed in curing, but if cheese is consumed before completely cured, these pathogens remain. Poor feeding and management of cattle also probably contributed by making undesirable organisms more abundant.[42]

Home cheesemaking in the nineteenth century, then, was an extremely exacting, touchy, and laborious process, demanding unremitting attention. E. H. Arr remarked that among Yankee farm people, cheesemaking was classed as "heavy labor" and thereby distinguished from spinning and weaving. Dairy manuals and journals often stressed that the dairywoman must rise at four A.M., and they sometimes outlined a timetable for the completion of various tasks during the cheesemaking day. The dairywoman had constantly to watch temperature and texture, and a huge variety of factors—the weather, cows' diet, the type and quality of rennet, and so on—could influence the process. Moreover, since the cheesemakers did not understand the role of enzymes and bacteria, they could never precisely control the outcome. One Herkimer County dairywoman summarized the work in 1859: "Thought and care are essential in all the various operations. Intense interest and anxiety are necessary *to do all these things well.*"[43]

A skillful cheesemaker could turn out a tasty, long-keeping, nutritious article. E. H. Arr remembered "such cheese as bulged your canvas sides" (she refers to a "cheese safe"), "prettily mottled with tansy or wholesome yarrow, and crumbling under the knife when cut. They had a toothsome way of dissolving in the mouth, and tickling the palate with a pleasant tingle." Others described good New York cheese as having a rich, mild flavor and a texture so buttery that the cheese melted in the mouth. Cheese was also, of course, nutritious; its advocates were fond of pointing out that a given amount of cheese contained twice as much "albuminous" matter—what we would call protein—as did beef, at a lower cost.[44]

A steady demand for well-made cheese provided incentive, and in keeping with the emphasis upon quality, women responded with a lively concern with improving the product, saving labor, and cutting costs. In the agricultural journals they vigorously debated the merits of various methods and tools, offered their opinions on proper milking technique, and recommended the best ways of handling milk and curd. Their often lengthy expositions evinced a keen sense for the business, as well as the art, of cheesemaking. One woman who made "over a tun" of cheese in 1860 calculated by experiment that cheese from unskimmed milk yielded more pounds of cheese per unit of milk, tasted better, lost less weight in curing, and fetched a better price than cheese made from milk partly skimmed.[45] Most of these women would probably have agreed with their colleague who insisted that the dairywoman's skill ranked high in importance to the farm's success. The peripatetic agricultural writer and reformer Solon Robinson was so impressed with Jefferson County's dairying families on a trip there in 1849 that he proclaimed, "The women and children here take more interest in agricultural improvement, and know more about it, than a majority of the men in some places."[46]

Early Tools

The basic process of cheesemaking would not change in its essentials until the century was nearly over. But the tools with which it was accomplished became more refined.

The tools of home cheesemaking are of interest for several reasons. They can offer concrete evidence of what it was like to practice a craft now utterly unknown to most people. Like the spinning wheel, flail, or scythe, they communicate lost knowledge. The vessels, presses, knives, and hoops of the nineteenth-century cheesemaker also tell a story about technological and cultural change. While cheese was still made at home, significant changes were introduced in several important elements of cheesemaking technology. New vats, curd cutters, and presses signified a greater reliance upon purchased equipment, and in turn a greater commitment to market production. They reduced physical toil. Though the new equipment was probably developed and adopted partly in response to the involvement of men in cheesemaking, it did not seem substantially to alter the household sexual division of labor; there are no signs that an increase in the proportion of male workers accompanied the adoption of this technology. Nor did new technology reduce the element of skill necessary to cheesemaking. It simply facilitated the work while preserving women's key role in production.

When the estate of Nicholas Gardiner, "late of Trenton in the county of Oneida," was inventoried in 1820, the recorders noted "15 earthen milk pans; 9 old tin pans and 1 skimmer; 1 tin pale [pail]; 1 tub; cheese ladder[,] basket[,] & hoops; 2 old cheese presses; 1 small copper kettle; 1 iron 4 pale kettle." With a few omissions, this list is a fair sampling of early-nineteenth-century cheesemaking equipment.[47]

E. H. Arr remembered such equipment in use in her childhood home in New England: "Twice a week, with much method and little bustle, quantities of butter and cheese were made ready for the market. The unctuous odor of these tasks comes back to me, and I still taste the all-pervading flavor of the cheese-room. I see the clumsy press, trickling with sour juices, the polished wooden bowls, the row of shining pans set out to scald in the sunshine, mistress and maid, in checked homespun aprons, shaping the golden butter or cutting the tender curd."[48]

Figure 4.2. In early home-based cheesemaking, women used unwieldy wooden tubs. *Progress of the Dairy,* 18, from the collections of the Henry Ford Museum and Greenfield Village.

Arr (Ellen Chapman Hobbs Rollins) waxed nostalgic about the old days, but the grinding everyday routine of early cheesemaking must have partaken little of the poetic. Cheesemakers had to curdle the milk using massive, unwieldy wooden tubs (figure 4.2). They heated milk or whey in a brass or iron kettle set over a fire. Then they poured the warmed liquid back into the tub. Lifting this heavy and awkward equipment was backbreaking work.[49] Early cheesemakers cut the curd with crude wooden knives, or "breakers." These implements cut a single, imprecise swath through the curd. When the whey was released and the curd sank to the bottom of the tub, the dairywoman dipped off the whey, put it into a kettle, and heated it. Then she returned it to the curd so that the hot whey would cook the curd. These operations again involved heavy lifting and pouring hot liquids. When the cooking had finished, the dairywoman further separated curds and whey by placing a cheese "ladder" across the tub or suspending in it a basket, sometimes called a "cinque" or "strainer." She put the curd into this strainer and let the whey drain back into the tub.[50] Finally, the curd went from the basket into a "hoop," a wooden mould lined with cloth, usually made by a local cooper. The hoop varied in size and shape according to the desired size and shape of the finished cheese.

As these descriptions suggest, working with early cheesemaking equipment was not only physically taxing but outright dangerous, as cheesemakers worked with scalding liquids in heavy containers. These dangers were vividly described in an 1843 broadside describing the gruesome death of fifteen-year-old Maria Hocrij, of Eaton, New York:

> As her folks were at Work in the dairy, one day,
> A scalding the curd, for the cheese, in the *whey*,
> They let down the *Kettle* by a *Windlass* or crank,
> Below the first floor, into a Caldron or tank

> The caldron was boiling, with Water, half full;
> They used it, sometimes, for their hogs and the fowl,
> To boil up their food and to fatten them well,
> Twas adjoining the place where those animals dwell
> The kettle being rais'd, she was steadying the same,
> When she slip'd, and into the caldron she came.
> Her father let go of the *Windlass* and *Crain*,
> To save his dear child from the scalding and pain

> Being in haste, he was careless, did not make them fast,
> And the kettle went down on this dear creature's breast
> Where it held her so fast, that two minutes or more,
> Elaps'd, before he his child could restore.
> In that liquid flame, What tortor she felt;
> Her cries would have made, e'en an adamant melt.
> Submerged in the boiling hot Water, she lay,
> Held down by the *Kettle* of hot scalding *Whey*

> Her face and her hands, they only escap'd
> This hot bath of fire, that she had to take;
> And she was so scalded that her flesh it gave way,
> In taking her out of the place where she lay.
> As they took off her clothes, the skin and the flesh
> Came off in large masses, *we* do here confess:
> Her blood turned inward, and so freely did flow;
> Out of the cavities made, it forced its way through.[51]

If possessed of the stomach to read further, a curious reader could learn in still more excruciating detail the physical and spiritual circumstances of Maria's lingering death. Here, of course, was another reason to exclude young children from the cheese dairy; but it is clear that anyone who worked over open fires, with boiling liquids, was in jeopardy.

The step after cooking—pressing—was less of a menace to bodily integrity but nonetheless an awkward and cumbersome procedure. Early presses were massive, crude contraptions, sometimes relying simply upon a stone for their power. In others a large screw supplied pressure; this could be tightened down upon a round, flat piece of wood (the "follower"), which fit into the hoop. "Beam" (or "lever") presses employed the principle of horizontal levers exerting vertical pressure via weights and fulcrums. Both lever and screw presses required physical strength, as well as continuous tinkering in order to ensure constant, even pressure. They also took up a considerable amount of space.

Changes in Cheesemaking Technology

On September 29, 1860, Maria Whitford went to her uncle Ezra's house to "learn how to make cheese in the *vat*," because she was going "to make cheese for Uncle Ezra's folks [i.e. his wife] while they are gone to the fair to Elmira." Maria Whitford's encounter with "the vat" typified the experience of many a home cheesemaker during the antebellum era. Cheese vats first appeared in the 1830s and gained popularity in the 1840s and 1850s. Local newspapers in the cheese dairying regions carried ads from local artisans willing to build vats. B. F. Usher of Boonville, a "practical Tin, Sheet-Iron, and Copper-Smith," advertised "Dairymen's Vats Cheese Hoops Cream Pails," "made to order at the lowest cash prices or short credit."[52] In central New York and Ohio several manufacturers began to produce specially designed patent vats. The Ohio firm of Roe and Co. advertised in the early 1860s that it had sold several thousand of its patent vats to home cheesemakers.[53] These vats replaced old-style kettles, wooden tubs, and in some cases strainers. The new devices consisted of rectangular containers, mounted on legs. The wooden jacket was lined with tin (or sometimes zinc), with a two-inch space between the liner and shell. Water (hot or cold) or steam could be introduced into this space. A removable stopper in the bottom of the unit allowed for whey drainage. Some vats could be tilted for draining whey and for cleaning.

The new vats facilitated home cheesemaking in several ways. Water could be let into the jacket with little lifting. The vats permitted evening milk to cool until morning; the cheesemaker need no longer "run a curd" in the evening.[54] Vats completely eliminated the necessity to pour steam, hot water, or whey from one open container into another. Moreover, the sealed firebox now

The Women's Work of Household Cheesemaking

Figure 4.3. Patent cheesemaking equipment. The wooden vat has metal liner, plug drain, and boiler below. The curd cutter and dipper processed curd more quickly. This apparatus made cheesemaking safer, easier, and more efficient. *Rural Affairs*, 1863, 261.

confined the hottest substances to a secure space, thus considerably reducing the dangers posed by open fires and by hanging, swinging, tilting containers. Steam or hot water also "cooked" the curd more evenly and thoroughly. The stopcock reduced the need for baskets and ladders.

There is ample evidence that cheesemakers quickly recognized the vat's advantages. One writer testified that his "Improved Dairy Apparatus" was designed to "lighten the labors of a much-loved mother," to avoid "clumsy, awkward, lifting, back-breaking, health-ruining tubs and kettles."[55] Mrs. M. J. Stephenson wrote the *American Agriculturist* in 1861 that "Roe's patent cheese vat is a great saving of labor. We had it last year for the first time, and now we would not be without it."[56] T. C. Peters and others observed dairywomen using cheese vats in the early 1860s, and all agreed that the new technology made the work easier.[57] Users also believed that the vat made possible a better quality of cheese, because with even heat and more precisely regulated temperatures, they obtained more curd from a given amount of milk.

New curd-cutting knives ("dairy knives") appeared at about the same time. These too varied in design but commonly consisted of several (usually four) parallel vertical blades, usually made of steel, set into a wooden handle. The tender curd demanded delicate treatment; if the curd was bruised or handled roughly in cutting, it would lose fats essential for flavor and texture. The steel knife cut the curd more smoothly than its cruder wooden predecessor had done. It also reduced labor because its multiple blades meant fewer cuts.

Patent presses were the third major innovation in cheesemaking technology in this era. It is not clear that these devices contributed to any improvement in quality; that press patents were more numerous than any others —they accounted for about three-quarters of patents relating to cheesemaking —suggests that no one had found a really superior answer to the problem of pressing. But the new presses at midcentury did seem to lessen the labor associated with pressing. This was primarily accomplished by exploiting the weight of the curd itself to exert the pressure; hence the term *self-acting* (figure 4.4).[58] These devices all allegedly made it possible to put a cheese into the press and leave it, instead of having to return frequently to tighten a screw or adjust a weight. But they still did not answer the need for the gradually increased pressure that most cheesemakers regarded as desirable. The "self-acting" presses therefore either eliminated only part of the work or resulted in a compromise on quality.

There is evidence that devices designed to allow a worker to turn a number of cheeses simultaneously were also being developed at this time. A notice in the 1835 *Albany Cultivator* publicized an invention of Henry Wilber of Richfield (Otsego County), who claimed that with his cheese turner a woman could "easily turn twenty-four heavy cheeses in a minute."[59] Years later a trial over an allegedly fraudulent attempt to claim patent rights for a turner revealed that improvised turners similar in principle had been used by cheesemakers, both male and female, throughout the Mohawk Valley during the 1840s and 1850s.[60]

All of these new technologies were actively used not only by men but also by women. Historians have recently raised provocative questions about relationships between women and technology. Generally, both industrialized and preindustrial or nonindustrial women's work, whether aided by technology or not, has historically been repetitive, interruptible, and manual. As capitalism consolidated, a complex process of industrialization took place in which housework acquired a more heavily gendered—and devalued—meaning. On twentieth-century farms in the Pacific Northwest, the tractor and

Figure 4.4. Kendall's patent press. Numerous patent presses required less frequent adjustment than older types. *Rural Affairs* 1846:246.

combine marginalized and further devalued women's work, because women were no longer needed to cook for huge harvest crews.[61]

The case of cheesemaking does not fit comfortably into these patterns, for several reasons. Cheesemaking was a skilled craft; indeed, its most distinctive feature perhaps was that it departed from the usual character of women's work. It was repetitive in its essentials, but daily variations inevitably occurred. It was certainly not interruptible, and it involved skill in a number of different operations rather than quickness in one.[62]

Cheesemaking technology did not affect women in the same way as later farm technology did, since the new tools indisputably saved women's labor but did not marginalize or displace women. These tools were acquired at about the same time as tools that eased men's work. Again the pattern differed from that in the twentieth-century Palouse of Oregon, where men acquired equipment first, thus reducing the need to feed harvest hands and making it harder for women to justify the acquisition of household technology. Why these differences? The different impact of farm technology on women in the nineteenth and twentieth centuries was rooted in economic and cultural circumstances peculiar to the times and places in question. The household structure of cheese dairying, combined with contemporary observations about women's domination over processing, provided the context in which dairying technology would serve rather than displace women. Sons were leaving to take paid employment or to farm elsewhere, while wives and daughters did the work.

Further, any available technology would have to accommodate women, because women possessed skill and knowledge not easily transferable. Scientific knowledge of cheesemaking was so rudimentary that everyday personal experience occupied a crucially important place in producing good-quality cheese. *Harper's New Monthly Magazine* characterized home cheesemaking as a "skill that was vested in intuition; it was the maiden's dower, the matron's pride." A cheese dairying farmer who had started out in northern New York in the 1830s attributed his success to his luck "in getting for a partner, a dairyman's daughter. The good, practical lessons learned from her mother, proved to be of much value." Finally, unlike reaping—a mechanical operation that could be performed by machine—no single cheesemaking technology available could substitute for this arduously acquired knowledge. To put it another way, at this point in time the skill was not amenable to either technological or scientific solutions, and since it was organized on a household basis, women could retain a significant measure of control.[63]

Buildings

A specialized architecture accommodated the household production of cheese, shaping and reflecting the nature of the work. Lavinia Mary Johnson remembered that the Dean family had built a "plank house . . . with a large pantry and milkroom on the east." Later they "had another kitchen built south and joined by a covered stoop and a cheese room by the kitchen."[64] The cheese room was placed near the kitchen, so that Mrs. Dean's various chores were spatially grouped together. E. H. Arr remembered that in New England "mistress and maid" made cheese in a "cheese room" near the kitchen. The *Ohio Cultivator* in 1851 visited Mr. Jacob Perkins's four-hundred-acre farm; the new house, the reporter said ambivalently, "does not indeed come within the five Grecian orders of architecture; but then it is picturesque, roomy, admirably ventilated, and perfectly convenient. The cheese room, with its apparatus, and presses, in a basement room, with stone walls and tile floor, on a level with the grounds around, was cool, and looked as though it would be so all summer." The article continued: "Under the wide projecting eaves of this house, an open space is left for the admission of air, for the ventilation of the upper western rooms, intended to be occupied for the cure and storing of the cheese. The whole arrangements of the house look to utility, and not show, and must be seen to be appreciated, in these days of dwarf cottages, and low and little rooms."[65]

These and many other documentary references show that in nineteenth-century dairy country, a landscape without cheese houses was as inconceivable as one without silos would be today. Yet for the most part these structures have disappeared without a trace. Just a few meager remains have survived the transformations of the past century. One of these is J. R. Hopson's cheese house in Herkimer County, New York. Hopson, a prominent local cheese producer, made substantial quantities in the middle third of the century. He built a one-and-one-half-story wing onto his house, constructed of rough board-and-batten siding on the exterior with lath and plaster inside. Only two windows, one on one side and one on the gable end, lighted the main story; a single gable-end window lighted the loft, which was probably used for storing cheese. A door in the south side provided the only entrance. Into this dark, low-ceilinged interior Hopson probably crammed several large tubs, plus presses, possibly later a vat, milk cans, and strainers. (We know that this is likely because the equipment, now in the Frisbie House Museum of the Salisbury Historical Society, Salisbury Center, New York, was only recently removed from this room.)

In West Winfield (Herkimer County) a more elaborate cheese house, also built early in the nineteenth century, still stands. This freestanding two-story building, 50 by 22 feet, is oriented perpendicular to both farmhouse and barn. The first story is constructed of thick cobblestone, while the second, balloon-framed level has wooden construction. Each floor was divided into two rooms, and each floor had a stove. Martin and Rhoda McKoon, its builders, had cleared their farm in 1796. They and their fifteen children made cheese early in the nineteenth century, and as the generations succeeded one another the farm developed into an extensive dairying operation with many cows and hired hands. Like the Hopsons, they made cheese on the ground floor and stored it on the second.[66]

Neither of these buildings possesses an intact interior, so it is necessary to turn to published plans in order to get an idea of how these spaces worked. Gurdon Evans's plan (figure 4.5), published in 1851, also gives insight into how cheesemakers rationalized the process. His building, like the two standing examples, had two levels—a cellar and an upper level over that. He recommended that the cellar should be

> settled about three or four feet below the surface, provided with a drain, emptying, if possible, into the slop tub in the cow house. The wall ... should be of stone or brick, and from eighteen to twenty-four inches thick; the bottom is best made of water lime, which will ... soon harden into a perfectly level and smooth surface, quite imperishable, and easily kept clean. The side and top walls of both stories, should be finished with plaster, by which means a uniform temperature, indispensable to curing cheese properly, is more easily secured.[67]

In the plan, at *1* was the room for making cheese (measuring 14 by 18 feet). At *2* was a storage closet for pans, pails, and the like. A carefully insulated icehouse (*3*) stored milk. In the "kitchen and general store room, 11 by 18" (*4*), a stove (*5*) generated steam for "warming milk, heating water, etc." Milk was warmed and curdled in a "tin cistern" (*7*) "surrounded by a wooden vat," where water hot or cold was introduced at *a* through a lead pipe. Other lead pipes (*c, d*) variously drained water out, channeling it into a cleaning tub and out of the building. The cheese was pressed at *8,* and then turned onto *9,* a wheeled table for finishing. This rolled the pressed cheese to *10,* an "elevator raised by cords, pulleys, and weights" to lift the cheese to the storage loft. Evans stressed that the wheeled dolly eliminated the need to lift the cheese "by main strength," a provision probably conceived with women workers in mind. Nonetheless, the heavy cheeses, often weighing seventy or more pounds, still had to be turned by hand. Workers could clean implements at *12* and drain slops out at *14;* they could pump water from outside at *11.* Above, the cheese arrived via the dumbwaiter, and workers again moved it on a dolly to the second-floor shelves. The cheese tenders would ascend a stairway (*6*) to check and turn.

What conclusions may we draw about these various spaces for cheesemaking? As far as mundane conditions of work are concerned, all of the plans and descriptions suggest a rather dim, possibly close environment, made so

The Women's Work of Household Cheesemaking

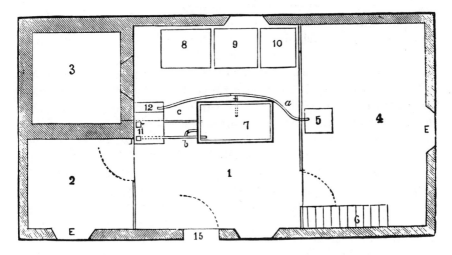

Figure 4.5. Plan for a cheese house. Cheese houses were planned for economy of motion and labor. *1*, make-room; *2*, storage closet; *3*, icehouse; *4*, kitchen; *5*, stove; *6*, stairway; *7*, vat; *8*, presses; *9*, finishing table; *10*, elevator; *11*, pump; *12*, cleaning table; *a–c*, pipes. Gurdon Evans, *The Dairyman's Manual* (Utica, 1851), 75–76.

by a basement location, low ceilings, and (often) a paucity of windows. In the crowded "make-room" itself, the heat would create a warm, humid environment.

Although men as well as women frequented the cheese houses, they were predominantly women's spaces. Mrs. L. B. Russell of Michigan indicated much about perceptions of the territory when she showed a *Michigan Farmer* correspondent "her" cheese house.[68] These areas made up a middle ground between the barn—increasingly dominated by men—and the house. In this sense cheese houses accurately mirrored the transitional nature of cheesemaking, from a domestic task involving traditional skills in food preparation to market production on a substantial scale. Their spatial organization underlined the emerging commercial character of the work: people organized cheese houses so as to make use of labor-saving technology and efficient layout, and thereby to make the most systematic possible use of labor.

In outward appearance the buildings blended into the vernacular complex of farm outbuildings including corncribs, hog houses, smokehouses, poultry houses, and icehouses. Like these others, the cheese house had its distinguishing features, most notably the slender stovepipe protruding from the roof;[69] but its fenestration, construction, and scale fit in with the building context. So in this regard as well, the cheese house simultaneously represented both the older and the emergent modern worlds of agriculture.

The spatial world cheesemaking families constructed was often compared with that of a factory. The *Michigan Farmer* used telling terminology in describing home cheesemaking: the cheese dairy's "operatives have the organs indicating order, neatness, and activity well-developed."[70] Significantly, the author chose a term that also applied to contemporary factory workers and described virtues associated with industrial discipline. Lewis Falley Allen used still more explicit language in the section on dairy buildings in his *Rural Architecture:* "The dairy is as much a *manufactory* as a cotton mill, and requires as much convenience in its own peculiar line."[71]

Yet in other ways this label was inappropriate. Though farm-based cheesemakers sold to national and world markets, their enterprise bore stronger resemblances to the traditions—already in decline in the nation's large cities—of craftwork. Farm families owned the land and their tools; they commanded specialized skills and controlled the work process from start to finish. No division and devaluation of labor was occurring here. Work rhythms too had much in common with preindustrial patterns. Daily weather variations made it difficult to impose any but the loosest "system." Sharp seasonal contrasts characterized the work; dairying families enjoyed comparative leisure during the slack winter months when the cows were dry, the fields snow-covered. The complexity of the system becomes apparent when it is compared with other, contemporary ways in which production was organized.

The Household Economy of Cheesemaking in Context

In the cheesemaking household we see a carefully tuned integration of agricultural and social systems. How did the household fit into the broader context of its day? As a setting in which women and their work commanded a central place, it has a great deal of interest and offers highly suggestive interpretive evidence concerning contemporary historiographical issues. John Mack Faragher has called for a history of rural women that examines them in the context of "family and kin relations and of gender." Faragher sets out a provocative, sweeping outline of change in rural America as it was shaped by, and affected, women. He argues that "the tendency of men to reject traditional culture in their struggle for a more commercial world created a gender dimension to the conflict between traditional and popular culture. 'Backward' farming was also the farming culture in which women's work was fully integrated as an essential part; 'progressive' farming might eliminate the old notions of 'reciprocity' without even a bow in the direction of women's roles in work and life."[72]

This paradigm fits well with women's experience in the Illinois frontier community of Sugar Creek that Faragher detailed in a compelling portrait. But it does not apply to dairying women, whether in central New York, New England, or the newly developing Western Reserve. Whether or not intensification in dairying was a choice in which women participated, cheese dairying in midcentury New York State of necessity fully integrated a tradition of "reciprocity" into a commercial production system. This suggests that region and crop specialty were important determinants of the economic organization of farm households and of women's place in that organization.

Men and women on antebellum cheese dairying farms fashioned gender relationships more closely resembling those of Nanticoke Valley (New York) dairy farms later in the century. Nancy Grey Osterud has found that these relationships stressed "mutuality." Though the exact division of labor varied with individual couples' preferences, men and women worked together. Couples reached a consensus on such issues as family limitation. Men and women shared "mixed modes of sociability" in work, leisure, religion, and community organizations. At the same time, women shared work among themselves. Neither the market's standards of value (associated mostly with men) nor the value placed on a "living" (associated more with women) dominated. Mutuality could empower rural women.[73] The demanding nature of cheesemaking work seems to have precluded extensive involvement in religious and community organizations, but otherwise the pattern of work and the hybrid nature of antebellum cheesemaking farms anticipated the characteristics of later dairy farms.

The nature of home cheesemaking comes into focus still more clearly if contrasted with some nonagricultural pursuits in their preindustrial stages. Shoemaking is an example. In preindustrial shoemaking, shoes were produced in the household and a gender division of labor prevailed. Men did the skilled work of attaching uppers to soles. Women sewed the uppers in a process called "binding," performing less skilled work that derived from the traditional association of women with needlework. Since they had to keep the men supplied with sewn uppers, shoe binders were unable to exert a great degree of control over their work, but they could integrate it with child care.[74]

The comparison with household shoemaking points up an important feature of cheesemaking: women did work that required at least as much skill as men's work, and probably more. That women were not displaced by new technology supports this observation. In this regard cheesemaking women were able to control their work to a much greater degree than women in other industries. But though women's work in cheesemaking was not predicated on men's work, its intrinsic demands placed limits on cheesemakers' latitude. The manifest conflict between cheesemaking and child care is a case in point.

Home cheesemaking developed about the same time as the New England putting-out system for making palm-leaf hats. In this cottage industry, also dominated by women, most households producing palm-leaf hats were poorer than the average farm household, and palm-leaf hatmaking enabled daughters to stay home rather than migrate to seek employment in industrial towns. But at the same time, outwork threatened the traditional semisubsistence and barter economy.[75]

There are some parallels and some contrasts between hatmaking and cheesemaking. As with shoemaking, raw materials for hatmaking were supplied not by the farm family but by coordinators outside the family economy, and hatmakers worked for wages. In this respect dairying men and women alike had more control over their work than did participants in the outwork system. Perhaps even more glaringly than hatmaking, cheesemaking embodied contradictions. It was (generally) larger in scale and took up more of the workers' time and energy than hatmaking did for New Englanders. It involved traditional women's skills, but the scale of the endeavor and the fact that it was exclusively for market meant that these traditional skills were employed in a context industrial in its organization, goals, tools, and space. Cheesemaking households were more prosperous than hatmaking households, and as we shall see later, this circumstance would be crucial as the family debated the ends to which its resources were put, and crucial in undermining the hybrid system.

In northern cities the artisan tradition supplied a base for resistance to capitalism; in the South the yeoman tradition in the upcountry, and the "commodity cultures" of the staple regions, served a similar function.[76] But among northern cheese dairying families, the household system did not function in this way. If anything, cheesemaking families often aligned themselves with the emergent quasi-industrial order rather than resisting it.

The primary reason for this difference was that the household system was riven along lines drawn by gender and generation. Taken together, household structure, work organization, and material culture contributed to a distinctive household dynamic that developed within these productive entities, which were at once families and manufacturing concerns. Most notably, the combination of the household structure—dominated by young adults in the status of children and wage workers—with intensive, large-scale production suggests that family and work relationships were intimately tied together, in a dynamic that combined the values of a "competency" with pursuit of profit, family emotional ties with market activity. It also involved balancing authority with autonomy—the needs and aspirations of parents and children, men and women, employers and employees. These elements have always been present to one degree or another in households, but because farm goals were multiple and because every household member's contribution was critical, nineteenth-century cheese dairying presented a particularly volatile mix.

5

The Social Dynamics of Household Dairying
Forces for Change

> If not to the rising fair generation, to whom shall we look for the hands that are to supply so important a portion of subsistence as the products of the dairy?—*Prairie Farmer,* 1841

Even as cheese dairying families established their tightly regulated, cooperative family economies, dynamic forces were developing that would undermine them. Spreading transportation and communication networks, economic diversification, and expanding educational facilities brought increased opportunities to people all over the rural North—opportunities to switch or even invent occupations, to acquire advanced education, to trade in goods from an ever wider variety. Rural people responded to these trends in a variety of ways. Within some families deep splits appeared, exposing the inequalities between the sexes and generations. In other cases family members together made concerted efforts to steer sons and daughters toward the new opportunities. However manifested within the household, the overall effect was to create a situation in which the household economy was increasingly untenable, and to ready the countryside for dairy centralization.

The Context: Cheesemaking Households as Centers of Negotiation

As we have seen, economic development—in our region, epitomized by the Erie Canal—stimulated the rise of dairying itself. It also brought wide-ranging changes to the region. These are well known. The relevant point for our purposes is that the canal helped promote phenomenally rapid urban growth and equally rapid expansion in trade, manufacturing, banking, and other occupations. The construction of the canal, and later its maintenance, opened up thousands of new jobs, directly and indirectly. Across the canal came new consumer goods and national publications. Simultaneously, the background of New England culture influenced the emergence of a vigorous educational system encompassing both public and private institutions. Farm families' participation in shaping this environment, and their responses to it, ultimately would prove a source of change.

To this point I have portrayed the household economy of cheese dairying as a smoothly functioning cooperative enterprise. Indeed, in many cases cheesemaking households functioned well, apparently affording to all participants a sense of satisfying, remunerative, common endeavor. Percy Wells Bidwell remarked of the "self-sufficient economy" that "it harnessed together in the productive process all the members of the family[;] . . . it concentrated attention upon the interests of the family group rather than upon the interests of its individual members."[1] The same could be said of the cheesemaking economy. Certainly there is ample evidence that men and women in dairying were positively engaged in their craft, actively seeking improvements, exchanging information, creatively experimenting, and sharing common goals.

Yet the very fact that individual "interests" were subordinated to family goals could cause problems. Gender and generational inequities increased the possibility that an individual's interests would diverge from the family's. It was precisely because sons and daughters were critical to the farm economy that it was so difficult to resolve conflicts between a family's economic goals—immediate and long-term—and children's aspirations for individual accomplishments, social life, and learning beyond the farm's confines. Like Connecticut Valley households earlier in the century, these cheesemaking households frequently became centers of negotiation, and in turn the outcomes propelled them into change.[2]

Quite possibly, the more a given farm was committed to large-scale cheese production, and the more critical close cooperation became, the more the physical and psychological pressures of work increased. In an environment where so much depended upon smooth-functioning personal relationships among family members, any tension could be magnified. Conflicts within families were potentially the most explosive disruptions to home cheesemaking. Disputes between sisters and brothers, children and parents, wives and husbands, highlighted serious differences about the goals of family dairying and divergences in interest based upon gender or generation.

These differences dramatically revealed themselves in passionate debates on three separate but related questions that arose in the dairying districts. One was the issue of whether women "ought to help with the milking." The second was whether cheesemaking was unacceptably debilitating labor harmful to women's health. The third was whether boarding-school education was appropriate for farmers' children, especially for girls. All three represented

Transforming Rural Life

disagreements within dairying families, and within the farming community, over the division of labor and the goals of the family farm.

The Division of Labor in Milking

The customary family economy of dairying was based partly on voluntary cooperation, to be sure, but there was also a strong element of patriarchal authority involved, as the Americanized image of the dairymaid as daughter suggested. E. H. Arr recalled that "extreme deference was exacted from children to parents, and from youth to old age." An 1838 poem in the *New England Farmer* expressed this well:

> *A farmer's daughter,* bright and fair;
> A faithful daughter too,
> As both her daily industry
> And her obedience show.
>
> She's versed in all the housewife gear,
> The labor and the care;
> And in domestic management
> She's sure to take her share.
>
> She loves to help her mother kind,
> 'Tis daily her employ;
> And why, to learn this handiwork,
> Should *any* girl be coy?
>
> She washes, and she irons too,
> She knits her father's hose;
> She cooks good rolls and johnny-cakes;
> And mends her children's clothes.
>
> She rises too at twilight dawn,
> And milks the bonny cows;
> She's ne'er afraid to give the kine
> Some hay from off the mows.
>
> To make good butter and good cheese
> Is her familiar work;
> Nor does she scruple e'er to bake
> A pot of beans and pork.[3]

Feminist scholars have criticized the assumption that the family can be studied as a single undifferentiated unit. They argue that families are loci of conflict and negotiation. This assumption gives rise to contentions that the idea of the "family farm" is inherently exploitative, or that patriarchal farm family structures in colonial New England denied women any claim to the products of their labor: gender interdependence was reciprocal, but not symmetrical.[4]

Household cheesemaking presents an intriguing variation on these points. It originated as a substitute for home textile manufacture, and thus had close ties to earlier patriarchal forms. Yet at the same time, the gendered organization of work assigned to women the most skilled labor, reversing the customary skills hierarchy in the artisan workplace. Thus "asymmetry" in the case of household cheesemaking lay not in the value assigned to the work but in an inequitable distribution of the *amount* of work. These circumstances led to tension but also potentially provided a means for women to mount an effective protest. By the 1840s and 1850s the ideal of unquestioning obedience was being seriously challenged in dairying families and communities.[5] Farmers' wives and daughters especially articulated their uneasiness that they had borne too great a share of the farm's burdens.

"Is It Right to Ask the Women Folks to Milk the Cows during the Busy Season?" The question, posed in 1857 by the *Genesee Farmer,* provoked such a voluminous response that the exchange extended over months and the editors could not print all of the letters received. One of the opening salvos captured vividly some issues at stake.

"E.A.B." of Oxford, Chenango County, contended that it was "undoubtedly . . . reasonable" to make such a request; but, he warned his male readers, "it is not safe to do it," even "ever so meekly," because such a misguided petitioner "will have a busy season, and a very warm one too, right away." "My wife," he continued, "is quite prepared to show the impropriety of the whole thing, and wishes to write out an argument at length, but is prevented from doing so, because she is confined to a single page in the *Genesee Farmer,* and once she begins to talk it up, she wont get through till after the 'cows have come,' and it would not be decided who would milk them." This correspondent adopted a humorous posture, but beneath the joking ran a bitter current of sarcasm. His resentment rose nearer to the surface as he went on. He described the rigors for men of the "busy season" and alleged that while he toiled in the field, "his wife [was] busy with company, his girls enjoying themselves about the house, and not one of them so much as giving so interesting a subject [milking] a single thought, and every one of them afraid of their shadows, made in the milking yard, should they by accident get there." (Here he was copying verbatim from an oft-reprinted lament that turned up frequently in the farm journals.) He concluded with a story about his brindle cow, which would only be milked by women but was neglected when "sewing, quilting, missionary and many other societies" claimed his daughters' attention. He reacted bitterly to their defiance; clearly they exercised a substantial degree of power within the household setting.[6]

"E.A.B." set the tone for the debate in the *Genesee Farmer*'s pages that year. While most of the printed responses (including those from women) replied in the affirmative (the editor maintained that "nearly all" of the "ladies" had approved of milking),[7] still there was deep disagreement, and both sides invoked potent arguments. Their letters reveal some of the emotional and social dynamics at work within dairying families.

Many of the women's letters voiced support for their menfolk. "E.N." wrote, "I think it is the duty of a farmer's wife . . . to see that the cows are milked. . . . Woman was intended to be a helpmeet for man, and I do not know in what *better* way a farmer's wife can assist him, than by taking charge of the dairy. Farmers have so many things to attend to, that I think they can very well dispense with milking the cows." A "Farmer's Wife" from Jamestown, New York, echoed this note of sympathy: "What farmer's wife, who has an interest in her husband's affairs, has not witnessed how perplexing and worrisome his business is at many times?" She cast milking as "a little kindness in doing a chore here and there," a sign of a daughter's love or a wife's devotion. Another respondent, "Cousin Alma," contrasted the work of milking with domestic pursuits she implicitly cast as frivolous: "In a farmer's family there should be no drones; and if there is more work out of doors than in, women should not object to milking cows, or otherwise lending a helping hand to whatever they can do. It only exercises the same muscles that are required to make lemon custards, or whipping Italian cream." But "Mrs. M.C.L." hedged around her *yes* with qualifications: "If favored with health, when man is actively engaged in their common good, woman should be willing to attend to her appropriate duties. If the barnyard is properly cleaned, it will neither injure her dress nor person in the least. It is no more beneath her dignity to milk, than to do the work afterward attending upon it, for surely it is not more laborious."[8]

It is worth paying some attention to the use of language in these opinions. The first two, for example, place men's work in a higher priority than women's and cast women's milking as a duty or obligation, in keeping with patriarchal authority patterns. The notion of a "drone," a parasite, was a scathing condemnation of family members who would, in this view, not contribute to the well-being of the whole; the borrowing of this term from natural-history lore is suggestive, for the implicit comparison is to the beehive, where individuals were subordinated and where all work was aimed toward a common goal. The last respondent located objections to milking in the low status of the work, and answered those objections in her reply, scoring a telling point by noting that processing the milk was as physically taxing as the milking itself.

The Social Dynamics of Household Dairying

Some advocates extended their arguments beyond the practical to analyze the implications of excluding women from the milking. A correspondent from Jefferson County believed that "right-minded, industrious women" rejected the "false notion . . . that women's appropriate labor is strictly confined to the inside of the farm house." This particular writer was a man; he maintained that in Jefferson County the women "help milk, without being asked, and consider it a shame to those women who refuse to maintain their right to share in the labors and toil, the joys and sorrows of their husbands, brothers, and lovers." Another, unfortunately anonymous, contributor retorted to a writer who had maintained that milking was outside woman's sphere: "I wonder if his heart exulted over the thought that he could *cage* his wife, and that although she might hop from base to dome to will, if she opened the *outer door* for any *useful* purpose she was *out of her sphere?*"[9]

The arguments in favor of women milking shared several characteristics. They commonly either ignored or explicitly rejected the association of either "masculine" or "feminine" qualities with the actual work of milking. Second, with one exception they implicitly compared household labor with the work of milking; they treated "indoor" and "outdoor" work as equivalent, all contributing to the maintenance of the farm, rather than contrasting them as qualitatively different. At the same time, however, they often implied that men's work was more important than women's, thus reflecting the continuing ambiguity accompanying attempts to define the value of work within a hybrid system suspended between precapitalist and fully capitalist economies.[10]

Finally, they blended practicality with sentiment in a powerful mix that suggests a dynamic of emotional obligation within the household labor system. "Mrs. M.L.B." expressed this succinctly: "LADIES, milk your own cows. It will improve your strength, increase your cash, improve your complexion, remove your pride, strengthen your digestion, and hopefully *relieve your consciences*" (emphasis added).[11] The Jefferson County farmer's appeal to "joys and sorrows" and the "Farmer's Wife's" appeal for a "little kindness" indicate the same dynamic at work. The ties between emotional and economic life, then, could promote or impede the functioning of the family farm; conflict could be especially intense because it directly affected not only the farm's future but also intimate emotional relationships among people.

The discussions over milking possessed a generational dimension. The older generation was consistently portrayed as favoring women milking. They often invoked the example of biblical women or of Puritan grandmothers, often in conjunction with one another: "Wonder if Mr. W. does not think that the maidens of the olden time, who drew water for the camels, as REBEKAH, or gleaned in the field, as RUTH, were rather low lived?" "Our Puritan grandmothers were worthy help-meets of the men. . . . They were dairymaids and cooks, as well as friends and sweet-hearts, in the good old primitive times." That younger farm women began to distance themselves from some farm work was suggested by Lavinia Mary Johnson (1801–91), who remembered that in the Dean family the patriarch's wife was "always at work" but "her son's wife was not quite so industrious."[12]

On the other side of the debate, objections to women's milking reveal intrafamily conflicts in their emotional, practical, and ideological dimensions. Among the 1857 responses to the question about women milking, those opposed doubted that a woman could milk "without neglecting her household work." One farm wife argued that once a woman agreed to milk, she would find herself doing it year round, "for industrious farmers find very little cessation in the labors of the farm." Further, she argued bitterly, "Admitting, for argument's sake, that there are 'busy seasons' for out-door workers, is it not correspondingly so in-doors? Most certainly it is. Then, of course, there is no more reason for women milking one season of the year than another." Her reasoning indicates the extremes to which an ad hoc approach to milking might be taken—in this woman's view, to an unfair and unequal division of labor. Another woman writing in the *Ohio Cultivator* for 1860 pursued more explicitly the idea that farmers' wives who milked endured severe exploitation by their husbands: "There is a class of farmers that would have a woman do all her housework, and milk seven or eight cows, tend the milk, and do the washing, and every thing."[13]

In contrast to those favoring women milking, objections often invoked the doctrine of spheres. One maintained that "women are out of their sphere with wild cows and other unruly cattle, and the [barn's] nasty and unpleasant covering." Another sketched a scenario of a woman milking: "She goes out, with a fluttering heart, among a drove of cows, and instantly prompts an example of perfect confusion. Here is proof that cows show by their actions that it is no place for a woman." This letter writer also held that women were physically too weak to milk, and that it would be equally "ridiculous" to ask men to help in the household busy season.[14]

An 1841 "Dialogue" in the *New England Farmer* captured in more detail the currents of thinking that reveal potential sources of strain within dairying families. Four young women persuade a young man, brother to one of them, to milk the cows so that the girls may take an evening stroll. When he first hears their request, "Charles" asks "Harriet," "Would you have us, after working hard in the field all day, milk at night?" Without hesitation Harriet replies, "Yes I would. I do think you ought not to work so hard that you cannot save time and strength to milk." A perplexed Charles responds, "Somehow I always thought it one of those things girls perform better than we, so that it is best for them to do it." This argument receives another swift retort: "Plausible reasoning, surely—telling us we do it *better*, will never make it easier for the poor *wrists*, or more within our sphere of duties." Her female companion joins in: "I always think those that say so, have an inveterate hatred to milking themselves." Before Charles has time to gather his wits, the young women pose another argument: "If we have to take care of the milk after it comes into the house, that is our full share of the work. Only think how much there is to be done to it[:] . . . to strain, skim, . . . do up for market, then the host of pans and pails to wash." Charles commits another blunder by wondering, "What else can you find to do? You find much more time for walking, visiting, etc., than we do." This is received with a sputter: "Idle creatures aint we! If so, why so great the want for female labor? . . . We take needlework with us when we visit; but when gentlemen visit, they leave their work behind." At the end of the dialogue it emerges that the young women have an ulterior object in view: they hope to persuade the young man to spare his future wife, who is another of their companions.[15]

In this exchange the young women vigorously challenge the idea that women are better milkers than men, exposing its potential for emotional manipulation and attacking it as a mere expedient. In its place they themselves employ a gender stereotype and the same kind of emotional appeal, arguing that women are weak and that milking is out of their sphere. Thus two opposite gender stereotypes become vehicles for expressing family arguments between men and women over the division of labor.

Yet at the same time, in the dialogue the young women express their discontent less with outdoor labor per se than with what they regard as an unevenly distributed burden of labor; one of them remarks, "If I had outdoor work to do, I should prefer raking, hoeing, weeding, or many other kinds of work, to milking."[16] They also decry men's inability to perceive the quantity and difficulty of women's indoor work—a protest, however implicit, at the devaluation of women's work. They seek not an exemption from labor but rather a fair sharing. In this sense they do not explicitly challenge the cooperative ethos but in fact invoke "mutuality" in order to adjust the balance.

But still others rejected milking not because it was excessive extra work, but because they regarded it as dirty and unrefined—in short, as unladylike. These objections brought forth especially sharp differences. The *Albany Cultivator* sharply criticized these young women, but in the process exposed emerging new values: "Half the young girls nowadays hardly know, at least they pretend it would be immodest and not at all ladylike to be presumed to know, whether the milk comes from the udder or the horns." The writer contrasted the daintily dressed belle, with "satin shoes, gilt hair-comb, and insect waist," with her predecessor, dressed in thick shoes and kerchief, and charged that the belle "would not recover from her fright for a week" if she glimpsed herself, in a mirror, so coarsely clad.[17]

A less critical perspective, presented from the viewpoint of the young women involved, appeared in a curious piece of doggerel reprinted in several farm journals around 1850. Signed with the pen name "Angelina Abigail," the poem was entitled "Tete-a-Tete of the Milkmaids." Its verses began conventionally enough, invoking romantic, idyllic rural imagery of the sunset, dew, and balmy breeze, in the style of popular British pastoral poetry. But these images were set in contrast with the interruptions of impatient cows. The girls hurry to chase down the wandering bovines, beating back their kicks and fighting off wet, swinging tails. One exclaims, "I fain would wander, Sally / To some green and Qui[e]t valley / Minus horned cattle." They conclude with a critical reflection on the intrusion of commercial values and pressures into their work: "Mortals now get milk and honey / Only by hard work and money!"[18] The form in which its author chose to present her complaint was particularly effective given its ironic possibilities; the poem suggests a self-conscious parody of the saccharine romantic rural poetry that appeared so regularly in the agricultural press. By caustically contrasting the bucolic idyll with the reality of hurried, difficult, dirty, unpleasant work, the author forcefully defended the position of young women who disliked the messy reality of milking.

Women's Overwork in Cheesemaking

A more general issue—that of overwork in cheesemaking—also received attention in the agricultural press. As we have seen, making cheese on a large scale was very hard work. Most women had to combine it with myriad other tasks. Some denunciations of overwork in cheesemaking echoed a theme of the milking debate: the unfair division of labor. The *Rome Sentinel* in 1851, for example, argued that butter and cheese production "made fresh demands upon the worthy housekeepers and their industrious daughters. The wheel and loom were relinquished, but the milk pail, the cheese tub and the butter ladle took their place and we could not help thinking they had made a poor exchange, as far as labor was concerned. They have contributed largely to the growing wealth of the farmers of this entire region, and have done their full share . . . towards placing the farmers of this and other dairying states, in comparative comfort and independence." The editorial condemned the "drudgery" "imposed" upon women, arguing that "farmers' wives may be ladies as well as others, and often better deserve the appellation."[19]

The Social Dynamics of Household Dairying

An 1861 commentator in *Moore's Rural New Yorker* reiterated the belief that a serious inequity existed in the sexual division of labor in cheese dairying. T. C. Peters, a well-known writer and agricultural reformer, after visiting farms in New York State, concluded, "I am the more inclined to the opinion [that the biblical Eve made cheese] from the fact that Adam saw a good chance to shirk the hard work. Even to the present day, by far the largest portion of the work is done by Eve's daughters." An Oswego County dairyman maintained that dairying relieved men of the "laborious tillage"; while grass farms needed "less of our own labor," they increased the amount of "female labor so indispensable to the neat dairy." Mrs. Eloise Abbott, who grew up in Jefferson County, New York, remembered that her mother was always late to bed and early to rise, but that the men enjoyed leisure time on rainy days, during the evenings, and on the Sabbath. Historians too have noted that pastoral farming involved less labor than arable farming, though they have not fully realized that this reduction applied mainly to men's work, not women's. Indeed, as in frontier Sugar Creek and turn-of-the-century Nebraska, women's work appears to have given men time to participate in public activities; local newspapers are filled with accounts of political ferment as the antislavery movement and later the Republican party gathered force.[20]

Those concerned about overwork also appealed to fears about women's health. Mrs. M. B. Bateham, editor of the women's page of the *Ohio Cultivator*, made a tour of the dairying regions for three months in 1848. She was particularly critical of cheesemaking as a drain upon women's health and energy: "As a general rule," she wrote, "cheese making and female labor are combined. Indeed the condition for women in dairies is little better than servitude, and in too many instances this is the lot of the mistress of the family. All of our readers know that we are an advocate of female industry, [but] giving a moderate share to all, rather than overtaxing a part. . . . The husband soon adopts the fashion of calling her old woman . . . a title not very flattering to female vanity." That some farm women agreed is evident in the diary of an anonymous Oneida County widow, who complained in·1861, "My natural protectors have been taken from me, they who should have sheltered me from the necessity of such heavy manual labor."[21] Maria Whitford of Almond, New York, kept a diary in 1859 that shows how wearing the continuous round could be: "I milked and made a curd to fix for cheese got along very well only was about sick. . . . Did not get up til half past seven. Sam'l got the potatoes on to cook. I made cheese and churned the cream, ironed the clothes."[22]

A "prize article" on dairying in the *American Agriculturist* for 1859 invoked the notion of women's weakness and physical vulnerability to argue for women's exclusion from the cheese dairy: "A man is better than a woman for a large dairy, for these reasons: men are stronger, enabling them to do heavy lifting, which is oftentimes required.... [It] is too severe for any woman but those of gigantic stature and strength.... We believe in the emancipation of women from the drudgery of a heavy dairy."[23] Concepts of overwork also showed a deep ambivalence about ambitions for material gain that seemed actually to generate more work, both in the dairy and in the home. A fictional "Journal of a Farmer's Wife" published in 1853 maintained that "we American women work beyond our strength.... [Why] should we toil so? What do we gain...? A grander house, perhaps, gaudier furniture, finer clothes, and more luxurious food. Are we the healthier and happier for these? No, indeed." The writer directly associated working "beyond our strength" with the worry and toil of cheesemaking: "Turned the cheese again. I declare I wish it was sold. I have had to turn and rub it twice as often this year as ever I did before, and yet I can't keep all the mold off." This writer seems to believe that she has crossed the boundary between "competency" and "superfluity" and strongly implies that the change has not been for the better.[24]

Education

These public exchanges illustrate that while men and women, children and parents, on dairying farms might accomplish their manifold tasks efficiently within their tightly organized households, the system also gave rise to serious conflicts, especially where the sexual division of labor was concerned. Nowhere did these tensions reach greater proportions than with the question of education beyond the primary-school level. These disagreements commonly took place between the sexes, but not always; sometimes the split occurred across generations, or between families. Nearly all of the participants in this extended discussion agreed that advanced education somehow changed the outlook and aspirations of farm children; they diverged most vehemently on whether the change—which tended to bring an orientation away from the farm—was desirable.

Though some attention was devoted to boys' education, by far the more controversial issue was schooling for girls. Midcentury New York State agricultural journals frequently printed articles and correspondence on the subject, much of it contributed by girls themselves. These dialogues reveal farm girls and women and their families struggling over a question that, like the milking and overwork issues, inspired strong feelings.

In 1850 a young woman identifying herself only as "Helen" wrote to the *Ohio Cultivator* with a dilemma. She very much desired instruction in the "higher branches of learning," but her parents had reservations: they were afraid she would return "too proud to work and too poor to live without."

The Social Dynamics of Household Dairying

Helen herself felt at a loss to know whether they were right. She respected her parents' judgment, but at the same time she could not understand what was inherently wrong with her aspirations to learning. Confused, she appealed to the *Cultivator*'s editors and readers for help.[25]

Indeed, Helen's quandary was a common one. An argument commonly advanced against educating farm girls beyond the common-school level was that an academy education, especially where it involved boarding away from home, led farm daughters to reject farm life and work. The 1845 *New England Farmer* printed a complaint from a widowed farmer whose academy-educated daughter, though well versed in Latin, French, mineralogy, botany, and needlework, knew nothing of the dairy or of spinning. The daughter, he complained, "cannot think of milking . . . [for] no young lady does that," and so the father was forced to buy his milk and cheese. Still more distressing to him was her reasoning that even if she knew how to spin and knit, it would be "much cheaper to buy stockings at the store."[26]

The farmer had good reason to see in his daughter's behavior a threat to his most deeply held values. In his view the attitudes that "fashionable" schools inculcated threatened the cherished notion of a competency, as the younger generation's market-influenced calculations clashed with parents' desire for a measure of self-reliance. The daughter also questioned the father's commitment to work as a worthy activity regardless of monetary return. The girl's position was the more stinging because it implicitly flouted the older generation's knowledge in favor of information sources originating outside the family and even beyond the community. And to add a final insult, the educated daughter scorned farm work as unrefined.[27]

The more extreme critics of farm girls' education developed a repertoire of vivid images to accentuate their point. Most popular were comparisons of the spinning wheel, milk stool, or cheese press with the piano, symbol of pretension and useless refinement. In a typical 1848 piece titled "Milk-Maids Turned Pianists," the anonymous author lamented the demise of the "rosy-cheeked" milkmaid, declaring that the charming and useful girl who by "sl[e]ight of hand . . . made each cow yield her lacteal treasures" had been replaced by a mediocre musician. A writer in the *New England Farmer* made a more explicit connection between girls' accomplishments and their aversion to dairying. Decrying the "fashionable" mania for boarding schools, the writer pointedly observed that the present generation's grandmothers had not sent *their* daughters to boarding school "to learn to make butter and cheese."[28]

Others echoed the fear that well-educated women would choose to abandon the dairy. A young woman correspondent of the *Michigan Farmer* (1853) expressed a familiar theme; having spent sixteen months at a New York State boarding school, she concluded that the "school girls ... are taught to look with contempt upon farmers" and that they regarded women's farm work as "drudgery." Similarly, a correspondent in the *Prairie Farmer* worried that young women would quit dairying because the academies taught them to despise it. The possible implications alarmed him: "If not to the rising fair generation, to whom shall we look for the hands that are to supply so important a portion of subsistence as the products of the dairy? ... [If] all this is expected of men, and not of women, how miserably shall we hereafter drop away in the produce of a most profitable and most useful article." T. C. Peters surveyed the developing market for dairy products in 1842 and hoped for the day when "our sons and daughters will think it quite as important to make good cheese ... as to play well upon the piano.... The time has gone by for the American woman to be the frivolous, useless toy that an erroneous system of education has made her.... Her destiny is a high one; for upon her, in a great degree, depends the future prosperity of our country and the perpetuity of the republic."[29]

Fears about the implications of daughters' education were not founded simply upon apprehension about its expense or utility. They exposed the possibility that daughters, because of fundamental inequities in their household status, possessed interests at variance with the farm's goals. Farm daughters, for example, could not expect to inherit land; thus their stake in maintaining the farm over generations was potentially less than that of their brothers.[30]

But criticism or outright condemnation of education for girls was far from universal among the farm population. Significantly, many defenses came from farm girls themselves. These young women upheld the intrinsic value of learning and accomplishment, and some took issue with the supposition that refinement and farm work were incompatible.[31]

A writer assuming the pen name "Annette" addressed critics' fears sympathetically while still asserting that farm girls deserved advanced education. Indeed, she heartily advocated acquiring "the more refined comforts of life; especially those intellectual enjoyments so indispensable to the happiness of a well cultivated mind." She took the schools to task for failing to attune their instruction to the needs and lives of farm people. She charged, for example, that teaching in the sciences was "superficial [and] uninteresting," and altogether inappropriate for people in everyday contact with nature. In another letter she also criticized farmers who, when they found "themselves in possession of a handsome competency," sent their children away to school but failed to welcome them back by making "home attractive"—for instance, by establishing home libraries, by enhancing decorative gardens, or indeed by cultivating their own minds so as to offer more stimulating companionship.[32]

The Social Dynamics of Household Dairying

A "Dairy-Maid" from Wayne County invoked religious and romantic rhetoric to justify musical "accomplishments," suggesting that far from representing a threat, music perfectly complemented rural life: "Would our friend S.W. suppose that farmers' daughters are insensible to the thrilling sounds of music, when the carol of a thousand sweet voices call[s] them forth to their tasks at early dawn. . . . The dairy is in order, breakfast over, and the city Miss is still locked in the arms of Morpheus; while we have an hour to devote to music, that calls forth all the finer feelings,—all that is good and lovely in nature and mind—and it inspires with a devotion that points heavenward."[33]

Other defenders of girls' education employed language less temperate. Several young women were moved to send retorts to the *Genesee Farmer* correspondent who had blasted the "Milk-Maids Turned Pianists." A "Milk-Maid" sarcastically thanked him "for the trouble he has taken to learn [farmers' daughters] something useful. But we should like to know if he would prefer to have us deaf and blind to all the wonders of the age."[34] This correspondent asserted a spirited determination to keep up with the times, a desire frequently voiced in the debate.

Family Strategies

John Mack Faragher has suggested that in the antebellum period rural women seldom openly resisted patriarchal domination, but instead accommodated by focusing upon "shaping the family order" and developing "strong female kin networks" to compensate for their disadvantages.[35] The debates over milking, dairying work, and education indicate that, to the contrary, many put up strong resistance to the patriarchal order, possibly using their valuable dairying skills as leverage in intrafamily negotiations.

Maria Whitford seems to have achieved a measure of success in this regard, though at considerable cost. Through her dairy work, poultry keeping, and spinning and weaving she accounted for a significant share of the farm's income, large enough that on July 20, 1859, she proudly wrote in her diary that she had "come around by *Mr. Shaw's and paid him forty dollars toward farm*" (her emphasis). That her husband, Samuel, often washed, cooked, churned, tended the garden, and mopped suggests that Maria was able to negotiate a measure of reciprocity in labor.[36]

In the 1841 *Genesee Farmer* a farmer's daughter described a more overt negotiation. Her mother and sisters bargained with her father to "let us have the butter and cheese we make this summer" for money to repaint the house. At first he balked, because he wanted a new wagon and harness, but he relented when the women pointed out that he could "have all the avails of the farm for that." In supporting the daughters' claim, the mother reasoned that unlike girls who spent their days "living and dressing upon the hard earning of someone," her daughters were educated in productive domestic skills.[37] This account is very suggestive in its portrayal of men and women negotiating over the proceeds from cheese sales, raising the possibility that some dairying women exerted a claim to use that income for their own purposes, and that in doing so they challenged the old ideas of competency, or extended its definition, in favor of "making the farm attractive." Like young women who made hats in New England, daughters in dairying families used their skills as bargaining chips within the family.[38] Like Pennsylvania farm women, New York State dairying women were beginning the slow process of "loosening the bonds" of patriarchal society.[39]

The acrimonious exchanges in the agricultural journals reflected differences within the household, among generations and between men and women. Yet there is also evidence that families responded to these differences with strategies that would contain them, if not resolve them. Total New York State production, for instance, did not increase significantly between about 1840 and 1860, even though the area of land under cultivation, and the number of farms, increased. While prices and labor supply, of course, probably influenced this trend, it is also likely that household priorities, and perhaps internal strains, also contributed to limiting expansion.

Moreover, a great many farm families in dairying country eventually adjusted their practices to accommodate the new generation. Evidently many delegated the milking to men, for example, and more significantly, they increasingly embraced advanced education for daughters and sons. As far as timing was concerned, dairying and schooling were not necessarily incompatible; children could work on the farm during the cheesemaking season and still attend school in the winter months. But whether or not it interfered with farm work, families more and more found ways to send their children to school. E. H. Arr noted that "it was thought to be generous in a farmer to let his daughter 'learn a trade,' thus freeing her from the heavier drudgeries of farm-work." As journal correspondent "Annette" put it, "Many farmers give their daughters a liberal education, with the natural expectation that it will have a tendency to increase their own happiness and the happiness of those around them."[40]

Historians of education have questioned the assumption that advanced schooling was rare in the nineteenth century. Rates of high-school attendance were significant in nineteenth-century communities where such institutions existed. It also seems that advanced schooling was not invariably prohibitively expensive. Establishing a "select school" often required relatively little in the way of resources, and such schools proliferated, even if they were usually short-lived. Moreover, advanced schooling was not solely an urban phenomenon.[41]

Indeed, Oneida County's experience suggests that a term or two at an academy or "select school" became part of the experience of many young people from farming families.[42] The county boasted a number of such academies and institutes; the New York State census listed sixty-seven in 1845—very likely this is an undercount—and many were located in rural districts, suggesting that among the farming population of the county there was a demand for such institutions, and that they flourished because they were responding to that demand. The schools therefore should be seen not solely as catering to bourgeois families, but as the product of agricultural conditions as well.

In Clinton the presence of Hamilton College made the town a natural location for educational ventures. The Clinton Liberal Institute (coeducational) and the Houghton Seminary, Domestic Seminary, and Home Cottage Seminary (all for girls) flourished there during the 1840s and 1850s. The coeducational Delancey Institute in the hamlet of Hampton (town of Westmoreland) accommodated 125 students in 1846. In centrally located Rome, a market and transportation hub, was the coeducational Rome Academy. Boonville boasted a Young Ladies School, and Vernon the Van Eps Institute, a girls' boarding school, as well as the Mt. Vernon Boarding School. There were academies in Remsen and in Prospect, a village in the town of Trenton.[43]

Many of the schools' annual catalogs listed all of the "ladies" and "gentlemen" of the student body with their place of residence. Most students came from the rural townships of the county—Marshall, Sangerfield, Westmoreland, all prominent dairy towns—and among them there were usually several from cheesemaking families. For example, in 1848 at least seven girls in a class of eighty came to the Domestic Seminary from cheesemaking families. At coeducational institutions both sons and daughters of cheesemakers attended (though not necessarily in the same classroom; parents in Rome voted against such an arrangement in 1851). The student population at the Home Cottage Seminary and the Delancey Institute also strongly represented butter dairying families. Moreover, dairying families were involved in establishing and running the schools; three cheese dairying farmers from Westmoreland were among the fifteen trustees of the Delancey Institute in 1846.[44]

The cheesemaking households that sent children to these schools do not seem to have varied significantly from the usual types. Since these conclusions are based on the small number of students who can be unmistakably identified with cheesemaking farms, they are tentative. Where farm value is concerned, the households represented all levels, from a minimum of $1,200 to a maximum of $20,000, with most ranging between $2,500 and $5,000. (The average value for cheesemaking farms in 1850 was $3,883; for all New York State farms, $3,250.) Thus academy education was not the exclusive domain of the wealthy. Though tuition at some academies was high, some boarding schools charged only $7 for an eleven-week term.[45] This reinforces the point made by "Annette" that daughters of the middling classes now enjoyed increased access to further education: "Nor is it daughters of the rich and fashionable alone, but many of the cultivators of the soil . . . begin to see the necessity of giving their daughters something more than a common school education."[46]

The one significant characteristic these households shared is that families who sent children to academies tended to fall into the "mixed-labor" category; that is, they had both live-in hired labor and children over the age of ten. Since the children are known to have attended boarding school at least part-time, this suggests a substitution of paid workers for children. Such a strategy would result in a double drain on family resources, via wages and tuition expenses, and would thus propel them further into the capitalist economy. Viewed from this perspective, the transition to capitalism takes on another dimension: families expanded their commitment to capitalistic enterprises in order to educate their children.

At these institutions girls and boys from rural Oneida County were exposed to young people from other backgrounds and places—students from other states and from New York City, and some whose missionary parents were posted in faraway spots such as Burma. They might even study together with fugitive slaves; the Young Ladies Domestic Seminary in Clinton was founded by the abolitionist Hiram Kellogg. At school pupils acquired education labeled both practical and polite, including "Reading, Spelling, Writing, Geography, Arithmetic, Grammar, Natural Philosophy, Chemistry, Geology, Mineralogy, Botany, Algebra, Rhetoric, Mental Philosophy, Moral Philosophy, Vocal Music, Latin, and Greek." For an additional fee they could take piano lessons, French, German, and drawing. Some of the schools boasted facilities for scientific experimentation.[47]

As always, school life involved more than formal instruction. Teenager Mary E. Root kept a diary between 1860 and 1863, when she was a student at one of the Clinton academies. Her activities illustrate not only the school's emphasis but also something of the opportunities school life afforded, and of her own responses. Mary heard a "lady preacher"; attended evening "Society" at the Reverend Mr. Bissell's home; enjoyed the girls' performance of "The Flower Queen"; responded with romantic sensibility to the local landscape; received an invitation to view a scientific exhibit at the Gentlemen's Department; read and enjoyed a popular women's magazine, the *Ladies' Repository;* contemplated her prospects for salvation; and took "gymnastics" from the famous Dio Lewis. The girls at the academy enjoyed an unbroken round of social activities not only among themselves but also with the gentlemen students, both at the other academies and at Hamilton College. Interspersed with Mary's entries relating to school were notes on home stays, during which she helped her mother clean, bake, churn, and wash wool.[48]

The diary of Sarah Burgess, another boarding-school student, corroborates Mary Root's experience. Sarah came from East Winfield, Herkimer County, in prime cheese dairying country, and she and other women in the family made cheese regularly. In May 1860, for example, she wrote that "Ruth is making cheese [and] Mother made some soap"; that day they also boiled sap for maple syrup and bought gooseberry plants to set out. This suggests that, in the pattern common to cheese-producing farms, the family's women were involved in a variety of commercial and subsistence activities. In the winter of 1859 Sarah, then aged sixteen, attended Temple Grove Seminary. She excelled academically, especially in composition and in French, and enjoyed the social aspects of school—parlor calls, meals, friendships, sleigh rides, and lectures. But she was always homesick and regarded the academic routine as monotonous. When she finally went home, she was happy to do so and wondered uneasily if her final departure from school meant that she was now one of the "Young Ladies." Sarah clearly felt uncomfortable thinking of herself as an accomplished young lady, perhaps because she understood the pejorative connotations the phrase carried in many farming circles.[49]

Academy students formed strong associations with their peers. An unsigned, undated letter in the Barnes family collection from Oneida County, probably written by young Mary Barnes, expressed yearnings for her school friends after she returned home. After a period of unhappiness, the girl wrote that she had regained "control over her feelings."[50]

For these young women education was a transforming experience. Schools such as the Troy Seminary and Mount Holyoke functioned as "incubators" for a new type of female identity; the students went on to establish careers and to educate others in turn (indeed one Troy graduate, Urania Sheldon, maintained a well-regarded girls' school in Utica). Moreover, seminary products married far less often than other women in their generation, and if married they bore fewer children.[51]

It is not possible to make an extensive study of Oneida County schools' students here, but from what we know of their experience, it is possible to make some tentative assessments of what it meant. The same process of generating new identities was being repeated, if on a more modest scale, in these rural schools. In the nineteenth century, women tended to forge identities in keeping with their world of dependency and limited personal choice; they often made deference a virtue.[52] Thus the conventional (usually masculine) meaning of the term *individualism* does not strictly apply to girls' experience. Girls in rural schools, for example, formed strong peer-group attachments. Yet within the context of the family farm, the "identity" they developed *operated* as individualism, because it stressed the pursuit of activities and personal interests unconnected with the farm's goals.

Though discussion in the farm press tended to treat boys' and girls' education separately, in everyday fact the divergences in their school experience were probably small compared with the common ground they shared. Indeed, the school's formative influence was probably just as powerful for boys. Male students could expect to receive a somewhat more extensive bill of academic offerings than girls. But they too could acquire "accomplishments," a reminder that the periodicals' almost exclusive association of music and art with women's education was inaccurate. The young Babcock boys, for example, went to the Fairfield (Herkimer County) Academy and also took piano lessons. Arthur C. Hackley, a farm boy from Bridgewater (Oneida County), filled his days with a variety of work and leisure activities; among the latter were music and drawing lessons. And the same aspirations that inspired families to seek education for girls informed their choices for boys; in the *Genesee Farmer* for 1840 a "farmer's son" advocated advanced schooling as a means of shedding "rusticity, coarseness, vulgarity, [and] impoliteness" for "good-breeding, politeness, . . . fine manners, and gentlemanly appearance."[53] Particularly when boys came from families with several sons, education might well offer more realistic alternatives for the future than would the prospects for inheriting land. Moreover, male and female pupils in the rural academies constantly intermingled; this reflected prevailing modes of socializing in rural communities, in which men and women participated together in many social activities.

The academies' transforming impact was not just personal but also social. Some families were able to negotiate the change successfully and to incorporate educated members who then made up the new generation of farm husbands and wives, helping to reshape rural communities later in the century. But far more important in terms of social change was that the academy environment presented alternatives to farm life, many of which potentially conflicted with agricultural values and work. The critics, however hyperbolic, were correct to associate further education with new values, and even with a tendency to abandon agriculture altogether. This explains why the issue provoked such an intense response in the countryside. Not only in public discourse but within dairying families, decisions about education automatically involved reevaluations of appropriate goals and activities for farm children, especially girls. By choosing to educate its sons and daughters beyond the common-school level, a family steered those children away from the farm.

In some instances the shift was so thoroughgoing that families literally reoriented themselves wholesale. Recent research has concluded that migrants to upstate New York cities were often well-to-do farmers and their families, moving with relative ease from life on the farm to occupations in business and artisanry. Similarly, in Rhode Island those who migrated to Pawtucket from "agricultural, nonindustrial towns . . . [often] took up white-collar employment."[54] This suggests that farm families often saw themselves as members of the middle classes, and also offers yet another reason why the transition to an urbanized and industrialized economy in the region was accomplished with relatively little conflict. Education was quite likely an important factor in facilitating this ease of movement.

New York's dominant New England background also influenced the nature of change. In New England the rural village was "the active site of innovation and enterprise," where a culture of "improvement," commerce, and consumerism pervaded local society. Rural people eagerly exploited this culture to acquire new levels of middle-class identity and sophistication—and ultimately to break away from agriculture. A "mobile marketplace of ideas," as disseminated by itinerant sellers of almanacs and other printed materials, exposed rural people to innovation and to new values—for example, those of work discipline.[55] In Yankee-dominated New York a similar atmosphere prevailed. We can see women's aspirations for broader learning when we read Maria Whitford's diary. Even during the busy summer months she managed to steal away from tending her chicks, cabbage plants, cherry trees, cheese vat, and raspberry canes to attend the Lyceum, Orophillian Society, and Ladies Literary Society. She heard pacifist Elihu Burritt and abolitionist Frederick Douglass, among others. Perhaps influenced by reformist thinking, she set about making a "bloomer dress" in the spring of 1859.[56]

More often, the shift was a generational one, as sons and daughters acquired knowledge, skills, and personal associations that would aid their departure from farming. For boys, of course, many nonfarming occupations beckoned. Catalog copy suggests that most "gentlemen" students were preparing for college, for careers in business, or for the professions. It was probably with well-founded apprehension that a *Genesee Farmer* correspondent from Trenton Falls (Oneida County) complained in 1861 that when the farm boy attends an academy, he "becomes associated with a class of lads from the city and large towns, who look upon labor as degrading and upon him as their inferior." Driven to "maintain his dignity," the farm boy is "finished for anything pertaining to farming."[57]

For girls the route out of farming would necessarily follow a different path. Academies not only functioned as centers for education; they also were marriage markets. Couples marrying from this environment would likely leave agriculture. In the next decades this tendency would show up in alumni newsletters, which would follow many women seminary students as they married professionals or businessmen and migrated away, often to urban areas. Lavinia Mary Johnson remained single, but a number of her childhood friends met their husbands at school and followed them away from rural areas.[58]

It is important to stress that farm daughters were not necessarily seeking a purely "domestic" future; though they might criticize overwork and reject dairying, their background in the ideology and practice of "mutuality" also exerted influence on the directions they took. Mary Root and other girls were exposed to a variety of cultural alternatives for women; besides their mothers and other female relatives, they observed women pursuing intellectual or literary careers and even serving as members of the professions. Though relatively few took it up, for example, missionary work was becoming an acceptable endeavor for women. The Judson wives' exploits in faraway Burma inspired young evangelical women all over the North.[59]

But most significantly, a seminary education, impractical as it seemed to those who had only farm work in mind, fitted girls for the newly "feminizing" occupations, and again alumni records show that former students entered these fields.[60] In the mid-nineteenth century the most significant new development that actually permitted some girls to act upon individual aspirations was the feminization of teaching. To be sure, girls' range of choices was still severely limited; they seldom could hope for complete economic or personal independence. But viewed against the even bleaker backdrop of the early nineteenth century, the scene had changed significantly. By 1860 New York State had over eight thousand women schoolteachers. As many as a quarter of the women in antebellum Massachusetts taught at some time in their lives, and the proportions in New York were probably not far from that. If public-school positions were not available, an aspiring teacher could start up a "select school" with relatively few resources. Teaching could offer a comparatively large measure of personal fulfillment and autonomy for young women.[61]

The example of Rosetta Hammond Bushnell shows how a young farm woman, even one possessed of limited resources, could move from farming to teaching. She lived in Fayette (formerly Guilford), in Chenango County. Rosetta worked on her family farm at times, milking cows (when her father was unable to milk), churning, raking hay, raising vegetables, spinning yarn, making "knitting sheaths," and cooking for the hired men during haying time. She also carried water for the men haying. In 1857 Rosetta spent the month of September picking hops in Oxford and made $4.09. Significantly, that is one of the entries in which she actually records a cash income, however small. The following winter she went to the "Academy," possibly the Oxford Academy, an old established private school in that town. At the same time, she worked on another farm as a "help," baking, churning, milking, and making cheese, for a dollar a day. "Wish I could live without work," she wrote tiredly in February 1858. Ultimately Rosetta obtained a "certificate," probably to teach, and taught school for a while. Her sequence shows how she was able to cobble together a living and also to acquire an education sufficient to allow her to enter teaching.[62]

The pattern in Oneida County was part of a broader movement in the antebellum education of young women. Early colleges and normal schools also attracted farm daughters in significant numbers. Over 40 percent of Massachusetts normal-school students were from farms, and their numbers were greatest at coeducational institutions, probably reflecting rural traditions of gender-mixed schooling and socializing. Most of Mount Holyoke's early students were from farms of below-average value, and from families with a preponderance of daughters. These families probably sent their daughters to the college out of a need to plan for a "gap" in the life cycle between adolescence and marriage, and also occasionally in order to gain the qualifications necessary to help support the family through teaching.[63]

By contrast, early women students at Alfred University came from middle-income households in which their labor was very important, since Allegany County, like so many in upstate New York, was a dairying area. In fact, "in almost all areas of production of goods involving women's labor, the families sending daughters to Alfred tended to produce more than those that did not." Oneida County's circumstances bore a closer resemblance to those in Allegany County than to those in Massachusetts; daughters were far from "superfluous," and dairying families possessed adequate resources. Allegany County farm families and daughters placed a premium upon education, seeing in it inherent value as well as practical, vocational preparation: "The desire of families to educate their daughters or of the daughters to educate themselves was fairly universal. . . . Moreover, Alfred seemed to offer its women students substantial opportunities to compete with men . . . [and] seems to have played much the same role for young women that the antebellum college played for young men, aiding in the transition from rural to urban life and moving women from farms into careers."[64]

Thus unlike the choices of those women who stayed on the farm, compensating for legal and cultural disabilities by fashioning strong networks of mutuality and kinship, the case of cheesemaking highlights the exploitative aspects of the household system and shows that many farm women responded by rejecting rural life and society altogether. For increasing numbers, schooling now served as a means of securing a future outside of agriculture, whether as a town or city wife or (in the shorter term) in the newly feminizing occupations. In her memoir *New England Bygones*, E. H. Arr hinted at the power of new values when she pointed out that the wives of "forehanded" farmers were "looked up to. This was because they enjoyed a partial exemption from toil." Arr's reminiscences celebrated women's work and its products, but even she believed that farm women worked too hard. These ideas informed the activities of families who, by investing in education, exploited the prosperity cheese sales brought to repudiate implicitly their lives of labor.[65]

In just a generation a dynamic innovativeness had replaced conservatism in family strategies. When families had entered dairying in the 1830s and 1840s, they had done so at least in part to preserve the patriarchal structure of authority. The next generation's alternative vision questioned that authority and even its agrarian basis. The gathering momentum in farm family strategies signaled a weakening commitment to the lineal family obligation—the desire to maintain the farm and pass it on to the next generation.[66] What we see in the actions of farm families who argued over the division of labor in milking, criticized women's excessive labor in dairying, or sent their children to boarding school is nothing less than a key source of a major nineteenth-century change—that momentous shift from a rural, agrarian society to an urban, industrial one. Like the young New England farm women who flocked to the manufactories of Lowell, central New York dairying girls found a way to avoid the fate their "worn-out mothers" had suffered. Given the disproportionate numbers of rural dwellers among antebellum Americans, perhaps the focus upon the cities themselves as loci of change has been misplaced. When the Civil War forced rapid shifts in economic conditions and labor availability, the household economy was already ripe for change.

6

The Rise of the Factory

> In many farm houses the dairy work loomed up every year, a mountain that it took all summer to scale. But the mountain is removed; it has been handed over to the cheese factory.—Mrs. E. P. Allerton, 1875

In 1851, in the Oneida County canal town of Rome, a well-to-do farmer named Jesse Williams erected the first modern-style cheese factory in the United States. Williams had been born into a Rome family of émigré Connecticut Yankees in 1798. In 1822 he married Amanda Wells, and the couple embarked on a prosperous farming career, concentrating almost from the beginning on dairying. In 1835 Williams's census return listed 265 improved acres with sixty-five cows, and the farm grew still larger after he inherited his father's farm in 1837. In 1850 he and Amanda were turning out twenty-five thousand pounds of high-quality cheese. Their household looked like many then operating in Oneida County, but Williams and his family came to local notice when they began to specialize to such an extent that they bought flour and, almost unthinkably, butter. In an apparent effort to improve their product, one year they rented out their farm and traveled throughout the dairy regions, visiting cheesemaking farmers to learn and observe. Thus by midcentury the Williamses were already well schooled in large-scale cheese manufacture.

The story of how the original factory was established centers around a generational difference within the founding family of cheese factory production. Trouble began when the couple's newly married son and daughter-in-law balked at making cheese at home. Eldest son George married in 1850. Some versions of the story imply that his bride lacked confidence in her cheesemaking ability or perhaps even simply refused to make cheese. Others speculate that it was George, not his wife, who was deficient in cheesemaking skills. In any case, in order to obtain his customarily high price for George's cheese as well as for his own, Jesse Williams persuaded his cheese factor to contract for both farms' cheese—from 160 cows. To process this quantity of milk was beyond the household's capacity, and so the Williamses decided to make cheese in a central location. By 1854 the Rome Cheese Manufacturing Association had ten patrons, including two of Jesse Williams's sons.[1]

Williams achieved instant local fame, but his innovation failed to catch on immediately; in 1860 cheese factories were just an interesting novelty, unknown even to most inhabitants of the dairying country beyond the town of Rome. But by the mid-1870s the crossroads cheese factory had become ubiquitous throughout Oneida County, often appearing next to the village schoolhouse at the center of the rural neighborhood; home cheesemaking had virtually disappeared. The rise of cheese factories signified a dramatic and far-reaching change—especially, of course, for the people involved in it, but also because in historical perspective it captures in one brief, concentrated process many of the developments integral to change in nineteenth-century rural America. The factory system helped to extend and consolidate changes begun earlier. It brought farming people more inextricably than ever into the orbit of the international market and of cash exchanges; it hastened the transition to wage labor; it introduced new pressures for conformity to uniform practices, both in processing and in farming; and it radically altered gender patterns of labor within the dairying household. Cheese factories in other respects represented considerable continuity with past patterns of economic and social organization, but to central New Yorkers the departures seemed more apparent.

An explanation of the rapid rise of cheese factories is thus essential to an understanding of the nature of change in the nineteenth-century rural America. Historians have found widely varying rural responses to industrialization and capitalism. Some view the "Great Transformation" primarily in terms of class conflict between yeomen, who defended precapitalist economic relations, and capitalists, who tried to control productive resources. Especially in the South, capitalist rationalization, mechanization, and government policies brought about massive displacement and migration. Even in other regions, industrialized agriculture was sometimes imposed upon unwilling rural people who people clung to their antiscientific, familocentric, locally oriented way of life in the face of pressure from urban interests such as Country Life reformers and land-grant institutions.[2]

But other scholars, mostly studying northern settings, find far less resistance. In Sugar Creek, Illinois, the transition to a market economy and society occurred as "conservative change"; farmers sought opportunities to raise crops for market, but did so in order to maintain stable communities in which kinship was central. Indeed, stability and consensus seem to have been typical in many late-nineteenth-century rural communities such as Chelsea, Vermont. In the Connecticut Valley the transition took place smoothly, as the pursuit of household goals led families into market activity and eventually away from agriculture to industrial production. This pattern also held in Oneida County, where cheesemaking families' middle-class status probably made it far easier for them to negotiate the passage out of home cheesemaking—or indeed, out of agriculture altogether—than it was for poorer, less well educated and less skilled farm people.[3]

Where dairy centralization specifically is concerned, explanations have relied upon economic models, locating the sources of change in prices, labor availability and wages, the press of competition, transportation, and markets, all part of the logic of capitalist development. Though these basic attributions are similar, scholars' evaluations differ. The earlier historians, writing from the 1940s to the 1960s, tended to see factory consolidation as a positive event for women, alleviating burdensome labor and accepting that when dairying became a "business," men became involved in it. More recent historians, benefiting from the insights generated by the rise of women's history, are more critical. They see in dairy factory centralization a loss for women, since men both took over control of production and claimed the income.[4]

The larger economic structures with which these historians have been occupied undeniably influenced change in nineteenth-century dairying. However, for a better understanding of why dairying centralized, it is necessary to take a closer look at the household itself as a source of centralization, because at the root of the factory system's success was the willingness of thousands of individual families to become factory "patrons." This approach affords a different perspective both on the impetus for dairy centralization and on its implications for women. It focuses upon the household and upon its complex interaction with the wider context, local and beyond. As in the Connecticut Valley, circumstances internal to the farm household were crucial in influencing the way the farm family interacted with the wider economic, cultural, and social setting. Tensions within the household system, especially its demands on women, helped to fuel the emergence of "rural capitalism."[5] While external events set the factory boom in motion, the factory system spread as rapidly as it did not because cheesemaking families always made more money when they became factory patrons, but because the factory system mitigated contradictions that had been brewing within cheesemaking households for some time. It relieved conflicts over the gender division of labor, and it accommodated the rising aspirations of individual household members to a new balance of work and leisure.

Precedents

By midcentury the people of Oneida County were well familiarized with centralized manufacturing. Relatively large textile-making operations had already been present for a generation or more, flourishing or declining with general economic cycles. In many an Oneida County village, smaller concerns used waterpower to turn out items from furniture to farm implements to housewares. Indeed, farmers often invested in stock from local manufacturing companies, so they actually had experience participating in industrialization and followed its fortunes.[6]

Where cheesemaking specifically was concerned, pooling the milk from several farms was not a new idea. It had numerous precedents in the distant past and in more immediate times and places. Collective cheesemaking customs of several European dairy regions had been publicized in America through the agricultural press and through the example and memory of immigrants. In Swiss villages, for example, families had customarily hired a man to make cheese, using the milk from several farmers' herds; the cheesemaker kept a tally of each farmer's contribution and reimbursed participants at the season's conclusion. A Swiss immigrant named Jacob Miller apparently operated a similar arrangement in the town of Westernville (Oneida County) in the 1850s. In bordering Lewis County, immigrant cheesemakers bought up milk from local farmers and made Limburgh cheese, also in the 1850s.[7]

Probably more influential were traditional arrangements practiced in the English cheese districts, especially in Cheshire and the West Country. When these regions developed their characteristic cheeses in the seventeenth and eighteenth centuries, the women cheesemakers in some villages joined their milk to make "great" cheeses.[8] Quite possibly New England settlers remembered these ancient customs when they founded Cheshire Counties in Massachusetts and Connecticut. Indeed, in Cheshire County, Massachusetts, local cheesemakers had pooled their milk to make a huge inaugural cheese for Thomas Jefferson.[9] This was a notable event more because of the size of the cheese than because the milk had been gathered from several farms; *Moore's Rural New Yorker* noted in 1866 that the factory was "but the complement" of an old New England custom, whereby farmers "changed" or "swopped" milk to make cheese. The wives would keep track of amounts and take turns making cheese "until the terms of the agreement were fulfilled," when accounts were settled.[10] The Oneida and Herkimer County families who migrated from New England may have brought at least the recollection of such customs with them. As the Yankee migration expanded further westward, cooperative dairying ideas went along; a Wisconsin man named J. G. Pickett claimed that his mother had operated a cheesemaking service in 1841 for six patrons, taking milk from their thirty cows and returning the cheese in proportion.[11]

Other activities helped to set the stage for the factory system in a less direct way. Under the "three-fifths" system, tenancy arrangements for cheese dairying frequently applied to large-scale operations. Tenants would rent substantial farms, some with as many as a hundred cows, and work them on shares, receiving three-fifths of the cheese.[12] In other cases prominent local dairymen tried to get farmers, especially those operating on a large scale, to produce cheese to uniform standards; Alonzo L. Fish, a leading Herkimer County dairyman, coordinated such an effort in the 1840s.[13]

Dairying people not only had memory and precedent to draw upon; they had also tried experiments with newer versions of cooperative cheesemaking. Goshen (Connecticut) cheese had a high reputation in the market; in 1860 *Moore's* attributed the quality of Goshen cheese to the "considerable number of local manufacturers who keep few or no cows themselves but buy the curds of neighboring farmers." A similar system, on a larger scale, was pursued in Ohio's Western Reserve in the late 1840s and early 1850s. Factories were set up to accommodate large quantities of curd.[14] But ultimately the curd factories proved unworkable because the curd was difficult to transport without spoiling and varied drastically in quality; anyway, farmers who would go to the trouble of making a curd generally preferred to manufacture their own cheese as well.

Credit for originating the idea of the modern cheese factory is usually given to Jesse Williams. His 1851 cheese factory in Rome became the prototype for the thousands that went up during the next decades. Williams combined the idea of pooling milk with that of organized factory production. The basic principle of the cheese factories set up on the Williams model was that farmers hauled milk to the factory a few miles away twice daily.[15] Fewer than a dozen "patrons," or as many as sixty or seventy, might send milk. The number of cows whose milk was collected also varied considerably, from 150 to over 1,000, but most commonly factories were set up to handle milk from between 300 and 600 cows. Most factories produced from one hundred thousand to two hundred thousand pounds of cheese per season, May to November, using the labor of from two to a half-dozen employees. Most early factories' collective owners paid a cheesemaker to process the milk, set charges for supplies (usually from 1 to 2 cents per pound), marketed the cheese, and returned earnings pro rata to patrons. As factories proliferated, total U.S. production shot up from 103 million pounds in 1860 to 243 million in 1880. New York State's production alone went from 50 million pounds in 1860 to 150 million in 1900.[16]

As businesses, cheese factories represented an increased role for capital in the industry, though amounts were modest compared with those needed for large-scale enterprises. Costs for setting up a cheese factory ranged from $1,500 to $10,000, depending on location, size, and equipment. The most popular way to "get up a factory," according to X. A. Willard, was to organize a joint-stock company, in which shares of $50 to $100 each were taken by patrons in proportion to the number of cows from which they expected to deliver milk; expenses and proceeds were divided pro rata. The stockholders elected officers, who managed the enterprise, and usually an individual or committee was elected to sell the cheese. Sometimes an individual established a factory, assumed ownership of the plant, and undertook to make the cheese, charging patrons by the pound. The cheese then belonged to the patrons, who selected a salesman from among their number. In 1870, for example, I. S. Weller bought a cheese factory in Boonville for $2,000. Every spring the local paper covered meetings at which Weller's patrons elected a salesman, a treasurer, and committees to examine the treasurer's books and to inspect the milk. In these first two forms, the factory worked on the principle of providing a service—making and selling cheese—for farmers. The third type operated on a different principle in that the proprietor actually bought the milk outright from the patrons and paid for it in cash. This last form, uncommon until late in the century, was a more speculative venture than the others, because the proprietor bore all of the risks and reaped benefits or absorbed losses.[17]

Advocates of the factory system supposed it to possess several advantages over home cheesemaking. They thought, for instance, that a single good cheesemaker would make better, more uniform cheese than would dozens of makers with varying degrees of skill. Factory cheese usually commanded a premium of 1 or 2 cents per pound over cheese from "private dairies." Savings resulted when factory supplies such as bandages and salt were ordered in bulk. And by centralizing cheese production, factories eliminated an enormous amount of work done on the farm.[18]

Contemporaries universally referred to these new institutions as "cheese factories," but it is important to recognize that in many respects they hardly deserved the appellation. Capital requirements were minimal compared with those of, for example, earlier textile-manufacturing concerns in New England and even in Oneida County itself, where the 1832 McLane Report had found several textile mills representing capital investment of as much as $40,000 or $50,000.[19] Owners and investors were local people, rather than capitalists from outside the area. Even the largest cheese "factories" employed only a half-dozen workers, compared with the dozens or even hundreds in the textile firms. These workers, moreover, possessed the same skills that their home-cheesemaking counterparts had wielded—the work was not divided into smaller, less skilled tasks. A further bar to characterizing this form of off-farm cheesemaking as fully "industrialized" is that only the processing was centralized; milk, the primary ingredient, was still produced on farms. And as we shall see in chapter 7, technology expanded in scale but did not fundamentally depart from that of home cheesemaking. From a comparative perspective, then, it might be more accurate to call these concerns "farmers' cooperatives" or "off-farm cheese houses." But from the participants' point of view, they were definitely "factories," for reasons that we shall explore in the next chapter, and so the historical terminology is used here.

The Impact of the Civil War

During the years immediately after Williams established his enterprise, only a score of cheese factories appeared on the New York State scene. The year 1863 was a turning point after which they multiplied rapidly. By 1864, it is estimated, 205 cheese factories made about a third of the cheese produced in the state.[20]

The triggering event for this burst of factory-building activity was the outbreak of the American Civil War. With the Civil War came several changes in market conditions. The southern outlet for New York State cheese, of course, was closed off. In its stead, the amount of cheese sent to Britain, always an important destination for American cheese exports, expanded dramatically. Five million pounds of American cheese were sent to Great Britain in 1859; that amount jumped to twenty-three million in 1860, to forty million in 1861, and to more than fifty million in 1863. Exports to Britain increased mainly because of a substantial price rise, caused by wartime economic conditions: English importers paid for cheese in gold, which in the United States was then exchanged for the inflated paper money the government issued in order to finance the war. Prices for cheese in some cases doubled in a few years.[21] The situation encouraged not only the diversion of homemade cheese from domestic markets to export but also the construction of factories to produce even more, adding milk from herds that would not ordinarily have contributed to cheese production.

Factories were the more imperative because the war also took women and men from home. Some dairying families lost female laborers as an indirect result of the war. The expansion of bureaucratic systems—for example, in government procurement or in business contracting for war supplies—combined with the scarcity of men to create opportunities for women in such areas as clerical work,[22] and possibly also in the rapidly growing canning and garment industries. The feminization of such occupations as teaching also accelerated. In New York State the swing was pronounced between 1860 and 1864: in 1860 there were 8,224 female teachers and 18,129 men teachers; by 1864 there were 21,181 women and only 4,452 men.[23]

One Oneida County family's correspondence offers an example of how the war disrupted family labor patterns and may have contributed to the spread of factories. Abijah Barnes of Boonville served with the local regiment in Virginia; in the winter of 1864 Abijah wrote home to inquire whether his sister Mary had returned to the farm. From the correspondence it is not completely clear where Mary had gone, but the context suggests that she was in Rochester. Abijah complained: "When is Mary coming what is she staying there so long for she told me she was earning a[l]most five dollars a week." Probably $5 a week was justification enough in Mary's mind. It is impossible to know whether Mary's departure forced the family to turn to the cheese factory to process milk; with both Abijah and Mary gone, it must have been difficult to manage. At any rate, by summer Mary was living at home again, helping to run the farm. Abijah approved: "It did not seem write [sic] to have her away." In Abijah's absence Mary and her parents shared the farm's work: "Mother is washing the milk cans and Father is planting corn. . . . Father has got his crops most all in[.] We milk 16 cows and our milk goes to the cheese factory."[24]

The military demands of the war, of course, presented the most significant drain upon labor resources. Civil War military needs were in general disproportionately met from rural regions, and Oneida County was no exception. An estimated four thousand Oneida County men served in the Union Army. Farmers and laborers accounted for far and away the largest occupational groups in wartime service from Oneida County. As husbands, sons, and hired hands left, the result was a serious dearth of male laborers on the farm and high wages for those who remained; a selected group of enumerators estimated for the New York State census in 1865 that Oneida County farm wages had risen 75 percent during the war, to an average of $30 per month. Even when veterans returned, many were disabled and could no longer perform farm work.[25]

There is some tantalizing evidence that the war experience also affected some men's attitudes toward work. Abijah Barnes gave a hint of such changes when he wrote to his family from the front in February 1864. He inquired regarding the possible sale of his parents' farm, remarking, "I suppose you do not feel much like working it." Then Abijah allowed that soldiering had made him "lazy." "You will hear a great grown if he [i.e. the soldier] haves to work," he warned. Though army life, of course, could be harsh and demanding, Union troops typically engaged in combat only sporadically and otherwise spent long periods in encampment—a leisurely existence compared with the daily toil of farm life, especially life on a dairy farm.

Whether returning veterans acted on these sentiments remains to be researched, but in any event they could not help with the farm while they were away. How, then, was the work of cheese dairying to be carried on? According to the usual division of labor on the cheese dairying farm, the absence of men would only occasionally create problems with the actual processing of the milk, but it would always mean a lack of manpower for haying, milking, foddering cattle, harvesting field crops, and the like.

In some cases machines supplied the need for labor; it is well established, for example, that during the Civil War the numbers of mowers and reapers used on northern farms increased rapidly, as high prices and labor shortages combined to provide favorable economic conditions for their adoption. By 1864 enough mowers had been produced in the North to supply three-quarters of the farms over one hundred acres, and implement manufacturing became a major industry in New York state, leading the nation even to 1870.[26] The mechanization of mowing proceeded in Oneida County as elsewhere; in August 1863 the *Black River Herald* observed that "the incessant rattle of the mowing machine, the busy activity of every man and boy that can wield a fork or rake, declare how important to the farmer is the present crisis."[27] Farmers felt all the more urgency because hay in March had brought as much as $20 per ton. But not all farm families acquired mowers immediately, and anyway the mower's use was limited to one specific, albeit crucial, task among the many that demanded daily and seasonal attention. Most farm work still required human manual labor.

The most logical people available to fill the gap were farm wives and daughters. By the spring of 1863 the *New England Farmer* was reporting that women were to be seen out in the fields wielding rakes and spades. The *American Agriculturist* too noted the next summer that "woman's labor in the fields" had formerly "been a rare sight," but that "the present scarcity of laborers leads many to employ women in field labors. During the month of June thousands have found profitable employment in weeding carrots and mangels, setting out cabbages, tobacco, etc., lending a hand in the hay field, and perhaps in the corn and potato fields too." The journal also published letters from women who claimed to be managing dairies alone. In the January 1865 issue a front-cover vignette entitled "Farmer Folks in Wartime" depicted farm women, children, and grandparents haying, sending off provisions, packing fruit, and slaughtering hogs for the winter. In the next issue the editors assured their readers that this scene "was no mere fancy sketch. Numerous letters received at the *Agriculturist* office show that the women of America are worthy descendants of their heroic grandmothers, who gave their husbands, sons and brothers to the country, and themselves filled the vacant places in the more peaceful, but not less important, fields at home." The editors included a sample of such correspondence, from a woman who harvested corn, hay, oats, sorghum, potatoes, and pumpkins while her husband was away in the army. "I hope I am serving my country as every patriotic woman should do," she wrote, "in trying to raise food for the thousands 'in the field' and the thousands more to go."[28]

The upstate-based farm journals also gave coverage to the efforts of farm women on the home front. In October 1862 *Moore's* ran a cover story entitled "War, Work, and Woman." New York State women, the article claimed, were effectively managing farms and businesses in their husbands' absence. It featured a Yates County woman named Margaret Livingston, who asserted that women were proud to exercise their "right to participate in the labors and sacrifices" of the war. She likened farmers' wives to a "home guard," furnishing hospital food and supplies beyond their own families' immediate needs. She even dared to hope that more women would acquire farms of their own as a result of the war. Another New York State family who attracted a good deal of attention in the farm press were the Roberts sisters of Pekin, Niagara County. *Moore's* followed their progress as they took over their family farm, doing all the harvesting, haying, and dairying. The sisters made enough cheese to require a substantial cheese house, so theirs was a case that paralleled the situation of many Oneida County farm families.[29]

That Oneida County women took up field work like their counterparts elsewhere across the state is suggested by a complaint from Rome's Jonathan Talcott. He grumbled in an 1863 letter to *Moore's* that he had "been cognizant of a number of American females who have performed farm work, such as raking and binding wheat in harvest, and other harvest labors." Talcott was deeply offended by the sight, because he believed that "continued out-door labor on the farm by a young girl, or woman, tends to lower her position in social life, not only in the eyes of those who witness it, but also in those that perform such labor."[30] That the sight of women performing field work elicited both excitement and revulsion suggests that the trend was unmistakable.

The war absorbed northern women's energies in other ways as well. As Margaret Livingston intimated, rural women as well as their urban counterparts organized relief societies for the soldiers at the front (and later for the freedpeople). This activity was an extension of their usual work at home, for they were essentially supplying food, clothing, and other supplies—for example, articles such as pillows, "comfortables," and lint—to the men of their families and neighborhood. Nearly every week the Oneida County papers printed notices of the societies' activities; in February 1863, for example, the *Black River Herald* reported that the Ladies of the Augusta Soldiers Aid Society sent bandages, stockings, mittens, lint, and dried fruit to the men with the Oneida County regiments at the front. Later the paper printed effusive letters of acknowledgment from the commanding officers.[31]

In the New York State cheese-producing districts, to these extra activities was added an increase in home cheesemaking. Despite the growth of factory output, fully one-third—about fifteen million pounds—of the state's increased cheese production during the war years came not from factories but from stepped-up production in home dairies.[32] While vats and cutters may have reduced the physical strain involved, the overall impression for the cheese dairying districts is of women's labor and time stretched exceedingly thin. Frank Metcalf, a Westmoreland army volunteer, captured the anxiety of rural men who worried about the burdens their absence had created for women. He wrote to his mother in 1863, imploring her not to volunteer her time for the war effort: "It will be well to beware doing *too* much. You cannot be forgetful of grand-mother's fate, and hard work shall never bring *you* to it as long as I have the power to prevent it."[33]

Thus the Civil War provided several conditions crucial to the rapid spread of cheese factories. The currency situation provided impetus to factory owners and investors. Centralization also supplied a solution to families plagued with labor problems, whether because of a lack of female or of male laborers. In households where women were absent, the household could continue contributing to cheese production without them. For households where women were carrying on in the absence of men, the factory relieved women of a major task and freed them for other pressing work—work that would ordinarily have been done by men.

Expansion in the Postwar Period, 1865–1875

Wartime conditions provided the original impetus to set up cheese factories, but the period of greatest expansion occurred in the decade after the war's close. Between 1865 and 1870 alone the number of factories rose from about 250 to more than 900. By 1875 one observer estimated that over 90 percent of New York State cheese came from factories.[34] Cheesemaking centralized first in Oneida and Herkimer Counties; in 1865 each already had more than one hundred factories, and by 1875 in Oneida County an overwhelming proportion of the milch cows were being milked for either factory cheese, homemade butter, or fluid milk.[35]

Though the shortage of labor eased somewhat, other circumstances that had developed before and during the war persisted afterward and helped to accelerate the trend toward centralization. The 1860s and especially the 1870s were a period of sustained demand in Britain; exports to Britain remained steady or were only slightly down from 1866 to 1870 but rose steadily between 1871 and 1881.[36] The British manufacturing classes, which consumed most of the American product, were increasingly able to afford American cheese, which was often lower-priced than British cheese. Moreover, the uniformity and therefore the predictability of exported American cheese constantly improved, aided also by improvements in transportation.[37] American dairy producers benefited as well from a series of disasters that struck British agriculture in the late 1860s. The dreaded rinderpest devastated many cattle herds, killing thousands. This calamity was accompanied by several years of severe drought. At the same time, though, the expansion of the rail network and the growth of cities also brought opportunity for British dairy farmers; they began to abandon cheese production in order to sell fluid milk. In turn, the market for imported cheese widened still further.[38]

Despite these favorable trends, the decade after 1865 was uncertain economically for New York State dairymen. It was not clear in 1865 that the British market would continue to expand, and in any case production increased so quickly—partly because of the growth of the dairy industry in Canada and the Midwest, especially Wisconsin—that even with an expanding market, prices did not stay high. Delight at prosperity in the early 1860s soon turned to gloom and apprehension, as deflation, price fluctuations, and depression in the postwar period generated fears that the factory system would result in chronic oversupply. Ominously low prices (see figure 2.1) shook cheesemakers' confidence in 1867, 1870, and 1875. According to *Moore's*, 1874 was "A Hard Year for Dairy Men," the first year since the boom began that total receipts actually dropped. During the following two years news items began to appear asking what to do about a cheese "glut," or how to dispose of "Our Surplus Cheese." No less an authority than X. A. Willard lamented that "if the past year [1867] is to be a sample, it will require a magnifying glass of more than ordinary power to see [profits]." Writers began to consider how the domestic market might better be developed. Cheese factories began to produce butter also, or even cheese filled with lard, in order to stay in business. The problem of oversupply grew so distracting that some even began to wonder if it might not be better to revive home cheese production. By 1878 *Moore's* could ask, "Is the Factory System a Failure?" and H. T. Stewart could publish an article in which he suggested that the variety inherent in decentralized cheese manufacture would be preferable to factory production, since "heavy lifting, slopping and other work which might be considered undesirable under our present popular notions might be avoided," and "a hundred heads and minds" would produce better results than a single one. His reasoning was the very inverse of the usual arguments in favor of a factory.[39]

Not only did factory owners worry; at the level of the individual farm there is little evidence that patronizing the factory was more profitable than making cheese at home, or, in the case of patrons who had not formerly made cheese at home, that milk selling represented a significant new source of profit. Contemporary champions of the factory system insisted that the individual farmer profited more from sending milk to the cheese factory than from making it into cheese at home. But the evidence shows much greater ambiguity than these claims would imply. In fact, no consensus existed.

Factory proponents, of course, tried to assure potential patrons of greater profits. They cited the successes of individual patrons. The Oneida County factory manager Gardner Weeks published figures in 1866 showing how patrons at his Verona factory had fared. George Benedict, with twenty cows, received $1,455.75; for milk from the same number of cows W. N. Peckham received $1,557.64. William Wyman, on the other hand, possessed only four cows and received $348.62. These figures represented only factory receipts and did not account for costs to the farmer of producing the milk in the first place. Others did try to calculate the margin of receipts over expenses; a contributor in the *New England Farmer* in 1867 figured an excess, to a farmer with thirty cows, of $271.00 for sending milk to the factory and of $225.60 for home manufacture. The critical factor in this argument was the higher price usually commanded by factory cheese, usually from 1 to 2 cents more per pound than for homemade cheese. X. A. Willard produced figures in 1863 to argue that factories could make cheese at a smaller cost per pound than home dairies. Not surprisingly, his figures showed that larger factories achieved lower unit costs. Lewis Falley Allen did not attempt any rigorous cost accounting but instead looked about him and associated the rise of western New York cheese factories with prosperity. Dairy farms, he thought, were "wonderfully improved," and dairy farmers no longer bumped to town in a "rough lumber wagon" but instead rode in comfort in "handsome, commodious carriages."[40]

But by others' calculations taking milk to the factory did not pay very well. A former patron wrote to the *Ohio Farmer* in 1866 to argue that dairying families knowledgeable about cheesemaking would be better off sticking to home dairying. He pointed out that high-quality homemade cheese fetched prices that were competitive with those for factory-made cheese, and judged the time lost in interrupting work to haul milk to be considerable and valuable. He pointed out that even if factory cheese did receive a higher price, factory charges—up to 2 cents per pound, sometimes more—would offset that advantage. Moreover, many factories kept the whey, and so the farmer had to find a substitute swine feed. Another factory critic seized on this latter issue to claim that factory patrons' farms suffered because soil-enriching hog manure was less plentiful than formerly, resulting in poorer crops and reduced profits. Throughout the 1870s farmers constantly complained about their "small returns." A patron from Oneida Lake wrote in 1872 that at prevailing prices, "despondency fills [patrons'] minds, and they have startling visions of johnny-cake, buckwheat and buttermilk." His words would strike a resonant note with rural people, who regarded these frontier-era foods as emblems of poverty.[41]

One contemporary analyst argued before the American Institute Farmers Club in 1870 that just two types of farmers now prospered in the dairy districts. The first kind, farmers possessing large farms (three hundred or more acres) and large herds, hired a first-class cheesemaker and erected a small factory on their own premises, with capacious vats, a good cheese room, and swine to fatten on the whey. These farmers left the milking and processing to the cheesemaker and concentrated on growing corn and hay. The second group was at the opposite end of the economic spectrum: poor farmers with large families, who could "rent a grass farm" and keep from thirty to fifty cows, using family labor to milk. Everyone could help, the Farmers Club analyst declared, from a young age: "A girl of twelve can be as much service as a man, and yet not miss a day at the district school, and the best men and women of the country have been reared under just such pressures."[42]

This observer's portrayal is notable as much for the people it omits as for those it mentions. Missing altogether are the people who in actuality comprised the bulk of factory patrons. As we shall see later, most milk-selling households fell into a middle range; these farms averaged about a hundred acres, about fifteen to twenty cows, and were operated with a mix of family and hired labor. The Farmers Club writer implied that because they lacked enough capital to invest in a plant but still operated on a large enough scale to pay hired labor, these families were less able to exploit the factory to significant financial advantage.

Given the wide variation in both farm and factory conditions, it is probable that profits also varied. We miss an essential point, however, if we assume that it is in the first place feasible, and in the second place of crucial importance, to reconstruct balance sheets. Few farmers could state their costs with any precision; most attempts to set out an accounting were hypothetical. X. A. Willard, speaking to the New York State Agricultural Society, urged farmers to calculate their costs so that they could see "whether the business is profitable, or conducted at a loss.... The majority of farmers keep no current farm accounts. They can not tell what a bushel of grain or potatoes costs, or in fact any other farm product."[43]

As an agricultural reformer, Willard criticized what he regarded as sloppy and unscientific methods, but he represented a mentality that many farmers did not share. Dairy farmers, with an eye to "competency," had not incorporated rigorous financial accounting into their outlook during the antebellum period, and they did not abruptly change with the coming of factory production. Though they watched prices carefully and knew generally whether they were prospering or doing poorly, they lived in a context where the all-encompassing economic rationalism Willard called for was not relevant, because they still relied extensively on family labor and noncash exchange. As important as prices and costs were social structures. The household labor system was under strain, and the desire to reduce tensions moved thousands of farm families to dispatch their cheese presses to the attic with alacrity and harness up the milk-cart team.

The Social Roots of Reorganization

In order fully to explain the continuing proliferation of cheese factories in the decade 1865–75, we must turn to the social background. The factory system spread so rapidly because it exploited the traditions of local cooperation and at the same time reduced some of the internal contradictions fostered under the home system.

Preexisting neighborhood links developed over generations established a firm basis for the cheese factory. Not only were factory patrons long accustomed to "changing works," they were also bound by other ties. Members of the same ethnic group often patronized the same factory, as the preponderance of Welsh surnames in the list of patrons from the Gang Mills factory in Trenton (Oneida County) suggests; Robertses, Williamses, Griffiths, and Humphreys appeared regularly.[44] County atlases also show emphatically that the crossroads factory was a neighborhood endeavor, as most farmers in the immediate vicinity of a cheese factory sent milk there. The Markham factory in Vernon (Oneida County), for example, developed out of prior exchange relationships between merchant Markham and his neighbors. The Herkimer County atlas shows that factories were often co-owned by several neighbors.

In the town of Manheim (Herkimer County) the Manheim Factory was owned by Rice, Broat, and Co., and within a few miles of the factory lived both H. Broat and J. Rice. Similarly, the Manheim Turnpike Association's Cheese Factory had as president M. P. Timmerman and as secretary Andrew Van Valkenburgh, both of whom lived within a mile of the factory. Kinship too probably played an important role. There were four other Van Valkenburghs and four Timmermans within a few miles of the factory. The factory system found ready acceptance partly because it was built upon these longstanding ties, adapting familiar neighborhood networks to new purposes.

Still more important in tapping a potent social source of centralization was the factories' promise of relieving the internal tensions that had been gathering during the period of household manufacture, persisting and possibly even intensifying in the postwar decade. Though they disagreed on issues of profitability, factory advocates invariably focused upon the gender division of labor in the farm family: the exhausting nature of women's work in home cheesemaking provided their central rationale. This justification addressed the tensions that had been developing during the 1840s, 1850s, and 1860s.

The concept of "drudgery" is a knotty one to sort out in historical perspective. That farm women worked hard is amply documented. At issue is how to evaluate that experience. Did the notion of drudgery erroneously apply middle-class standards of propriety, labor, and anxieties about health? Or did farm women indeed sacrifice their health through overwork? Or did they use the rhetoric of drudgery to their own purposes? Historical studies to date seem to indicate that rural women's experience varied depending on the stage of development in their region, the type of agriculture they practiced, and the time period they lived in. In the Illinois pioneering community of Sugar Creek, the heavy labor of development work combined with the strains of incessant childbearing and lactation exacted a heavy toll. Similarly, agrarian women in Nebraska suffered bitterly. By contrast, midwestern farm women in the early twentieth century criticized reformers' perceptions of "drudgery" and questioned middle-class values of leisure and consumption (as opposed to saving). Nevertheless, there was always potential for conflict pitting women's interests against men's ultimate control of farm resources and labor. In New York State's Nanticoke Valley in the late nineteenth century, farm women were legally and politically subordinated but were able to develop "strategies of mutuality" that helped to offset these disabilities. Nancy Grey Osterud contends that middle-class denunciations of farm women's "drudgery" were "ideological" in nature, and that in reality farm women valued hard labor and took pride in it and did not always view themselves as exploited. But Osterud also observes that *both* the romantic view of farm women's work as imparting economic authority *and* the idea that farm women suffered nothing but unacknowledged drudgery are in error: in fact, the agrarian gender system was filled with contradictions.[45]

The internal contradictions of the farm household gender systems are very apparent in the discussion of "drudgery" that was so crucial to informing centralization in cheesemaking. In the case of cheese, the success of the argument from drudgery was based upon both real and ideological factors. Home cheesemaking made unusually harsh demands on women's physical capacities, demands that increased markedly in the Civil War era. By the 1870s a new generation of farm women were more amenable to the language of middle-class values than their mothers had been. This combination of circumstances made for a receptive audience. Participants in the discussion, moreover, were able to draw upon traditions of both cooperation and separateness in gender relations, turning inconsistent ideas to the same use. This overriding preoccupation with the gender division of work, expressed in terms familiar to all dairying families wrestling with the issue, explains why cheese factories held out such a strong appeal to dairy farming families searching for a resolution to the tensions plaguing them. In fact, if we look closely at Oneida County patterns of organizing factories, we find that while merchants and bankers appear, the overwhelming majority of owners and investors came from the ranks of local well-to-do farmers, many of whom operated substantial home cheese dairies before they decided to build factories. These were the very people who were probably most susceptible to arguments revolving around the gender division of labor.

Discussions of dairy centralization were lively and complex, and worth a close examination. Some factory advocates outlined their arguments in terms that implicitly or explicitly discarded economic rationality as grounds for patronizing factories. One such statement weighed pecuniary gain against labor saving, claiming that if the press of cheesemaking labor upon women was taken into account, it was better to take milk to the factory and receive a net price of 8 cents per pound than to produce farm-made cheese at 8½ cents. The author implied that a lowered income was preferable to continued overwork for women. In 1869 a column in *Moore's* maintained that the primary goal of cheese factories was to relieve families, not to establish private businesses, since milk was not sold outright but processed for a fee. This argument was a bit disingenuous, since its object was to exempt cheese factories from taxation. But the opinion did not altogether misrepresent the factories' purpose. In this respect the factory advocates ironically used anticommercial values to push for a system that ultimately would bring about a rural society more enmeshed in capitalistic ways.[46]

X. A. Willard emerged as a leading publicist for the theme of drudgery. He believed that in cheese dairying, more than in any other branch of agriculture, women grew "prematurely aged and broken [in] health" and suffered from "overtaxed muscle [and] incessant care." At the convention of the New York State Cheesemakers Association Willard proclaimed that "the flesh and blood of our wives and daughters are of too much consequence . . . to be worn out by this ceaseless toil," and he urged dairying families to abandon their cheese presses as an earlier generation had discarded the spinning wheel. The *Oneida Weekly Herald* employed more dramatic terms: "Already whole hecatombs of lovely ones have passed away as fleeting ghosts by constant, endless, exhausting toil and care over butter and cheese." It urged, "For fat cheese, . . . fat farms . . . , fat purses, [and] . . . fat women, put up the factories."[47]

Cheesemaking women added their voices to those of the men who advocated factories. The *Country Gentleman* in 1868 noted that the cheese factory enjoyed great "popularity with our dairywomen, as it relieves them to a great extent of the confinement and drudgery of cheesemaking." At the American Dairymen's Association 1871 meeting in Utica a reporter from the *New England Farmer* observed that central New York farmers, their wives, and their daughters had "dairying on the brain." This was because the new factories promised to "relieve the indoor and outdoor drudgery of the old system, and . . . make farm life more attractive." He added that even for those who had not made cheese at home, dairy centralization would reduce buttermaking work.[48]

A more detailed and colorful statement of a dairying woman's perspective came from a Wisconsin farm wife, Mrs. E. P. Allerton, in an 1875 essay entitled "Dairy Factory System—A Blessing to the Farmer's Wife." Allerton began with a retrospective on the old days:

> Formerly dairy work was a private enterprise, a family affair, which nobody outside the domestic circle had any business with. There may be some here tonight who do not know how it was done. Under the new improvements old methods lie buried out of sight, but it is well to dig them up now and then, if only that we may enjoy the contrast.
>
> There were various ways of turning out home-made cheese. The simplest and most primitive was after this fashion: The milk and whey were heated in the kitchen boiler; the curd was set in a tub, and drained off through a concern that was called a "cheese ladder." The press was a lever wedged under a beam of the wood-shed, with a big stone on the end for the power, and the cheese for a fulcrum. . . . [The] object would appear to be the prying up of the wood-shed, but no, it was to press the cheese.
>
> It was not usually managed in that way, nor after so small a pattern. In many farm houses the dairy work loomed up every year, a mountain that it

took all summer to scale. But the mountain is removed; it has been handed over to the cheese factory, and let us be thankful time does not hang heavy on the hands of the farmer's wife now that it is gone. She does not need the dairy work for recreation.

Mrs. Allerton continued with a sarcastic barb at what she regarded as nostalgically naive portrayals of the old-fashioned dairy room:

> It is customary, or rather it was, for writers upon country life to expatiate upon the cool dairy room of the farm house, with its rows of nice cheeses, smooth and fragrant, . . . and the womanly pride of the housewife in showing them. This is the poetic side of the picture. But turn it over once more. . . . Behold her, as I have in more than one instance, performing the various functions of milk maid, house maid and dairy maid—yes, and of nurse maid, also, at the same time. Hear her say at dark that she had not sat down a moment during the day, except to take her meals, and that every bone in her body ached—that is prose, of the hardest sort.

She concluded with a critique of the idolatrous "worship" of work for its own sake:

> Hearty, honest work is a good thing for us all; but how much of it? that is the question. For my part, I think a little rest—a little blessed idleness now and then—is good, for a change. I hate to hear it said of a woman that "she is always at work." If she can't help it, she is to be pitied, and if she can help it, she is to blame. . . . A wife should not forget that she has something else to keep clear of rubbish than the house she lives in. If there must be cobwebs anywhere, it is better they should lurk in dark corners of the room, than in her heart and brain. . . . If [the farmer] wants a companion, she must have a little leisure now and then in order to be companionable. . . . All work and no recreation of any kind,—what does that make of a woman? A machine, a dumb automaton; but she will wear out some day. Lucky for her if she goes to pieces all at once and all in a heap, like the famous One-Horse Shay.[49]

Mrs. Allerton echoed a growing strain of thought then commonly expressed in the farm press when she criticized women who "always" worked; her articulate comments revealed an important dimension of factory supporters' thinking when she expressed hopes for a new balance between work and leisure.[50] Among the positive benefits to women, factory advocates invariably envisioned "comfort, leisure, and culture," "mental cultivation and social enjoyment." Some even found in worldly sophistication a worthy goal; it is no coincidence that to illustrate a contrast to "gentility," an author writing in the *Prairie Farmer* chose the "countrified" pursuit of making cheese. In central New York A. A. Hopkins, editor of the Rochester-based *American Rural Home*, spoke to local audiences condemning the work that rendered farm women "less and less a fraction of the married unit"; he hoped for conditions that would permit wives more "personal comfort, personal culture, and personal growth." That Hopkins and many other male factory advocates mentioned ideas such as these suggests that men were beginning to see advantages for themselves in centralization. If farm wives were relieved of the burden of cheesemaking, they would then be able to provide enhanced services for their menfolk. The possibility that factories would benefit *both* men and women helped to accelerate centralization.[51]

The expressions of women's viewpoint on dairy centralization placed some value on material comfort and on improved living standards, but the elements stressing intellectual, cultural, and social stimulation are much more prominent.[52] Even the criticisms of cheesemaking as "confining," while at the simplest level connoting a value set on physical freedom, also implied freedom from mental and social narrowness. These ideas suggest resistance to the unequivocal subordination of women's aspirations to family goals and refer implicitly to the conflict between the two.

Notably, motherhood was not prominent in these pronouncements. To be sure, Allerton placed woman at the "center" of the home and mentioned her function as "nurse maid," but she did not contend that the cheese factory would allow women to spend more time with their children. This was probably partly a function of the place home cheesemaking occupied in the farm family life cycle; child-rearing concerns would not be paramount for families with teenaged and young-adult children. But the thrust of Allerton's message is nonetheless significant: she placed heavy emphasis on the woman's capacity as a wife, anticipating that leisure would make the wife more "companionable." This emphasis reinforces the fact that the concept of "domesticity" or "separate spheres" as conventionally styled by historians was not necessarily a model for the farming community.

Though it addressed household conflict, the terminology Allerton used and the way she framed her conceptualization were shaped by the language of mutuality. Here the preceding decades of experience in the agrarian household—the day-in, day-out ideology and practice of cooperation between men and women workers—assumed great influence. In the 1860s and 1870s dairying women redefined cooperation in the realm of labor as the ideal of a "companionable" emotional and marital relationship.[53] This ideal fit compatibly with the individualism expressed in the justifications for the factory, because individual cultivation was regarded as a prerequisite for the companionable marriage relationship. The earlier conflict between women's individual aspirations and the cooperative work ethos was thus eased.

Critics of home cheesemaking who derived their reasoning from the ethos of mutuality did not necessarily view women as inherently incapable of cheese dairying work; many assumed that factory work would be less taxing than cheesemaking under the home system and envisioned that women would staff the new factories: "John Bull objects to our American cheese factories because they will deprive his wife and daughters of an honorable and healthful employment, and by relieving them from labor, lead to idleness and extravagance. Could they not possibly work in the cheese factory, or do anything else but make cheese in the old way?"[54] The cheese factory therefore mitigated conflict not by eliminating *all* women from cheesemaking, but by separating cheesemaking from other work. This explains why the early factory labor force included large numbers of women: many women did not reject cheesemaking itself, but rather the inequitable burdens they had to carry under the household system.

The language of reciprocity, of fairness in the sexual division of burdensome labor, also framed a popular analogy in which supporters compared the factory with other recent technological developments affecting men's work. The *American Agriculturist* in 1878 drew a revealing parallel: "What the mower, reaper, and thrashing machine have done for the farmer, the cheese factory has done for their wives and daughters." Rather than contrasting men's and women's situations, thus implicitly assuming a fundamental and qualitative difference, this idea *compared* men's and women's work. It is worth noting that in general the trend was increasingly in the opposite direction; the farm press regularly carried pieces complaining that men's farm work, but not women's, was mechanized. Cheesemaking thus provides an important exception to the prevailing pattern. The reorganization of the women's work of cheesemaking may have proceeded quickly partly because overlap between men's and women's work was so highly developed that dairymen understood the work well.[55]

Ironically, though it arose from circumstances of cooperation, the language of drudgery also invoked notions of gender difference. The work of cheesemaking was highly visible not only because men knew about it but also because it generated cash income. Its association with market activity encouraged the cultural assumption that the work was more properly suited to men than to women, thus revealing the ideological dimension to critiques of farm women's overwork.[56]

The rhetoric of drudgery assumed gender difference in the idea that women's physiology and "constitution" differed from men's. X. A. Willard especially, and to an extent Mrs. Allerton, believed along with many contemporaries that female physiology, especially reproductive physiology, rendered women vulnerable to ill health and to both physical and mental strain. Hence the numerous correlations of overwork with physical and mental collapse. Some historians have argued that concerns over health arose less from objective circumstances than from deep-seated anxieties about the direction of social change. In this sense, though the role of objective circumstances was real enough, the concern over drudgery was part of a more generalized cultural anxiety about women's health. For example, some nineteenth-century thinkers borrowed from the law of conservation of energy to argue that women were born with a finite amount of energy, and that most of it was taken up in reproduction; they used a vocabulary that suggested women paid a physiological "penalty" or "tax" for work beyond these narrow limits.[57]

Thus the justifications for the cheese factory combined two intertwining sets of ideas of gender relations, those of mutuality and those of separateness. Though they drew upon quite different bases, the argument for reciprocity and the argument based on women's vulnerability reinforced one another: if women were more susceptible to overwork than men, then fairness demanded that women be relieved.

These arguments possessed a powerful appeal to a generation of farm women raised to understand both the language of mutuality and that of separateness. The women taking their places on central New York dairy farms in the 1860s and 1870s were the same women who as young people had argued with their parents and grandparents over farm work and education. As we have seen, conflicts between the sexes and generations harbored explosive potential, both in emotional terms and in terms of the farm's direction. Rationalizations for dairy centralization reworked the themes of earlier conflicts. For example, denunciations of drudgery focused upon resolving the tensions among individual aspirations, middle-class values, and family goals—the same issues that had been involved in the debates over whether women and girls should "do the milking."

The experience of many in this generation included exposure to new values and ideas at the numerous schools in the region. As we have seen, the educations acquired at these institutions encompassed a wide range of possibilities, but girls who attended were exposed to values of intellectual endeavor and cultural achievement—the very values stressed in the rhetoric of factory advocates. Moreover, we know that in many Oneida County schools girl pupils were introduced to prevalent ideas about women's physiology; the health reformer Dio Lewis, for example, taught classes at at least one of the county's private academies. Young women whose training encompassed these influences were likely to come to farming with expectations different from those their mothers had brought. They also seemed inclined to act on those expectations: Mary Barnes, for example, seemed to be showing initiative to pursue her own goals even under family protest. Thus young men and women stepped up the pressure for change. To this extent the prosperity of home cheesemaking helped them to create new lives for themselves and to participate actively in the shift to factory production.

The cultural dimension of industrialization in cheesemaking is thus evident in a number of areas, from the repertoire of models for gender relations to the influence of schools. Another important element in women's response was ancestry: Oneida County factory patrons came mostly from Anglo-American backgrounds. By 1870 the few remaining home cheesemakers in Oneida County were disproportionately immigrants from Switzerland, Germany, and France. Their traditions and gender systems were different, and they did not abandon home cheesemaking as readily. This ethnic division overlapped to some extent, though not completely, with class lines.

Women's Choices in Historical Context

Cheese factories had replaced the household business with astonishing speed. Women took a prominent role in the process. In contrast to the part women had played in the rise of home dairying,[58] however, through their activity in promoting dairy centralization they contributed to what historians have rather clumsily labeled "defeminization" in agriculture. Bengt Angkarloo, in a 1979 essay, maintained that one general "direction of change" in the West from 1700 to 1900 was "toward a more pronounced preference for males in the agrarian labor force." He associated the preference for males with "capitalist production for the market," maintaining that "opportunities culturally defined as women's work" disappeared along with diversified subsistence peasant production. The result, Angkarloo argued, was disproportionate out-migration by women, a "defeminization" of agriculture. Other historians have since followed Angkarloo's lead and examined specific instances of women's exclusion from various farm tasks, in various countries. In Sweden, cultural shifts in the gendering of work underlay preferential treatment given to men by both government and business in the early twentieth century, leading to women's exclusion. In Canadian dairying at about the same time, the increased importance of capital in dairying, combined with patriarchal household structure, allowed men to assert control.[59]

All of these accounts document the disappearance of women from agricultural production, a process that took place at varying paces in different settings. All also associate the "defeminization" process with the rise of capitalist agriculture and with shifting cultural concepts of men's and women's work. In assessing this trend most of these historians stress the evidence of a loss for women—a loss of skill, of occupational opportunity, and of economic power.

But to emphasize the element of decline in this story is to minimize women's involvement in the process, and to deny the legitimacy of their reasons for wanting to abandon home cheesemaking. In nineteenth-century New York these women were not helpless onlookers, stripped of an occupation they wanted to retain. Nor were they grateful recipients of patriarchal benevolence, as earlier historians tended to assume. Rather, they were active participants in their own withdrawal from home cheese dairying, motivated by contradictory ideas—by stereotypes of female vulnerability, but also by their belief in an equitable distribution of labor. They believed, with some justification, that they were making a trade-off that benefited them: they hoped to exchange hard, disproportionate labor for a greater latitude in the conduct of their everyday lives.[60]

The evidence from cheese dairying regions thus suggests that rather than resisting, many women actively embraced the new order. Their behavior and ideals bear out historian Allan Kulikoff's speculation that "by underestimating struggle within the household, social historians exaggerate resistance to capitalist economic relations."[61] Though there is evidence that Nanticoke Valley women helped to resist capitalism, we must consider that they were among those who remained on the farm, perhaps precisely because they had employed strategies of mutuality with some success. In order fully to explain the dynamics of rural change, it is also necessary to tell the story of those huge numbers of women who left, for whom mutuality failed; this story is a critical measure of discontent and inequity. In short, capturing women in transition points up the limits of mutuality. Ironically, while men's craft traditions often served as a basis for resistance to industrialization, this female-dominated craft tradition tended to function in the opposite way. Though they were able to control their day-to-day work to a significant degree, women cheesemakers wanted to withdraw because they lacked control over broader circumstances. Thus rural women's relationship to capitalism did not take just one form, but rather spanned a spectrum from resistance to enthusiasm, depending upon individuals' experience, background, and options.[62]

That dairying women were able to turn centralization to their advantage shows not only their struggles within the household, but also the importance of circumstances. In contrast to Nebraska farm women, eastern and midwestern dairying women had several advantages. They did not live in isolation like their plains-dwelling counterparts, and so they could maintain contact and exchanges with kin and neighbors. They participated intimately in a cash-generating activity that was widely recognized as important, whereas

Nebraska women's egg and cream work, though objectively as important to family income and subsistence, was perceived as a "sideline" because it fell more squarely within women's realm of responsibility. Though dairying women drew upon resources outside of the nuclear family to improve their position, the evidence shows that the nuclear family itself was often also a source of support. In this regard the exploitativeness of the nuclear family was not so thoroughgoing in dairy country as it seems to have been in Nebraska.[63]

While traditions of both separateness and mutuality justified women's withdrawal from home cheesemaking, once centralization was established, dairymen and factory owners would deemphasize mutuality in favor of gender stereotypes: they began to recast the work more exclusively in terms of gender qualities, rather than thinking in terms of an equal distribution of work. Ultimately dairy centralization would bring unanticipated pressures to exclude women entirely.

Thus political and economic events precipitated the transformation of cheese manufacture, but circumstances within rural households sustained it. In turn, as factory owners, managers, workers, and patrons shaped new procedures and work patterns, the factory system would give rise to a new set of social arrangements.

7

The Social Organization of Factory Work
Cheesemaking Becomes Men's Work

> I am only getting 73 and 7400 lbs of milk a day only making 12 chees they only wey from 70 to 80 lb a piece have only got out 519 chees have a nough to do to give me an apatit.—H. W. Smith, 1871

Early in the summer of 1870 a young man named Harvey Day left his home in Ypsilanti, Michigan, to learn the cheesemaking trade at a nearby factory, boarding with the factory manager, Mr. Smith, and his family. Day had an imposing task ahead of him: he hoped not only to master the intricacies of an unfamiliar and difficult craft but also to live up to the expectations of a demanding mother. Harvey had barely settled into work when a letter from Hannah Day arrived:

> I hope this will find you well and enjoying yourself in a strange place and at a strange work hope you will be conten[t]ed have good luck and learn your trade it will take courage and patience to do the work you will remember it is the one that holds out to the end that winds the first do not git home sick and leave till you get your trad leart but perseverence in faith will do the work and make the Man.[1]

Day evidently could not get home for the first holiday, the fourth of July, after he commenced his job, for on the sixth Hannah was writing, "I looked for you to be up on the fourth but didnt see you hope you had a good time if you spent the day in making chees that was not spending your money and time to no good use."[2] Hannah tried to make sure that Harvey's devotion to work never wavered:

> You must put your mind rite to work and not let those nice young Lady excite and draw your attension from your work you must make cheese when the milk comes the Lady will always be about so you will see that think work and plenty of Money is the first and the Ladys second A good worker is never at a

loss for friends be strickly honest and that is worth more than money be calm and resulute with patience and you will gain a good deal by it. . . . be a Man every time.³

Inspired—perhaps *prodded* is a more accurate term—by his mother's solicitude, Harvey Day plodded along at the factory. Hannah was not invariably stern; she rejoiced that her son "was not homesick," and she sympathized when Harvey "had to work so hard all the time and get no more for it then if you only made one cheese."⁴ But she found consolation in the prospect that "if you keep well you will learn something and it is not so hard for you as if on the farm and you will not wear out your clothes so fast." She was glad too that Harvey had a proper home; she felt reassured that "it is a good place that is the folks will bee kind to you and if you are sick they will care much for you and that is worth a grate deal when we are from home to know we have friends if needed."⁵ While he served his informal apprenticeship, Harvey formed close ties with the Smith family; the next season, when he had taken up a cheesemaking post elsewhere, H. W. Smith, the manager's son, wrote Harvey, "I miss your company very much" and "Your place is missed very much at the table."⁶

Like home cheesemaking, factory cheesemaking was grueling physical labor. Harvey frequently heard from colleagues in the business, most often from the younger Smith. Smith worked seven days a week. At peak times he was confined to the factory as milk flooded in: "We are having an increase of Milk we have 3847 lbs a day. . . . we have the vat as full as we can work we have the hoops as full as we can press them handily. Mr. Plank is making a new vat I expect to hafto send some [milk] back before we get the vat done we expect four or five thousand lbs of milk, it will make us lots of work."⁷ By September the flow of milk had abated, but by then, Smith said, "We have nearly 600 chees a day to turn." The following season Smith ironically reported, "I am only getting 73 and 7400 lbs of milk a day only making 12 chees they only wey from 70 to 80 lb a piece have only got out 519 chees have a nough to do to give me an apatit."⁸

To the sheer physical burden of the work was added pressure to get along with recalcitrant patrons. When Harvey suspected that some of the milk he received was half-spoiled, he faced the unpleasant prospect of having to turn away a patron. Day's friends urged him to return "changed" milk if he detected spoilage. Smith had his problems with patrons too; he left the factory early to celebrate the Fourth in 1871, only to learn that four patrons had dropped off milk while he was gone. He returned to discover that "they had all gone from the Factory weighed their Milk and run it off except 136 lbs that they left in the can they didnt put any names down nor figures so we had to guess on the Rennet and Salt."⁹

Factory cheesemakers did not always work alone; they needed assistants, and so Smith was able to write "from the Factory" in 1871 that "Herb Tiffney is helping me this Summer he wants to learn the trade Mr Plank payes him $8 dollars a month he is quite a good hand." The following season Smith wrote to Harvey:

> I am not quite so lonesome as what I was I have some company now. I have a Woman to help this summer. She is pretty good help, but after all I have to do all the lifting in of Milk and turning of the Chees. . . . We took in over 4100 lbs of milk this morning made 8 large Chees. . . . I put the milk in the No 2 vat first when I get half in it I let the Woman set it and then when the rest of the Milk is in I set mine, by that time she can go to work in hers and as soon as hers is done I let her work mine and I go to taking the C out of the Press and weigh out the salt and by that time the first Vat is ready to take out, and as soon as we get it salted I let her go stiring the other Vat and I put it in the press, then that is ready to take out.[10]

At different times several of Harvey Day's female friends worked as assistants in neighborhood cheese factories. In 1871 Ella Comstock wrote to him, "I have been up to the cheese factory for a week and a half I don't expect to stay long we have good times working together." A bit later, she "came home from the factory last Wednesday night they did not really need me any longer and I was ready to come home any time." When Harvey had been ill she offered to come and help him "if they are willing and you need the rest." The next season Ella noted that "they have lots of work to do in the factory. Ella Lord has gone and Clara Fisher is there now. They wanted I should come, but I did not see how I could while our folks were gone." Her letters make it plain that Ella combined periodic factory employment with farm work. In July 1871 she told Harvey, "Our folks have away with Mary Janes today and Lydia and I have had the afternoon all to ourselves I have milked the cow and Lydia is feeding the pigs we are going to have all things done before they get home." Ella enjoyed the variety in her work, but she knew that for a single woman, her position afforded little security; she fretted about becoming an old maid and concluded that if she met such a fate she would "be good natured so folks will like to have me stay with them."[11]

As the season advanced and the cheeses accumulated in the curing house, makers watched them anxiously for signs of "huffing" (bloating), leakiness, skippers (insect pests), or any of the other defects that could indicate a poorly made, possibly even unsalable article. The most conscientious application was no guarantee that the cheese would turn out well. "Harvey I forgot to speak about the Rawsonville factory," Smith wrote in the fall of 1870. "They have had bad luck lost a good many cheese—we heard they had lost 5 ton." Later, Day's associates shared his bewilderment and consternation when a batch of his own cheese went bad: Smith, perplexed, exclaimed, "I never had any chees that leaked whey when pricked." All he could suggest was that Day "see some of them other Factory makers" to try to find out what went wrong. In the meantime his friend Ella Comstock counseled, "I would not throw away any more cheese would sell them for a little less . . . make them harder and let them get colder before you put them in the press." This problem was the more perplexing because Day had produced excellent cheeses the year before; Smith had thought Harvey's cheese looked "splendid" and reported that another colleague thought it "beet mine all out." [12]

When at the end of the cheesemaking season the cheeses were finally safely off to market, Day received offers of other positions. Hannah's response to her son's success is not recorded in his papers, but her later letters are filled with resentment of what she regarded as Harvey's inattention. Whether or not he tired of her admonitions, he learned his business well, eventually becoming a prosperous dairy farmer and leader of the Holstein-Friesian Association. He also found time for the "Ladys"; he eventually celebrated a fiftieth wedding anniversary with his wife and co-worker, Lenore ("Nonie"). In a poem he composed for the occasion, Day fell to reminiscing about the cheese factory days, by this time—probably the mid-1910s—long gone:

> Somehow the old place and factory
> Comes to my memory today;
> Then we had cheese a-plenty
> But not present comforts, I say.
>
> We spent many years with the cheeses
> We worked hard and had our fun, too;
> But this here is the kind of living
> I hope we always may do
>
> Out with the cows in the morning,
> Getting the foaming whey,
> Good food a-plenty at mealtime,
> And rest at the end of the day.[13]

Harvey Day left an unusually vivid record, but his experience was shared all around the dairy districts, from the Midwest to the eastern seaboard. By the 1870s and 1880s thousands of factory hands, patrons, and managers were working together to turn out tons of cheese annually. Where everyday operations were concerned, various new arrangements obtained. Sometimes the cheesemaker was paid by the pound—in which case he or she usually hired assistants—sometimes with a salary. Whey was variously redistributed to patrons, fed to hogs, made into whey butter, or simply flushed away. Patrons supplied rennets. Charges for cheesemaking ranged from 1 cent to more than 2 cents per pound. Costs for bandages, salt, boxes, coloring, and the like were sometimes included in this charge and sometimes figured extra, in proportion to each patron's contribution of milk. Provision was always made for inspection and for fines in the event that patrons violated provisions mandating clean, whole milk. As the cheese was sold, patrons were paid by check, in proportion to the amount of milk they contributed.[14]

More significant for the social historian are the emerging patterns of work, social organization, and material culture under the new factory system. These patterns represented not so much the complete abandonment of older arrangements as their adaptation—not radical changes but incremental ones. Factory cheesemaking's dramatically enlarged scale, capitalization, new technology, and highly visible architecture obscured strong forces for continuity in social organization, forces that for a time would coexist with trends toward change. For example, to a surprising extent cheesemaking continued to be socially organized on a household basis, in the sense that factory owners, managers, and workers often shared a common residence, family ties, or both. Moreover, men and women still worked together to make cheese. Women sometimes continued to occupy the positions of skill and authority they had possessed under the home system.

Along with these established patterns, however, new ones were developing. As cheesemaking was recast into a quasi-industrial process, workers' relationship to the enterprise also began to change. Most significantly, the basis was being laid for excluding those women who had chosen to continue cheesemaking in the factory setting. The "defeminization" of cheesemaking was accomplished through shifts in the organizational structure, rhetorical basis, and financial foundations of the industry.

Buildings and Equipment

Certainly external appearances communicated a message of dramatic change. An entirely new institution had appeared on the rural scene, offering an important new focus for economic and social life. It is no wonder that rural residents saw in the erection of these structures a revolution. To begin with, the cheese factory introduced a recognizably new building type into the landscape. Arranged in single buildings or in complexes of two or three, they stood out in several respects. Most obviously, by their location, scale, and distinctive features they defined new industrial, commercial, and social spaces.

Most often factories stood independently in open-country neighborhoods, introducing a prominent new feature into the agricultural landscape. Many occupied a crossroads along with other important neighborhood institutions, the village school and church. Still others formed part of small-scale commercial and industrial complexes—stores, blacksmith shops, wagon shops, cider and grist mills—at village centers. A reporter from *Appletons' Journal* noted in 1878 that "warm dusty roads" led through the fields to the factory, past herds of "indolent cattle." The factory, a "large, freshly painted, neat looking building of stone and wood," was sited near an "old farmhouse and orchard."[15]

No matter what the context, cheese factories stood as visible signs that farming—and together with it, rural life—was undergoing deep transformations. Indeed, some factories' very origins signaled departures from conventional building customs, because they were designed and built by people who advertised factory designs, thus bypassing local neighborhood collaboration. William Blanding of Broome County, for example, advertised in the 1877 *Report* of the American Dairymen's Association that he was an "Architect and Builder of Cheese and Butter Factories."[16] More often, factories were locally designed and built, and it was their physical features that reminded rural residents of the factories' function.

Sometimes the sheer dimensions themselves departed from the common rural building context. The Fairfield factory in Herkimer County, for example, though a conventional 38 feet wide, stretched out to 148 feet in length and rose to three stories. The Willow Grove Factory in Oneida County consisted of an elaborate complex: make-room, 30 by 28 feet; press room, 14 by 26; and, at right angles to the first two, a curing room, 30 by 100. In the Whitesboro factory, also in Oneida County, the curing room alone measured 30 by 104. Regularly spaced rows of windows on both the second and third levels reflected the emphasis on uniformity and efficiency in production and hinted at a large storage space (figure 7.1). Chimneys, often tall and prominent, marked where heaters or boilers operated, and served further to distinguish factories from agricultural buildings (figure 7.2). Often too the smokestack revealed the position of a substantial engine room. The factory's most distinctive feature was the delivery window (figure 7.2); often covered over, this opening was cut so as to permit easy unloading from wagon height. It was usually above the factory floor, so that milk could be easily poured into vats below. As the point of contact between patron and factory, the delivery window served as visual symbol and as intermediary space. It separated patrons from the pro-

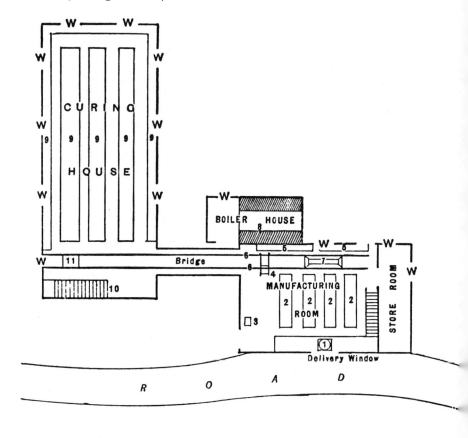

Figure 7.1. Plan for the Whitesboro factory, Oneida County. The basic layout of the home era is extended and elaborated. *1*, weighing machine; *2*, milk vats; *3*, engine; *4*, curd mill; *5*, presses; *6*, rails for sink; *7*, sink; *8*, boiler; *9*, cheese tables; *10*, stairs to upper room; *11*, cheese wagon. J. P. Sheldon, ed. *Dairy Farming* (London, c. 1880), 475.

cess, enforcing an air of exclusiveness and reminding them of management's potential power to reject milk and inflict public humiliation. Sometimes this message was reinforced by an arrangement whereby the window obscured the patron's view into the factory (figure 7.3).

Siting further associated cheese factories with other types of industry. Dairy advisers regarded a steady, preferably running supply of "pure spring water" as essential in siting, because water was indispensable in cooling and

Figure 7.2. Sanborn factory. With its two-story arrangement, delivery window, and smokestacks, this was a typical early cheese factory. X. A. Willard, *Practical Dairy Husbandry* (New York, 1877), 369.

cleaning.[17] A letter to *Moore's Rural New Yorker* in 1861 described the newly erected Ridge factory, near Rome, as sitting "at the bottom of a small hill, from which gushes one of the coldest and clearest springs in the county of Oneida."[18] Water was also useful for waste disposal; if the whey was not fed to hogs or made into butter, it had to go somewhere. X. A. Willard explained why: "It is usual to have a considerable stream of water passing under the manufacturing room, so as to carry off the drippings of whey and refuse slop, so that there be no accumulation of filth or taint of acidity hanging about the premises. Where whey and slop are allowed to collect from day to day about the milk room, the stench at times becomes intolerable."[19] Then as now, the resulting pollution became the problem of some unfortunate further downstream.

Inside too the factory presented novel vistas. In some, interior spaces were remarkably open and light. To reduce the number of obstructive posts, some

Figure 7.3. Factory delivery window. The arrangement betokened both efficiency and control. *Harper's New Monthly Magazine* (Nov. 1875), 818.

factory designers suspended the second-story ceiling from cast-iron trusses so as to reduce the number of supports needed below; L. L. Wight of Oneida County achieved an unusually open effect by using "iron rods suspended from bridges in the attic."[20] Rows of evenly arranged vats and presses spoke to the enlarged scale of cheesemaking. Sometimes spaces were connected by rail carts, vivid reminders of the transportation revolution. The curing room attested even more emphatically to the arrival of mass production (figure 7.4).

The equipment itself must have impressed visitors. Though none of it was so entirely foreign as to be unrecognizable, it surely all connoted a significantly enlarged scale. The emergence of cheese factories had stimulated a boom in technology and in patent activity. An 1877 article in *Moore's* listed basic equipment as including a three-horsepower steam boiler, six-hundred-gallon vats, gang press, curd sinks, scale, knives, hoist, milk agitator, and dippers.[21] Some factories used self-heating vats, but in many, large boilers supplied steam for power and for heating the vats, cleaning equipment, and so on; these supplanted smaller home types.[22]

The boilers required their own space, for safety and to keep room tem-

Figure 7.4. Curing house, Whitesboro factory. Mass production contrasts with the smaller scale of home-based cheesemaking. Sheldon, *Dairy Farming*, 476.

peratures down (figure 7.5). Moreover, their operation and maintenance demanded technical skill and knowledge. The shift to steam power was only just under way in American manufacturing, so for the contemporary observer these boilers inescapably linked even the humble crossroads factory to up-to-date modern trends. The huge vats commonly held from two hundred to six hundred gallons of milk at a time, quantities beyond the experience of most former home cheesemakers (figure 7.6).[23] The *Black River Herald* marveled in 1863 at an enormous new vat under construction by hardware men Riggs and Colton.[24] Measuring 14½ by 4½ by 20 feet, it held six hundred gallons. Even though he lived right at the gateway to the "North Woods," Mr. Colton had to travel far in order to find lumber big enough for the job. He lined the vats with "heaviest 4 cross tin." Vats continued to be made by local craftsmen—Mr. Plank made one for Smith's factory—but most were purchased from dairy supply companies. They thus represented links with the wider economy not only through their function but through their origin.

"Gang presses," marvels of efficiency compared with the old tools, pressed a number of cheeses simultaneously (figure 7.7). Finally, a plethora of com-

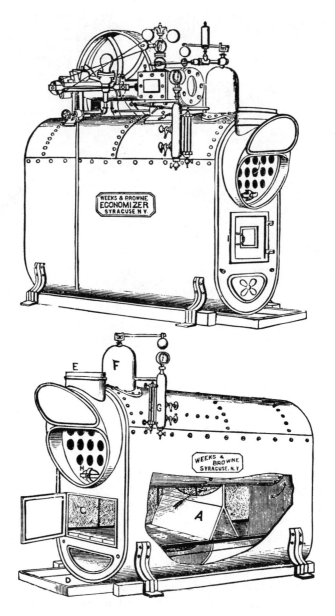

Figure 7.5. "Economizer" boiler, showing the incorporation of steam power into cheese manufacture. Lauren B. Arnold, *American Dairying: A Manual for Butter and Cheese Makers* (Rochester, 1876), 318.

Figure 7.6. Factory vats, c. 1875, resembled older equipment in most respects except for scale. *Harper's* (Nov. 1875), 821.

mercially produced accessories—curd knives, curd mills, milk cans, strainers, boxes, borers (for sampling cheese), and curd sinks—supplanted homemade ones. At some factories winches or pulleys aided in lifting heavy cans and cheeses. Some manufacturers introduced steam-powered agitators to stir the milk in order to keep it cool, and also to prevent the cream from rising.[25]

Yet though observers might notice their more novel features, cheese factories also borrowed heavily from familiar vernacular building conventions. The structures only minimally disrupted the regional vernacular repertory. Particularly after the initial phase of factory building, newly erected factories usually shared the common scale of rural buildings;[26] manufacturers found that smaller factories, serving a smaller area, produced a better product. In materials and exterior features too factory builders followed local convention.

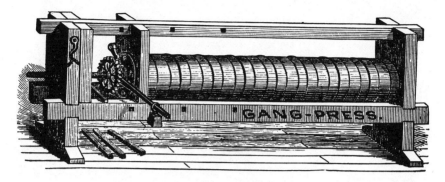

Figure 7.7. Gang presses accommodated larger-scale production. X. A. Willard, *Practical Dairy Husbandry* (New York, 1877), 403.

The much-publicized factory in Sanborn, New York (figure 7.2), replicated the masonry basement story characteristic of barns in the dairy regions. Vertical siding also recalled a popular covering for farm outbuildings. Ventilators astride the roof ridge were familiar sights on country barns. Windows too were a common type—six-over-six lights—and occasionally shuttered. Even the delivery window recalled a familiar local pattern in barn building: throughout central New York one can still see barns with small, covered entranceways built on earth banks so as to permit access to the barn at an upper level.[27]

Common construction techniques predominated. Factories consisting of several single-story buildings were usually of balloon-frame construction, sometimes—as in the case of curing houses—raised off the ground to keep pests out and help air to circulate freely. Those that housed all operations in a single multilevel structure were usually built with timber frames. In 1862, for example, Gottlieb Merry of Verona, New York, built a factory with traditional timber framing. The "industrial" building materials in Merry's factory—iron trusswork supporting the upper story—were hidden under a conventional-looking skin, much as tall, iron-framed urban buildings of the day were clad in stone facades. Thus the exterior features that distinguished the nineteenth-century cheese factory from a barn or other farm building were really very few.

In plan, factories were fundamentally enlarged versions of the private cheese dairy, divided into spaces for making, pressing, and curing cheese, distributed either horizontally or vertically. In the Sanborn factory (figure 7.2), for example, manufacturing and pressing took place in adjoining spaces, while the boiler stood in an ell. Cheese cured in a separate room on the second story. These arrangements recalled the private dairy plans of Gurdon Evans, from the early 1850s. In the Whitesboro (figure 7.1) and Willow Grove factories, make-room, press room, and curing room were spread out horizontally, linked together by passageways with rail carts to facilitate moving curd and cheese. These arrangements also developed ideas contained in designs of the 1840s and 1850s.[28]

Similarly, most devices used in producing factory-made cheese were larger and more sophisticated, but not fundamentally different, versions of home equipment. As we have seen, home producers were already familiar with vats, albeit on a smaller scale. Curd knives and sinks too differed little from earlier versions. Presses still employed the same principles—and encountered the same problems. Patents for these still tried to solve the ever-present problem of even pressure. Farmers were surely well acquainted with pulleys and winches.

Social Arrangements

The observer acquainted with home cheesemaking would also note familiar social arrangements. The age configuration carried over from home cheesemaking: as home cheesemakers had relied extensively upon the labor of older sons and daughters and of young hired hands, so factory hands (like Harvey Day) were primarily young people in their teens and twenties. Factory owners and managers were usually older, as it took time to accumulate capital and experience.

Moreover, for many cheese factory employees work was still inextricably tied to household and family. There is evidence from the Harvey Day papers, for instance, that cheese factory workers worked on managers' or owners' farms during the off-season. Indeed, the marked seasonality of cheesemaking continued old patterns; the season was extended, but still the work was far from year-round. Another indication that the cultural connection between cheesemaking and family persisted into the factory era is that manufacturing facilities often contained family quarters.[29]

Occasionally—there is no way of knowing how consistently—the census taker noted when someone "worked in a cheese factory" or considered himself or herself a "Cheese Maker." In the seven Oneida County towns' population census for 1870 we find a number of persons involved in factory cheesemaking. This information can tell us something about how factory cheesemaking was socially organized. Factory employees still frequently lived under family governance as children. In Trenton, William Wheeler owned a cheese factory erected on his farm premises. On the estate in 1870 Wheeler farmed his 250 acres, sending 30,000 gallons (about 258,000 pounds) of milk to the factory from his sixty cows. His daughter Esther (31) kept house, while his son George (29) was the "Cheese Manufacturer." Two teenaged sons and one infant lived at home. Boonville's Isaac Weller appears in the 1875 census as a cheesemaker. His son Horace (21) also listed himself as a cheesemaker. Weller's wife is listed as without an occupation, yet we know from newspaper accounts that she collaborated in running the family's factory. Also at home lived a daughter (19) and Isaac's mother. In Steuben, William Hunt (56) owned a 180-acre farm and kept nineteen cows; his wife, Sarah (42), kept house, while his son Lowrie (21) was a "Cheese Maker." Besides Lowrie, the Hunts had seven other children. Another household in which cheesemaking was still very much a family affair was that of Charles Bussey (46) of Verona. Charles, who owned no real estate, called himself a "Cheese Maker." His wife, Louisa (48), and daughter Helen (18) both "Work[ed] in [the] Cheese Factory," while Mary (17) kept house.

Sometimes cheese factory workers lived in situations that paralleled the earlier pattern in which farmers boarded their unrelated live-in hands. Harvey Day lived with his boss. Oneida County factory cheesemakers show the same tendency. In Verona, Gottlieb Merry was already a "Cheese Manufacturer" at age 27. In his household (besides his wife, Sarah, 21) lived cheese factory employees Ellen McGann (26) and Lettie Roberts (16). Next-door lived another cheese factory worker and his wife. Similarly, John Van Vlack, a "Cheese Maker," boarded cheese factory worker Ebenezer Ely. "Cheese-maker" Carrie Ferris (24) and cheese factory worker Adam Goodbread (23) both lived in farmer Richard Davies's household. Davies probably owned or patronized a factory, since he kept fifty cows and produced fourteen thousand gallons (about ninety-five thousand pounds) of milk. Finally, Franklin Foster's household included both family members and employees who worked in the factory. Foster's son Nathaniel (38) was a "Cheese Manufacturer"; his wife, Mary (35), and young son lived in with Franklin, as did farm laborer Edwin Stoddard (23), domestic servant Sarah Clements (25), and cheese factory employee Rebecca Smith (27).[30]

This and abundant other evidence shows clearly another, perhaps surprising, element of continuity: many factory employees—during the early decades of factory production, over half the labor force—were female. In the 133 factories reporting to the New York State census in 1865, 713 employees were men and 794 were women (figure 7.8). Why did some women seek work in factory cheesemaking at a time when the great majority of dairying women supported their own departure from home cheesemaking? An explanation lies

Figure 7.8. Men and women continued to work together during the first few decades of factory production. *Appletons' Journal* 4 (Apr. 1878), 300.

partly in the ideology and practice of mutuality. To be sure, mutuality contributed to the rationale for cheese factories, but it also left open the possibility that women would work in cheese factories under presumably more favorable and equitable conditions than at home. One factory supporter referred implicitly to more equitable conditions when he asked, "Could [women] not possibly work in the cheese factory, or do anything else but make cheese in the old way?"[31] The fact that female employees could count on their male co-workers to help with the heavy lifting encouraged the notion that factory cheesemaking would be easier than home cheesemaking, thus addressing the complaints of an unfair and uneven gender distribution of labor in the household. And probably women employees had fewer other responsibilities. While the dream of easier work for women seems to have been imperfectly realized, Harvey Day's papers do suggest that in factories the men were expected to do all the heavy lifting in addition to performing other aspects of the cheesemaking routine; his friend Smith at one point implicitly complained that his work was harder than the work of his woman assistant. Ella Comstock, by contrast, seemed to enjoy the work.

Women employees frequently also had other compelling reasons to continue in cheesemaking, grounded in their personal circumstances. Factory employment was attractive for young single women, for widows with families to support, and for wives who could not depend upon their husbands (or conversely, for wives who could expect significant cooperation from their husbands). They possessed the necessary skill and knowledge to move into factory cheesemaking. Moreover, daughters especially would have the opportunity to earn cash wages, instead of watching the profits of their labor return to the general farm enterprise. Further, they could do so without migrating away from kin and community networks. Ella Comstock is a case in point.

A closer examination of women cheese factory employees as a group helps to elaborate. Some women took advantage of continued access to cheesemaking work in order to make a living when men were either absent or perhaps inadequate providers. One intriguing pair who turned up in the census consisted of Levi Holmes (45) and his wife, Caroline (41). Levi, who listed no property at all, gave his occupation as "peddler of Pictures"; his wife worked in a cheese factory and owned $600 worth of personal property. Did Caroline find it necessary to work while her husband was away? We can only speculate, but it seems likely. In other cases women provided support to sustain female-headed households. In Trenton, Jane Conway, probably a widow, lived with her seven children between the ages of 12 and 30; of these, Sarah (26) worked in a cheese factory and Libbie (30) taught school. In Augusta, Emily Peebles (60) lived with schoolteacher Sarah Nye (38), Sarah's two sons (12 and 10), and Caroline Peebles (27), who worked in a cheese factory. Emily Peebles and Sarah Nye were both probably widows, mother and daughter. Thus cheesemaking provided women with an opportunity to support themselves and also, as before, may have permitted daughters to stay at home.

For some, cheese factory management offered satisfying and remunerative opportunity. A Miss Sternberg, manager of a cheese factory in Chenango

County, commanded an excellent reputation among her patrons. Trained at the well-known Eagle factory in Otsego County, by 1873 she had been running the Plymouth factory for nine years. Nearby, the South Plymouth factory was managed by a woman who, the *Utica Weekly Herald* reported, "evidently understands her business." Mrs. Margaret Freeman ran a factory in Salisbury Center, Herkimer County; Laura Kent of Oneida County was recruited to go out to Illinois to manage a factory because she was so well regarded as a cheesemaker. The 1875 Oneida County atlas identified a factory in the town of Ava as "Mrs. Dorn's." A "Cheese Factory Item" in the *Black River Herald* for 1867 described a visit to a local establishment owned by some of the town's leading dairymen. There the milk of over six hundred cows was "made into the most excellent cheese by Mrs. Rogers, whose experience and skill enables her to operate for the best interests of her employers." Women managers could command $100 per month in wages; more commonly they made in the range of $12 per week. These were high wages compared with other opportunities open to women.[32]

Moreover, unlike most employments open to women in the nineteenth century, the work afforded a degree of authority. Women workers in cheese factories occupied a far more public position than they had as cheesemakers at home. The term *private dairy,* referring to the home dairy, was not just a label for marketing purposes; it implicitly underlined the public nature of factory cheesemaking. To be sure, to managers fell the responsibility for maintaining the factory's reputation with patrons and with consumers, and they were constantly monitored and reminded of their charge. But conversely, they possessed the authority to discipline patrons.

More often, the woman would be part of a married couple who acted as a managerial team. Harvey Day and his wife evidently worked together. In Boonville, I. S. Weller and his wife ran the factory that he had purchased in 1870. Mr. and Mrs. J. C. Smith made the cheese at the Ridge factory in Rome in 1861. In at least one case the wife operated one factory and the husband another. In 1875 a reporter for *Harper's* was impressed that among such couples "there are many women who can manipulate the milk as well as men. The one possessing the skill wields the sceptre and gains the high wages; the other submissively works under orders." What the reporter interpreted as "submissiveness" was more likely a situation in which men and women cooperated. Nathan Huntington, who claimed to have opened the first cheese factory in western New York (1863, in Collins), recalled that he and his wife, Lydia, "were the whole thing. President, Secretary, Treasurer, and Salesman.... Lydia was an experienced cheesemaker but before we commenced she went to Ohio [to Anson Bartlett's] and worked at his factory two or three weeks and I went there and staid a short time to get the details of management etc." These couples may have been those for whom the ideal of mutuality was realized; Nathan and Lydia Huntington clearly transferred the ethos of mutuality into the factory setting.[33]

Where women joined their husbands in managing cheese factories, these arrangements continued, in a new context, the old established pattern of cooperation between farm wives and husbands. As they worked in this capacity, women factory managers continued in the role filled by wives in large cheese dairying farms, following the precedent established by women like Cornelia Babcock. If anything, the need for close cooperation may even have increased. Harvey Day's papers show how H. W. Smith and his woman helper divided the work routine in a way that plainly required careful timing. The same would have been true of husband-and-wife teams. While women's jobs might have involved less physically demanding toil than under the home system, the pressures were no less intense. On any given day they made hundreds of pounds of cheese, and even a small mistake could be very costly. The *Appletons'* reporter noted that the manager "knows that mistakes will not be tolerated."[34]

Most often, women worked as assistants, as had Harvey Day's friend Ella Comstock and her female friends. *Harper's,* on a tour of the dairy country in Oneida and Herkimer Counties in 1875, reported that "here we may find, perhaps, the traditional dairy-maid transmuted into the Jehu of the milk-wagon, for very often we find the pride of the dairy-man's family at the receiving door of the cheese factory." In Oneida County, local newspapers always kept an eye out for who was working at whose factory. The *Rome Sentinel* noted in 1870 that "Dan Yourdon, who runs our factory [in Western] has secured the services of those excellent cheese-makers, Miss Wood and Miss Dorrity, as assistants." The Vernon correspondent for the *Utica Weekly Herald* gossiped in 1874, "We are glad to know that Mr. Edgar Hills will keep the same cheese-maker this coming summer that he employed last year. She must be a first-class hand, for he has engaged her for the length of his natural life."[35]

Ella Comstock's remarks to Harvey Day may give us some idea how these women approached factory work. Unlike Harvey, who was formally committed for the season, Ella came and went as circumstances, personal and seasonal, dictated. She "helped." The term as used here indicates partly a subordinate position but also an attitude. Ella's stints at the cheese factory were consistent with the neighborly customs upon which most women's work in household cheesemaking was based. They came about through informal means, such as a request from a neighbor; probably more often than not the factory manager was indeed a well-known neighbor. Ella apparently stayed only for peak periods, then returned home. Her family exerted the first claim on her labor; she did not go to the factory when "the folks" needed her on the farm. Even the way Ella talked about going to the factory evoked the notion of work as a social event: "We have good times together." This was an outlook absorbed while living in a setting where work and sociability were intertwined, and this interrelationship continued in rural women's work.

But important forces for change would eventually transform factory work and social organization. Most significantly, despite the early importance of women in the factory labor force, the organization, language, and social makeup of factory cheesemaking were shifting decisively toward men. The very financial basis of centralized cheesemaking contributed to men's control of the business, since factory stockholders and owners were almost invariably men.[36] But even under household cheesemaking, men had possessed ultimate control over economic resources; the financial reorganization of cheese production merely consolidated and formalized men's involvement in the industry. The significant departure under factory dairying was less in the shift in economic power than in the redefinition of cheesemaking as men's work.

Cheesemaking as Men's Work

In practical terms men gained knowledge of cheesemaking in several ways. As we have seen, while women dominated under home cheesemaking, men could acquire cheesemaking skill too, as the work was often shared and not so extremely gender-marked as in other societies. Thus there formed a small cadre of men trained in the art well before the factory system developed. This exchange continued into the factory era; Harvey Day's story, for example, shows that he gathered information not only from men but also from his women friends. Since each factory replaced dozens of workers, relatively few men could rather quickly come to control a significant proportion of total production. As the factories spread, trade knowledge was more often transmitted by men to other men, hence weakening women's command of this exotic skill. Harvey Day, for instance, learned from Mr. Smith, who also taught other young men. Gardner Weeks, a Syracuse "Dealer in Cheese Factory and Dairy Supplies," also advertised that he would provide cheesemakers. This shift to domination by men was also reflected in published literature on cheesemaking. Factory reports, journal articles, dairying manuals, and the like were now produced mostly by men. Women still published their methods for home cheesemaking, but these addressed a fast-diminishing audience and often appeared on the ladies' page rather than in the dairying section.[37]

The industry's very reorganization as a centralized, capitalized, rationalized industry—as opposed to a farm-based craft—also helped to accomplish masculinization. Unlike other industries, cheese factories did not depend on expensive machinery tended by low-wage, often female, workers; rather, they relied on a significant investment in labor and on capital inputs, and so attracted men because factory cheesemaking was a relatively autonomous endeavor, in which managers exerted a good deal of control. Indeed, manager and owner were often the same person.

Accompanying the rising numerical proportion of men in the work was a shifting cultural perception of it. Cheese manufacturers came to compare their endeavor with other industries, whether agricultural or not. Industry spokesmen attributed to centralized cheesemaking qualities of system, science, and order. They scornfully contrasted home cheesemaking—and implicitly its women workers—as operating "by guess," as haphazard, and as slovenly. Both explicitly and implicitly they argued that the new system required qualities culturally associated with masculinity. As early as 1859 a writer in the *American Agriculturist* had maintained that men were "from education and habit of thought and investigation, better able to judge chemically and experimentally the various conditions of the milk, curds, and other ingredients composing its parts." By 1863, when factories were increasingly common, a reporter from the *Utica Morning Herald* visited the Union Cheese Factory in Oriskany, New York, and reported that "it does one good to witness the difference between the order and cleanliness of this model institution, and the suspicious and slatternly surroundings of some home dairies." A *Genesee Farmer* article, written by a man, denounced home cheesemaking as slovenly, unsystematic, and lacking in technical sophistication; he urged farmers to elevate the occupation to "the dignity of a particular science." *Harper's* contrasted the "science" of factory cheesemaking with the "rude curdling and ruder pressing in which our grandmothers' [cheese] achieved its gossipy reputation." These images implicitly associated allegedly unscientific, inferior cheesemaking with women; the dictionary definition of the word *slattern* is "a slovenly woman." Conversely, factory men were described as "scientific, enterprising, [and] commercial." Soon the notion that factory cheesemaking was a "science" was so widespread that even women cheesemakers accepted it; a woman in the *Prairie Farmer* modestly contrasted her home cheesemaking with a "scientific plan with complicated machinery," and others lamented that women could make cheese if they "only felt competent" to do so.[38]

It is important to note that this characterization partook much more of image-making than of reality. Men were gaining control of cheesemaking through organization, not necessarily through superior skill or knowledge, nor through science. In fact, cheesemaking had changed but little. Method and quality varied wildly from one factory to another, and from season to season,

Cheesemaking Becomes Men's Work

as Harvey Day's experience amply demonstrates. Individual factories could, and did, produce large batches of relatively uniform cheese. But from one factory to another no such uniformity prevailed; a glance at individual factory reports to the New York State Cheese Manufacturers' Association shows that, as in home cheesemaking, the temperature for "setting," the time for "cooking," and the time and amount of pressure all varied. Some factory makers added sour whey to "make it work faster";[39] others did not. Some adopted variations on the English "cheddaring" method.[40] Regardless of method, day-to-day adjustments were still dictated by the season of the year, the condition of the milk, and other variables. No matter how they made cheese, factory manufacturers possessed no better an understanding than their predecessors' of the scientific basis for cheese production. As *Harper's* put it in 1875, "The cheese-maker's agency is beyond his understanding. The most advanced students of animal chemistry can not explain this digestive process. They have been able to imitate some of its transformations, but the philosophy of the process is beyond them."[41]

Factory cheese probably did possess qualities different from those of its homemade predecessor, partly because so much of it was destined for export. Factory critic John Chapman, who lived in Oneida Lake (Madison County), argued that the factory system was incapable of producing high-quality cheese of "rich, nutty flavor." He maintained that the milk mingled in factory vats varied too much in quality, it was handled too much, and it was produced with an eye to shipping quality rather than to flavor. Chapman was correct. Under the factory system it was *inherently* more difficult to make good cheese, because milk from many farms and cows was mixed together and the cheese-maker had little control over its quality; indeed, as we shall see, there were incentives aplenty to adulterate milk, and few effective punishments. True, early factory cheese was predictable—but mainly in its mediocrity. According to observers, factory cheese not infrequently lacked the "rich, crumby, creamy" quality of homemade cheese. Gardner Weeks, a prominent Oneida County factory man, conceded that factory cheese occasionally even turned out to be rank and "mildly poisonous."

Finally, the masculinization of cheesemaking extended to ascribing "masculine" qualities not only to the work itself but also to the men who performed that work. Harvey Day's letters offer valuable insight into the personal manifestations of the masculinizing process. As in industrializing America generally, qualities valued for mass production were applied to personal conduct and to manhood. Day's own attitude toward the work signals the change, especially when contrasted with past attitudes and with his female co-workers' perceptions. Even if Harvey had been possessed of a less serious temperament, his formal, careerist orientation contributed to an outlook that emphasized pressure, competition, anxiety, and ambition over sociability. Harvey's mother's exhortations linked his work still more clearly to masculinity: "do not git home sick and leave till you get your trad leart but perseverence in faith will do the work and make the Man," she counseled; "be a Man every time." [42] For Day, social interaction represented not part of work but a hindrance. This outlook departed significantly from Ella Comstock's, and also from earlier attitudes. To be sure, pressure and anxiety had certainly hounded home cheesemakers, but the focus upon cooperation and "competency" had inhibited tendencies toward competition. Financial success, self-discipline, and competitiveness would contribute to Harvey's establishment as a "Man."

While enormous boilers and rows of six-hundred-gallon vats might seem to epitomize the most revolutionary changes wrought by dairy centralization, less visible changes would ultimately carry greater impact. The way was being paved for women's complete exclusion from cheesemaking. To be sure, most women were voluntarily withdrawing from household cheesemaking, but those who chose to continue cheesemaking in factories were also getting squeezed out. Even as they employed women, factory men seized upon gendered notions of work to shape a twofold process that simultaneously "defeminized" and "masculinized" the work. By 1900 the U.S. census reported that out of 11,700 employees in factories for "Cheese, Butter, and Condensed Milk," only 1,049 were female. In New York State only 345 of 2,430 employees were women.[43]

This again points up the limitations of "mutuality." While the ethos of reciprocity enabled women to circumvent some of the restrictions on their lives, it was less effective as a tool for maintaining access to prestigious jobs and for presenting an alternative to the gendered meaning of cheese factory work that developed so quickly. Practical problems also arose, since married women would have greater difficulty combining household obligations with cheesemaking work when it disappeared from the home. The process of defeminization was thus a two-edged sword. On the one hand, women had a hand in driving the process, since it had positive consequences for those who wanted to withdraw from home processing. But for women cheese factory employees, defeminization meant exclusion and loss of opportunity; it adhered more closely to the pattern that was developing in Western dairying generally and in individual countries such as Canada, Sweden, and Ireland.[44]

Adding gender to the analysis thus has implications for the continuing scholarly examination of the capitalist transformation of the countryside. In asking how this transformation took place, historians have focused upon issues of accommodation and resistance, market penetration, and class conflict. Few have investigated how relations between the sexes figured in the great transformation. The centralization of cheese dairying suggests that gender played an important role in transforming rural society and economy, that the "transformation" should be cast as much in terms of a reallocation of labor between the sexes as of a realignment of social classes.[45] Under the household system a basic inequity in the status and power of the sexes carried over to an unequal sexual division of labor. Conflict erupted over the division of labor, and these household tensions helped give impetus to the factory system. While some women accepted the ideological masculinization of cheesemaking, others hoped for a fairer balance in the gender division of labor, rather than for a redefinition of the work. But even though women comprised an essential element in the early factory work force, they were increasingly marginalized as men came to command the language, resources, and processes involved.

Participants in this transition anticipated in it both social benefits and costs. In its wake they set about forging new ways of living and working.

The Dairy Zone Embattled
Factory-Era Agriculture

> I shall not dispute that we may doctor up our lands to produce any desired crop, but to do so is expensive, and will often require more science and skill than are common in the country.
> —X. A. Willard, 1877

In 1877 X. A. Willard surveyed the extent of American dairying: "The great American dairy belt lies between the fortieth and forty-fifth parallels of latitude. It stretches from the Atlantic to the Mississippi, and possibly to the Pacific. Within its limits are New England, New York, Pennsylvania, the northern parts of Ohio, Illinois, and Indiana, the greater portion of Michigan, Wisconsin, Iowa, and Minnesota, and a part of the Canadas." The old "Dairy Zone" had expanded so that the concept was rendered essentially meaningless. Willard's comment signified an intellectual and practical transformation with profound implications for dairy farming.[1]

This reevaluation of the Dairy Zone occurred largely because westward-moving migrants, untroubled by thoughts that they might have passed beyond some mysterious boundary, had discovered that they could profitably make milk, butter, and cheese. Suitable climate and herbage, together with low land prices, favored dairymen in what is now the Midwest, and the factory system also worked to their advantage. Willard thoughtfully observed that "cheese dairying is no longer a privileged business, narrowed down to a few places where high skill in manufacture has built up an enviable reputation. It is opened up to many localities. [Westerners] come [to New York] and pick off your best cheesemakers. They erect their factories and meet you in the market on an equality." Willard's apprehensions were well founded. In 1880 New York State's production stood at 129 million pounds, Wisconsin's at 23 million; by 1890 New York's had fallen slightly, to 124 million, while Wisconsin's had more than doubled, to 54 million. But during the 1890s Wisconsin passed New York State with a phenomenal burst, leaving New York far behind: by 1900, even though New York had picked up the pace to 150 million pounds, Wisconsin cheese factories were turning out an astounding 393 million pounds.[2]

Before the factory era, consumers and sellers alike had sometimes identified cheese by its area of origin: Goshen cheese, Herkimer County cheese. Some believed that the unique combinations of soil and herbage in a given region resulted in a distinctive cheese. But the regional identities of various American cheeses essentially disappeared once factory cheese came to predominate. In an article on Colorado dairies, for example, *Moore's Rural New Yorker* argued that the "old notion concerning dairy soils" was scotched.[3] In the new thinking, "skill" rather than "soil" accounted for good cheese. Just two new categories superseded the old ones: "factory" and "private dairy" cheese. Neither possessed geographical associations.

By drawing attention away from local conditions and focusing instead on technology and process, cheese factories further contributed to the erosion of the Dairy Zone. With centralized cheesemaking came a focus on what happened to the milk after it arrived at the factory door. Cheese manufacturers devoted their meetings mainly to discussing the merits of various technologies and methods rather than to considering the impact of soil and grass on milk. Of course, factory owners worried about spoilage, but even then they concentrated as much on ways to alter the manufacturing process to accommodate poor milk as on enforcing sanitation and transportation rules.

Especially for eastern dairymen, the Dairy Zone's demise was ominous. In 1879 a reporter for the *American Agriculturist* recorded easterners' shocked realization that a new climate for dairying had developed. "Here in the State of New York, Eastern Pennsylvania, and New England and New Jersey, we have all along felt so sure of *Our Milk Crop,* that we laughed at the thought of competition.... As to butter and cheese, with our cheese factories and creameries we were beating the world.... All on a sudden we find ourselves *Beaten By the West!* The Western dairymen stand forward as equals in cheesemaking."[4] This disturbing discovery, occurring as it did in the aftermath of civil war, must have banished forever any remaining echoes of the naive antebellum rhetoric that had praised the unifying tendencies of regional specialization.

Neither did the prewar, republican emphasis upon independent proprietors' "competency" suit the new conditions, in which market competition ruled. Uncertainty and dwindling confidence pervaded postwar assessments of dairying's prospects. This mood reflected not only the wartime experience but also newly difficult economic circumstances. Nationally, the elements of economic change that led to these new conditions are familiar. New York dairy farmers were brought into competition with midwesterners—and, increasingly, with Canadians as well—as the national and global economies became increasingly integrated. In this regard changes accelerated that had first been felt with some force in the 1850s. Transportation linkages resulted in enlarged regional and national markets; industrial mass production grew phenomenally; and capitalist enterprises also increased their scale and reach. As exchange partners were increasingly separated, cash became more important as an exchange medium. The Civil War had dramatically accelerated all of these trends. Taxation, mechanization, and the premium on labor all presented increased demands for capital.

These changes reached Oneida County dairy farming families most decisively through the factory system. The cheese factory fever induced ever greater demand for milk, which combined with the generally escalating need for cash to stimulate changes in the dairy farming system that were to render it quite different from that followed under cheesemaking at home. In the newly uneasy climate, dairying families responded to the changing conditions they confronted.

Each of four Westmoreland (Oneida County) farmers visited in 1870 by the *Rome Sentinel*'s roving correspondent followed practices that in some way epitomized the new order. Ephraim Bessee cut seventy-five tons of hay for his twenty-five milch cows. William Watson made sure that his exemplary grass farm was "well stocked with thoroughbreds of various grades." "Practical dairyman" Lawton Goodsell kept a hop yard, in addition to his cows; and Samuel Isabell, an "energetic and scientific farmer," carried on an extensive lumber trade, supplying villagers with cordwood and building materials.[5]

William Watson, Lawton Goodsell, and their neighbors had made decisions that collectively resulted in an overall shift in Oneida County's agricultural profile. The total number of milch cows in the county had risen since 1850 by about ten thousand, to fifty-nine thousand. Moreover, cheesemaking occupied an increasing proportion of the herd: by 1870 cheese factories collected milk from 55 percent of Oneida County's milch cows. In some towns, such as Trenton, the percentage was as high as 80; under home cheesemaking it had more usually stayed around 25 percent and rarely exceeded 40 percent.[6] This activity resulted in an astounding increase in the county's cheese production, to over 8 million pounds (in 1875), and in fluid-milk production, to 600,000 gallons (from 184,000), while butter output remained about the same at 3.9 million pounds. From each cow Oneida County farmers coaxed more milk, about 2,850 pounds in 1875 as compared with about 2,400 in 1855. To support a larger, more productive herd, Oneida County dairymen also extended their grasslands; the number of improved acres and the number of acres in grass both increased. Dairy cows probably got most of the benefit from the larger expanse of grassland, since the numbers of beef cattle remained about the same while numbers of hogs and sheep declined.[7]

Home cheesemaking had required a certain minimum in milk, equipment, and skilled labor; but any farmer with even a few cows could take milk to the factory. This basic difference in the nature of the two industries enabled the factories to pull more farms into cheese production than had been involved in home processing; 475 farms in the seven Oneida County towns had produced cheese in 1850, while 632 sent milk to the factory in 1870. Milk-selling farms[8] were small compared with cheesemaking farms; the average size was smaller overall (126 acres versus 132), though the improved acreage was greater (104 versus 92). Milk-producing farms raised fewer milch cows (14.6) than cheesemaking farms did (16.0), but they put their dairying resources more exclusively into milk production, making far less butter than cheesemaking farms had made in 1850 (337 pounds versus 751), and less than farms that did not send milk to factories. While they did not achieve the productivity of cheesemaking farms (3,800 pounds of milk per cow versus 4,000), milk-selling farms still got far above the county average (2,850 pounds). They reflected the general tendency in Oneida County in that they supported fewer nonmilking animals than cheesemaking farms had supported in 1850: fewer neat cattle by about one (3.5 in 1870 as opposed to 4.8 in 1850), five fewer sheep (3.3 versus 8.4), and half as many hogs (3.2 versus 6.3). They raised about the same amounts of the major field crops—corn, oats, hay, and potatoes. They also continued to mine the forests and to tend orchards. Two products—hops and poultry—increased significantly in importance.[9]

New Cows

In 1878 the roving reporter for *Appletons' Journal* described the cattle that roamed Oneida County pastures: the hills were "dappled with the red, brown, and black, and fawn-color of grazing cattle."[10] The cows giving milk for the central New York factories, like the "natives" that had supplied milk for farm-made cheese, evidently were motley, multicolored, and varied in size and shape. But subtle changes were at work in the way Oneida County, and northern farmers generally, perceived and managed the animals they increasingly depended upon for their livelihood. Though natives continued to dominate, thoroughbreds or their "grades" received increasing attention and comprised a growing, if indeterminate, portion of the local herd.

The native cow still had her champions. In 1864 members of the New York State Cheese Manufacturers' Association listened to Leander Wetherell of the *Boston Cultivator* maintain that "there was no breed so profitable for the dairy as the native cow." Wetherell repeated an old formula: "A well-known farmer in Massachusetts . . . had tried the Hereford, the Ayrshire, and the Alderney, to test the matter and determine which was best; but there was no cow on the farm so profitable as the native cow."[11] The war between proponents of purebreds and of natives still raged in Oneida County too: in 1875 the *Boonville Herald*, with an undertone of smugness, announced the death of the fourth "Dutchess of Oneida," a high-priced purebred. Her demise, the editor chortled, would "dampen the ardor for purchases of fancy cattle."[12] Several well-known thoroughbred enthusiasts had establishments in New York Mills, just outside of Utica. In 1873 Oneida County farmers looked on incredulously at an auction there to see "how much money one man would throw away on one cow. . . . Our country farmers, who buy very good cows, could hardly believe their senses when bids . . . go up by thousands of dollars . . . [and] the auctioneer himself lost his equanimity."[13]

If farmers wisely avoided sinking huge sums into cows of unproven worth, they were no longer averse to judicious, low-risk investment in grade cattle for the dairy. Blooded cattle were on the ascendant in dairy country. The indefatigable X. A. Willard toured Oneida, Herkimer, and Otsego Counties in 1864 and 1865, reporting back to the *Utica Weekly Herald* on local crops and agricultural practices. As he passed through villages and hamlets, local farmers invited Willard to visit and showed him around their buildings and fields. He noted that "the whole region is probably better adapted to grass than grain, and contains a large number of highly cultivated farms on which are herds sprinkled with a goodly proportion of grade Ayrshire, Short Horn, and Devon cattle."[14] At Hosea Hall's farm in Annsville, Willard found "thirty-nine cows, mostly grade cattle being a cross of Devon and Short Horn." Among George Clark's twenty-four cows were "natives and grades of Short Horn, Alderney, Ayrshire, and Devons."[15] None of the farmers Willard encountered claimed to possess solely native stock, as had prizewinners in earlier New York State Agricultural Society competitions. Indeed, the *Utica Weekly Herald* contended that "crosses . . . are now abundant" in Oneida County, and the British observer J. P. Sheldon wrote that good grade cattle were more common in the dairy districts than in other regions.[16]

Still other reports suggest that grades were accessible even to moderately prosperous milk sellers, and that more and more farmers were raising their own cows rather than making major purchases from drovers. The *Rome Sentinel* reported in 1870 that William Watson possessed one of Westmoreland's best grass farms, "which he keeps well stocked with thoroughbreds of various grades."[17] According to the agricultural census, Watson's farm was not far above the average for milk producers. He owned eighty-four improved acres, twelve unimproved; milked twenty-five cows; and got from them eighty-seven hundred gallons of milk (about three thousand pounds per cow), actually below the average for milk-selling farms generally. He valued his farm at $7,000. We may infer, then, that many local farmers could probably afford grades. The affordability of grades was probably enhanced when, as the *Utica Weekly Herald* reported, local farmers "clubbed together" to buy blooded bulls.[18] More and more the *Herald* mentioned farmers who were raising their own cattle; a Mr. Budlong from Herkimer County was commended for raising his own herd of twenty-six cows of mixed grade and native stock, as was William H. Storrs of Oneida County.[19] That year (1870) at the local fair the dairy cattle display was "beyond former years," with Devons, Ayrshires, Jerseys, and Alderneys making their appearance.[20]

The most significant breed to make its appearance in the factory era, however, was none of those just mentioned but rather a relative newcomer to America: the Holstein. During the mid-1870s cows from this breed began to attract notice in the dairy districts as promising milk producers. "The Holsteins," declared the *American Agriculturist* cover story in 1871, "are preeminently dairy animals." They were apparently familiar in Oneida County as the factory system consolidated; they were exhibited at the county fair in 1873, and the *Utica Weekly Herald* even claimed in 1876 that locally the Holstein was "so well known that its representatives are easily recognized from sight." Local dairymen were instrumental in the formation of the Holstein-Friesian Association in 1885. The Utica paper further commented that although these docile cows had a reputation for consuming large amounts of feed, their milk traveled well—an obvious advantage for farmers jogging along to the factory with clumsy carts. The Holstein also reputedly thrived in the American climate.[21]

The productivity of breeds earlier touted for the dairy—the Channel Islands cattle, Shorthorns, and the like—had been debated, but the once the figures on Holsteins were widely published and once American farmers witnessed them perform, few disputed their capacity to outproduce other breeds. Lewis Falley Allen, in his widely circulated text *American Cattle* (1868), recognized Holsteins as good milkers, and by about 1880 the Holstein's superior milking capability was well established. Importations of Holstein cattle from Europe increased in the 1870s and especially in the early 1880s, attracting wide attention.[22]

Why did dairy farmers begin to shift their emphasis away from the natives and to show increasingly serious interest in high-yielding breeds? The changing cultural and social climate may have helped to pave the way for the new mentality. Class lines, and justifications for social hierarchy, were more apparent and more widely accepted than before. A new and repressive system of race relations was evolving, buttressed by equally rigid racial ideologies. In the face of unprecedented immigration, nativist rhetoric also emphasized purity, heredity, and hierarchy. Even though the new cattle were, of course, immigrants themselves, the mentality that led to their acceptance was quite different from the democratic defense of native cattle prevalent earlier. To replace democratic rationalizations for the native cow with concerns for heredity and explicit ranking was consistent with the new social context.[23]

More immediately and directly, the nature of production for the factory brought tremendous influence to bear on the way farmers selected cows. Farmers were paid according to how much milk they delivered to the factory, and so quite suddenly there arose a powerful incentive to increase production. The quality of the cheese was now someone else's business. This was a highly significant change from the home system, in which quality was at least as important a consideration, and in any case the small scale of the operation had limited the amount a single household could produce. This aspect of production for cheese factories governed farmers' responses and in turn contributed to the profound alterations in farming systems that occurred in the late nineteenth century.[24]

New Shelter

Though Holsteins were enormously important in the long run, they probably did not in the short term exert any significant influence on the increase in average yields, since their numbers at this point were still relatively small. Fred Bateman has concluded that improved shelter and feeding probably accounted for most of the increases in dairy cow yields in the late nineteenth century.[25] Indeed, life for dairy cows had improved quite dramatically. Because they no longer shivered outdoors in cold weather, and because they ate well, late-nineteenth-century cows usually got through the harsh New York State winters in better shape than their less fortunate ancestors. In turn, more controlled shelter probably encouraged farmers to expand their investment in cattle, since they could reasonably expect their stock to survive and thrive.

In 1866 an observer writing in the *Transactions* of the New York State Agricultural Society declared that "palaces for dairy cows" were replacing the messy, makeshift yards of an earlier day.[26] Indeed, beginning in the mid-1850s and continuing into the prosperous Civil War years, a barn-building boom transformed animal shelter. By the late 1860s and early 1870s many Oneida County dairy farms boasted new or rebuilt barns to accommodate their valuable stock. In 1875 the reporter from *Harper's* observed that "every dairy farm has a dairy barn of greater or less excellence."²⁷ Though the new barns, of course, varied in size, they shared common features that marked them as an evolving type.

In 1869 Benjamin O. Jackson put the finishing touches on a 130- by 40-foot barn just outside of Boonville. It was, stated the *Boonville Herald,* "one of the finest and most convenient barns we remember to have seen." The barn had room for fifty cows and their hay. Its basement walls, of "solid masonry," rose eight feet high, pierced with windows for light and ventilation. The basement floor had been paved, then "covered with cement and planks," and sloped so as to allow liquid manure to flow to a tank outside. Solid manure was removed and spread daily. The barn had a "pointed" (probably gabled) roof, "terminating in two huge ventilators." At the north end a "causeway" led to the attic level, so that workers could draw hay in by wagon, pitch it downward to mows, and then slide it via chutes to the waiting cows below, "so that in every movement of the hay the motion is *downward* by aid of its own gravity—an aid duly appreciated by all ease-loving herdsmen." Stanchions of "improved construction" secured the cattle. In summer, its owner boasted, forty cows were milked in the barn twice a day, with no odor to offend the fastidious. The barn also stored Jackson's mower and tedder, which plied his fields "like a huge grasshopper at exercise." "Truly the millennium for the farmers in this fertile dairy region is dawning," the chronicler rhapsodized.²⁸

Two years later, another Jackson, Conkling P., put up another barn just outside the village limits. In a departure from the tradition of communal "raising," Jackson hired at least some of the labor to build. His barn was much smaller than Benjamin's, just 40 feet by 60, but it contained the same features: a ventilated stone basement for the cows, with drive floor above, to which hay was conveyed by incline. The correspondent from the *Boonville Herald* admired the barn's red paint, white trim, and bracketed cornices. But its crowning glory, quite literally, was a weather vane in the silhouette of a cow. The observer balked at this final extravagance, wondering if the "golden calf" would "pay," but conceded that its owner would probably "enjoy it."²⁹

These two barns represented a new and increasingly common type. Barns of this "improved pattern" probably looked on the outside something like William H. Comstock's barn in Trenton, pictured in the 1875 Oneida County atlas (figure 8.1). Inside, according to X. A. Willard, the "modern barn now being erected in Oneida and Herkimer Counties" had a dual-purpose basement: "The stables for milking in summer are those in which the cows are kept in winter." Many had manure cellars beneath. Sometimes the basement was dug into a "side-hill," so that the stock entered on the ground level, usually through a gable-end door. The stables were usually arranged in two inward-

Figure 8.1. William Comstock's farm, in Trenton (Oneida County), boasted an up-to-date northern basement barn. Oneida County atlas (1875).

facing rows with a central aisle for ease of feeding. The cows stood about four feet apart, fastened by chains or in stanchions.

"The leading feature of the barns now being built in the dairy region," Willard continued, "is to have the drive-floor and bay above the stables." Often the upper level was arranged so that several loads of hay could be hauled in and emptied consecutively; horse stables and carriage houses often adjoined; and a loft was frequently built over the drive floor to store corn prior to husking.[30] Significantly, the term *dairy barn* was now attached to these buildings, indicating a specialized function (figure 8.2).

Henry Glassie, in a thorough field study of barns in the neighboring dairy county of Otsego, has identified the major features of this type. He labels it the "northern basement barn" and associates its rise with the introduction of dairying in the mid-nineteenth century. Glassie shows that the tripartite, one-level "English barn" gave way to a two-level structure that in concept was a "one-level barn built up on [a] basement." The lower level was used for "stabling" and the upper for hay storage. A ramp or a natural hill permitted

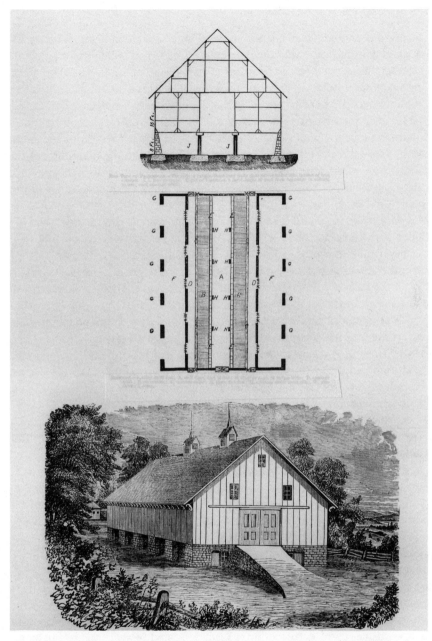

Figure 8.2. In the newly named dairy barn, the spatial arrangement facilitated the milking and tending of the cows. *Basement:* A, alley (8 × 70 feet); B, stall floor (4½ × 70 feet); C, ditch, or drop; D, space, or walk; E, stanchions; F, manure cellar; G, piers (2 × 4 feet); J, columns under cross sills; W, windows; I, doors. X. A. Willard, *Practical Dairy Husbandry* (New York, 1877), 518–19.

access to the upper level. Glassie argues that the northern basement barn developed within a folk tradition that allowed traditional builders to manipulate abstract "base concepts" in accordance with shared cultural "rules" concerning structural limits and aesthetic preferences. In this case, of course, the two "base concepts" were respectively the "English barn" and the "basement." The latter, Glassie asserts, was "variable according to use, size, and physical environment: attendant rules governed the possibilities for relating the depth of the barn and the terrain on which it was situated to the placement of the doors, determining the work flow through the barn, and, in turn, the practical traditional options for the placement and arrangement of stalls, stanchions, pens, and bins."[31]

That the northern basement barn increasingly dominated in late-nineteenth-century central New York Glassie has established convincingly. As his description of the basement's function is somewhat brief, the analysis here amplifies his interpretation by examining in more detail how the northern basement barn accommodated dairying, and also to suggest some additional conceptual sources of its "basement" component.

The northern basement barn had a well-developed functional antecedent in the milking sheds and barns that had appeared during the 1840s and 1850s. In these midcentury buildings milking had been spatially regularized in efficient parallel rows, arranged along the barn's lengthwise axis. The cows, immobilized in stanchions, faced each other toward the center feeding aisle. This arrangement simplified milking and manure collection.

Published barn plans can show more vividly how these barns worked.[32] They represented both actual and ideal barns that were usually more elaborate and expensive than the ordinary northern basement barn erected in central New York. Yet there are good reasons to regard these plans and descriptions as valuable indicators of what was probably happening on a more modest scale in typical barns. The plans, though on a larger scale, often shared the same basic designs that prevailed across the dairy region. Moreover, published plans can suggest where farmers stored and used their machinery; this is valuable information because surviving buildings in the field have been altered many times over as needs have changed.

Two plans from farm journals show how the milking-barn arrangement was incorporated into the northern basement barn. A "Dairy and Farm Barn" from Otsego County appeared in *Moore's* in 1871 and again in 1874; in this northern basement barn the lower level consisted of two longitudinally arranged rows of stanchions with central feed alley. In this case the manure area was on the same level as the cows, at either side.[33] The "cattle floor" in George Waring's barn, published in the 1869 *American Agriculturist*, featured the latest technology of the day: facing the central alley, cows were fed from a "rail way." The barn also accommodated young calves and working stock. As with Wheeler's plan, a cellar below served to collect manure.[34]

The full significance of the lower level lies in its relationship with levels above and (in some cases) below. In this sense the combination of "concepts" was not simply additive but integrative, requiring innovation as well as expansion. The new multilevel barns revolved around a single focal point, the dairy herd. In this regard they were creations of the market economy rather than of a particular ethnic group, since from New York to Wisconsin, dairy farmers of every nationality built them.[35]

They reflected specialization and a commercial outlook that valued sustained, concentrated, rationalized labor patterns over bursts of intense seasonal activity; the northern basement barn supplied a year-round workplace for the farmer and his hands. The point of having the "drive floor" and bay above the stables was so that feed processing, storage, and distribution could proceed efficiently. The most fundamental advantage gained, of course, was that of gravity: nearly all multilevel barns allowed for feed to be hauled into the barn on a slight incline and distributed simply by dropping it through chutes or openings to the cows waiting below, or to carts. The upper level—whether arranged lengthwise or crosswise—also functioned, as before, as a central space for processing feed. Grain could be threshed there; corn husked; hay chopped.

Even if the upper level housed functions superficially similar to those it had sheltered for generations, by the later nineteenth century these tasks were more often mechanized. We have seen that mechanical mowers and rakers aided in the single most important task, haying, beginning as early as the 1860s. Another implement important in haying, the horse fork, was coming into use as well. This device used horsepower to raise the hay into the barn.[36]

The appearance of machinery for feed processing, threshing, and the like on Oneida County dairy farms is difficult to document directly, but a few indirect clues hint that farmers were beginning to acquire such equipment in the postwar period. Directories and newspaper advertisements show that Oneida County supported a number of farm machinery dealers and makers. Census figures also suggest that individuals owned more tools. In Trenton, for example, where in 1870 a full 80 percent of the milch cows were milked for factory cheese, milk-selling farms possessed an average of $232 worth of farm equipment, well above the county average of $184. Local publications contained more references to processing feed in some way, usually by cutting, mixing, or (occasionally) cooking. The *Utica Weekly Herald* noted almost jovially in January 1873 that Jerome Steere of New Hartford had "recently left his thumb in a straw-cutter."[37] Other farmers boiled the feed, presumably using steamers. Daniel Spencer advertised a farm in 1873 that included a barn with running water.[38] In 1866 a visitor to William H. Storrs's farm in Trenton came upon him working at a "one-horse threshing machine," threshing barley. Storrs told his guest he used the machine also to cut straw and even to saw wood; he considered it "a great institution, doubly paying the interest of the expenditure." Perhaps Storrs had followed advice in contemporary agricultural journals, which frequently argued that horse-powered fodder cutters could double the farmer's income from forage crops.[39]

Again, barn plans offer more concrete detail about how planners accommodated the mechanical revolution. Writing in the 1877 *American Agriculturist,* one planner specified that he had left the center of the upper level "open and free for cutting and mixing the feed. Here should be a fodder-cutter and a large mixing-box, in the side of which should be a spout to carry the feed to the car on the floor below. If the food is steamed the boiler can be kept in a rear building . . . the steam being carried to an engine, which would work the fodder-cutter, and the steamer, both on the upper floor."[40] A "large stock barn" that appeared in *Moore's* in 1871 exaggerated all of these features to an unusual but revealing extent.[41] To this three-level barn actually adjoined a "factory" containing both steam-powered[42] and horse-powered machinery, for processing grain and apples and for running the "hay cutter, threshing machine and corn sheller" housed on the third level. Feed was dropped below to cars on "tramways" for easy distribution.[43]

Others tied the storage and use of farm machines explicitly to the improved dairy barn plan. Dairy barn designer D. W. Clark, publishing in *Moore's* in 1879, kept "horse power machinery for cutting feed" in his barn on the second level.[44] An 1879 "Farm Dairy Barn" erected by E. C. Ross of Paulding County, Ohio, was "designed to contain a root-cutting machine" in its root cellar, adjacent to a mixing box. Above, on the second level, was a "cutting room." Another plan, also for a dairy barn, showed a fodder cutter on the second level. A "Big Giant feed-mill is kept outside the stable, below this door, for grinding grain." Since the barn lacked a ramp, when it came time actually to store the grain "a hoist wheel [was] placed at the door near the feed bins, for drawing up bags of feed."[45]

Factory-era dairy barns thus consolidated the design trends of an earlier era. They also further developed the industrial imagery earlier adumbrated. The presence and operation of horse-powered machinery reinforced the notion of the barn as factory; the rhythmic clatter of horse-powered hay forks, fodder cutters, and threshers must have sounded very unlike the hand flailing and threshing of an earlier day. Human voices too were missing; machines probably appeared the more prominent because they replaced sociable parties of people with a solitary worker. Though the total number of workers on milk-selling farms did not decrease from the earlier era, postfactory farm work was spread more evenly over the year rather than concentrated in seasonal bursts. This accounts for why it *appeared* more solitary.

Many designers articulated the industrial analogy explicitly. In 1865 an article in the *Ohio Farmer* made the connection unequivocal, calling the farmer's barn "his Bank, his Treasury, his Factory, his Laboratory, the Headquarters of his Outdoor Life." This author described two barns he regarded as fulfilling these qualities; both made use of gravity power and of manure spreaders, horse-powered separators, corn shellers, and feed cutters.[46]

About the same time, an equally revealing article criticized what the first had celebrated. Its author condemned farmers who forsook beauty for mere money in their barn building. The "picturesque" barn, low and rambling, had been informed by "human sentiment" and offered a haven for hens or for children at play. By contrast, the "commercial" barn, tall, bare, and purely functional, barred all but those on business. The descriptions strongly suggested that the author was thinking respectively of the old New England barn and the northern basement barn. The writer's memory may have been clouded by nostalgia, but he pointed accurately enough to the shift in emphasis.

Thus as the notion of a geographically bounded Dairy Zone dissolved, farmers began to challenge previous definitions of the "natural" limits to dairying practice; the northern basement barn carried further the earlier attempts to provide cows with an environment that would enable them to extend the milking season. In some cases seasonality disappeared altogether as farmers milked their cows year round. The farm press began to carry more letters and articles from farmers who had experimented successfully with "winter dairying." Under this scheme cows were bred to calve in the fall, then were fed liberally over the winter. The practice originated in Illinois, where cheap grain was plentiful, but it was increasingly recommended for other areas on the grounds that milk kept better in cool weather and that butter and fluid-milk prices were higher then. "Is It Profitable to Milk Cows in the Winter?" Members of a New England farmers' club debated this question in 1868. Most spoke against the practice, generally on the grounds that it was unnatural and detrimental to cows' health; they invoked concepts that had dominated before the factory era, when dairymen like Alonzo Fish had believed that the natural order dictated spring calving only. But now a strong cadre of dissenters argued that those who milk till within four weeks of calving "make more butter per cow in a year."[47]

In fact, Fish had grasped an important insight. A longer period of milking, combined with poor sanitation, ignorance of bacteria, and the greater impetus to maximize production, put the entire herd at risk for various ills. In 1883 an article in the *American Agriculturist* identified "milk fever" as a severe hazard among "cows that secrete milk abundantly." It was "induced by high feeding, neglect of proper care, and insufficient exercise"—in short, by conditions frequently present as feeding intensified and as cows were confined longer.[48] Milk fever today is treated through dietary means, but at that time knowledge of bovine health was too rudimentary to offer solutions to the problem.

The risk of disease also increased simply because the stock was crowded together. Contagious abortion (brucellosis) had always been a problem but became a serious concern in the postwar decades. Willard worried that it had "become epidemic." Sudden "storms" of the disorder afflicted herds late in the century. Scientists today know that this dangerous disease is caused by a bacterium, *Brucella abortis*. Bovine tuberculosis, pleuropneumonia, rinderpest, and mastitis also brought troublesome problems. Crowding and lack of sanitation created prime conditions for the microbe that caused these diseases to spread and infect the cattle population. American veterinary medicine was still in its infancy; few professionally trained veterinarians served the rural population, so farmers still had to rely upon publications and local farriers.[49]

To farmers and their neighbors, then, the northern basement barn clearly expressed prevailing trends in dairying. Nearly every space in the northern basement barn was devoted in some way to feeding, milking, and housing the dairy herd. As more capital was invested in the barn itself and housed within it, the building also came to represent an increasing emphasis upon commercial agriculture. The new form carried implications for both bovine and human occupants: dairy cows surely benefited from the warmth their new homes provided, but they also were placed at greater risk for some diseases. As we shall see, farmers and farm workers used barn space to develop an industrial discipline in tune with factory demands, and to shape a more pronounced sexual division of labor. Altogether, while it was based upon traditional models and design processes, the northern basement barn represented an innovative response to changing conditions.

The Crisis in Dairy Lands

The substantial barns gave the countryside an air of prosperity, but an undercurrent of worry swirled beneath the excitement over flush times. Dairy farmers needed every little improvement they could manage, because they were apparently less able to count on harvesting large quantities of grass for their valuable cattle. Since hay and pasture grass still accounted for the bulk of the cows' diet, falling meadow and pasture productivity presented a serious concern. In 1863 hay was actually imported into some New York State districts, an occurrence previously unimaginable.[50]

The agricultural economist T. E. Lamont used several measures to conclude that statewide hay yields remained steady from 1866 to 1905.[51] Even so, central New York grassland yields did not meet growers' expectations during this period, when they sought not steady productivity but increases. Moreover, local yields varied enormously, probably leading observers to perceive declines in some cases. No reliable measures exist at all for pasture productivity, but pastures probably suffered in the postwar period. Census figures hint that the modest increase in milk yield was likely due not to an improvement in grassland management but simply to extra feeding, since milk-selling farms harvested the same amount of hay as had cheesemaking farms, but fed it to fewer cows.

Hay and pasture quality became more difficult to sustain partly because milk-selling farms were smaller and probably occupied more marginal land than cheesemaking farms. Also, new lands converted into pasture and meadow after 1870 were likely less fertile than the grasslands that had been used during the county's earlier stage of development, when settlers established farms on the best lands. The result was that although the average milk-selling farm had more improved acres than cheesemaking farms had had in 1850, the ratio of improved acres to each milch cow was greater in 1870 (7:1) than in 1850 (5.75:1). In this regard dairy farming had actually become *less* intensive. The *American Agriculturist* in 1874 estimated that typical dairy farms took up six acres for each cow, and it criticized this figure as excessive.[52] On Oneida County milk-producing farms the overall crop and livestock mix may have contributed in turn to a decline in grasslands: less stock meant less manure, stretched even further by competition from an increasingly popular but voracious crop, hops.[53]

A chorus of nervous public pronouncements reinforces what ambiguous statistics only adumbrate: across central New York, grasslands, especially pastures, were failing. "The question of the treatment of pasture lands is now one of the uppermost in dairy discussions," declared *Harper's* in 1875. In 1871 Anson Bartlett, a prominent cheese manufacturer from Ohio, confronted the American Dairymen's Association convention in Utica with troubling thoughts about the apparent decline in carrying capacity. Bartlett compared dairying with that most notoriously soil-depleting, nutrient-robbing crop, tobacco. He reasoned that every ton of cheese derived from grass removed hundreds of pounds of nutrients from the soil, and he recommended that farmers husband their manures even more carefully and ease the pressure on pastureland by raising more green fodder such as corn and sorghum.[54]

Bartlett was not alone in his concern. During the 1860s and 1870s agricultural leaders, writers, and practitioners themselves discussed the issue in the press and at various farmers' associations. The Central New York Farmers' Club convened in 1873 to discuss how to improve grasslands, in an atmosphere of anxiety. One speaker gloomily recalled that in his childhood two or even three tons of hay to the acre were common, but now he and his neighbors could barely wrest a single ton from an acre. "Hay is an exhaustive crop," he concluded, warning that it would be "only a question of time how long it will take to sell an Oneida County farm poor by selling the hay and not returning an equivalent for the mineral and other ingredients taken away." Dairy cows removed nutrients too, until "the grasses [are] literally starved out . . . [and] nature covers up the nakedness with mosses and motherworts; then the land is said to be 'worn out'." Unpalatable wire grass and orchard grass multiplied.[55]

Concern about grasslands was heightened because a changed context made soil fertility a more important issue than ever. Western competition from farmers exploiting fresh, fertile lands cast "worn-out" eastern farms into relief. In addition, factory cheesemaking was a very public enterprise. Just as participants kept close watch over cheesemakers, so they also turned their attention to how the milk was procured. One farmer's lax ways were everybody's problem. The farmer's everyday activities became public in ways he might not have imagined a decade before.

As usual, the most comprehensive and lucid explication of the problem came from X. A. Willard, in his popular text *Practical Dairy Husbandry*. In his chapter on the management of grasslands Willard cast a critical eye over New York State dairying systems. His analysis exposed the complexity of the issues involved and the limitations of nineteenth-century practice. He first focused upon the evils of overstocking and weeds, relating the two. Not only did too-close grazing result in feeble growth, it let in competing weeds and inedible species, which to eradicate would require enormous labor or even plowing up the pasture. But this latter course, Willard believed, was undesirable, first because it necessitated temporary reductions in the herd, and second because reseeding with "artificial" grasses produced a stand inferior to indigenous, self-sown grass:

> When nature furnishes the conditions for producing grasses that give the best results in milk, and when these grasses become firmly established in the soil, are we not pursuing a suicidal policy in destroying them under the impression that our pastures can be renewed at any time by plowing and re-seeding? . . . I

shall not dispute that we may doctor up our lands to produce any desired crop, but to do so is expensive, and will often require more science and skill than are common in the country.[56]

Willard was right. "Artificial" grass, both for hay and for pasture, was indeed problematic—not necessarily because it was deliberately planted, but because the plant on which farmers relied most heavily was highly vulnerable. The *Utica Weekly Herald* commented in 1873 on the local tendency to plant timothy "here, there, and everywhere," and other evidence indicates that farmers in the dairy regions were following this pattern.[57] This was unfortunate, because timothy did not return nitrogen to the soil as did clover and other legumes. Its shallow roots rendered it vulnerable to cold and to close grazing. Because of the preponderance of this single variety and the resultant lack of intermixed tall and short, deep- and shallow-rooted plants, weeds sprang up in the interstices. Insect pests also could multiply happily in a uniform stand. Even without all these problems timothy is ordinarily a short-lived perennial and requires continual replenishment.

As with the increasing interest in blooded cattle, probably the foremost reason for the excessive reliance on timothy had to do with the shift to factory production. With the premium set on quantity, timothy enjoyed even more popularity than before, because it made high-yielding hay; in turn, the improved dairy cows demanded more feed, reinforcing the tendency. The home cheesemaker's desire to include clover as an ingredient in high-quality cheese was less and less a factor in selecting seed for pasture and meadow. In the heavy emphasis upon timothy, the Anglo-American monocrop mentality reasserted itself.[58]

Farmers therefore sought solutions in keeping with their evolving system, consistent with prevailing values and assumptions. They approached the crisis by following several strategies. One technique well established by the postwar era was to cut hay earlier and to use machines. Lauren B. Arnold observed in 1876 that farmers were cutting their hay several weeks earlier than they had done in the 1840s.[59] Nutritional quality was thereby improved because the plants had not yet flowered. Cutting by machine reduced weather damage, and thereby resulted in better-quality hay.

More and more, farmers supplemented hay and grass with other food grown on the farm. Home cheesemakers, of course, had also fed crops other than hay and grass, but under the factory system the practice became more widespread and more intensified. Lewis Falley Allen warned that farmers who wanted to feed only hay over the winter, as many formerly had done, "had best retire from the business . . . if they intend to compete." The discussions of cattle feeding grew more sophisticated as farmers recognized the complex relationship among breed, feed, and environment.⁶⁰

Corn, fed to meet the late-summer shortage, was the most popular fodder crop. Census figures show that the acreage of fodder corn in Oneida County doubled between 1865 and 1875, from 1,132 to 2,064. By the turn of the century it would come to 11,000.⁶¹ A reliable and familiar crop, corn also fit into the new system because it effectively boosted milk production. Oats, shorts (a feed mix containing wheat bran), pea meal, and various root crops also were produced on the farm for feed. In 1876, after an abundant harvest, Boonville farmers fed potatoes.⁶²

Close to the city, where land was at a premium, the emphasis on grains and roots was greatest. Cheese manufacturer and farmer Lyman Wight, located in the Utica suburb of Whitestown, developed a system he called "half soiling." The term *soiling* referred to the practice of feeding cows in the barn year round instead of using pasture; Wight pastured his cows part of the year but fed intensively the other. He grew feed crops on his arable land and thus kept a relatively large number of cows on just half an acre per year per cow. He added that "cows give much more milk when soiled, and yield it more regularly than when pastured; and the waste of time, and the worrying of cows in driving them to distant fields in warm weather, are avoided." ⁶³

A second strategy was to purchase feed. An Oneida County miller rejoiced in 1872 that the cheese factories were providing a "great stimulus to liberal feeding"; he had benefited through increased feed sales, especially of corn and bran. Early in the 1880s the farm ledger of the Barnes family, from Boonville, showed regular purchases of animal feed for cows and other stock. The Nye family, of Vernon (Oneida County), bought middlings and shorts for their dairy cows throughout the 1870s. At a discussion on cheese factory patronage in 1867 one speaker said that "most of the Herkimer County farmers purchase their flour, and considerable quantities of grain fed to stock." Another agreed: "The great bulk of flour, of coarse grains for stock . . . came from other sections." These products probably came from the West.⁶⁴

Oneida County farmers showed an increasing interest too in purchasing fertilizer for their "worn-out lands." They had long been accustomed to making or buying lime locally, and of course had scrupulously saved manure. Now they began to consider other sources, because animal manures often fell short of the farm's needs; the Central New York Farmers' Club in 1876 discussed what sorts of fertilizers farmers could buy when their farmyard manure ran out. The *Black River Herald* declared in 1869 that "dairy farmers hereabouts need some active fertilizer on their worn-out pastures and meadows," and it endorsed "Raw-Bone Super-Phosphates," printing testimonials.

Factory-Era Agriculture

The *Utica Weekly Herald* in 1873 ran advertisements for "double refined poudrette" hauled from New York City at a steep $25 per ton. Westmoreland farmer William Brill told the Utica paper in 1865 that he purchased more than one hundred bushels of ashes each year for his pastures, at 10 to 15 cents each. By 1867 sample farm accounts published in the *Transactions* of the New York State Agricultural Society included expenditures on fertilizers.[65]

But "commercial fertilizers" were always dubious prospects. Dairy farmers joined in the excitement over fertilizer fraud that agitated the agricultural community in the later nineteenth century. Local newspapers published warnings from the agricultural chemist Augustus Voelcker about falsely advertised fertilizers; when bones and "sulphates" were scarce and expensive, he thought it doubly important that farmers get their money's worth. The scientist assumed that there was a strong demand for commercial fertilizer, and he believed that suppliers exploited farmers' fears. But scientific discussion of the issues was confused and inconclusive, mirroring the rudimentary state of knowledge at the time.[66]

In short, Oneida County farmers relied increasingly on feed and fertilizer purchased from off the farm; these materials were often imported from afar. Even the Holstein symbolized this trend, because it signified a shift away from locally available native cattle to more costly imported breeds. Barn building too more often involved purchased labor and materials. In a mutually reinforcing cycle, farmers then grew more crops to produce the cash income to buy feed, fertilizer, cattle, machinery, and barns.[67]

The most important new cash crop was hops. Hops, used in beer making, were raised extensively in Oneida County by 1870, as central New York became the country's leading hop-raising area. Acreage planted in Oneida County jumped from twenty-five hundred in 1865 to sixty-six hundred in 1875, and the harvest increased from 1.25 million pounds to over 3 million in the same period. During the boom in prices in the 1870s even a few acres could produce a tidy income. On farms with "hop yards," a few hundred to several thousand pounds were grown, all of which would be sold off the farm for cash.

Though they brought prosperity to the region, hops exacted a serious toll in the long run, because they were notoriously greedy. The *Utica Morning Herald* commentator worried about what he regarded as the destructive tendencies of hops:

> Where hops are grown, on a farm nearly every branch of business conducted on the farm is made subservient to that crop. Hop-growing requires large quantities of manure, the crop returns nothing to the soil, but is a continual

source of exhaustion, hence stock must be kept.... The hop crop eats up the larger share of manures made on the farm.... So, too, with the disposition of labor.... [This crop] is apt to run down the farm so that it will carry less and less stock and produce smaller quantities of grain year by year. Hop-growing we suppose must be profitable, since the apparent wealth or thrift of farmers who raise the crop is a noticeable feature.... But however this may be, hop-growing ... has a tendency prejudicial to a general system of high farming.[68]

Poultry raising also increased in importance. In Oneida County the reported value of both poultry and eggs nearly doubled between 1865—the first year statistics on this were published—and 1875. In 1865 Oneida County farms produced $23,000 worth of poultry and $37,800 of eggs; ten years later, $44,500 of poultry and $70,652 of eggs. Quite possibly these figures understate the amount of poultry raising going on, because this was conventionally women's work. The New York State manuscript census indicates that milk-producing farms were more likely to report significant chicken and egg production than their cheesemaking counterparts had been. Account books confirm that the poultry business was an important facet of milk-producing families' operations. Mrs. Susan Nye of Vernon, New York, regularly bought hens, bought feed for them, and sold their eggs. In bordering Madison County B. F. Cloyes raised hens for eating, and his wife sold eggs. The Oneida County town of Westmoreland became a poultry center, where enterprising businessmen farmers raised eggs for hatching and in turn sold them to farm wives. Articles on poultry production began to appear in local papers.[69]

Timber for lumber, bark for tanneries, and cordwood also produced cash income. Of course, timber had always been a crucial source of income, but now the emphasis shifted away from the potash and salts of pioneer days to building materials, fuel, and leather goods destined for burgeoning cities. Orchard products also found a market in cities and villages. In Annsville the developing canning industry provided an outlet for sweet corn during the 1860s and 1870s.[70]

Thus by the mid-1870s central New York dairy farms were "diversified" in the sense that they produced a variety of cash crops. Perhaps by a business standard these enterprises were well balanced because they spread economic risk. But where ecological and family sustainability were concerned, they comprised a farming system considerably less flexible than the one it replaced.

The new type of diversification on milk-producing farms resulted in a growing detachment from the old subsistence mooring. Under the home cheesemaking system, auxiliary products had lent themselves to numerous uses—for family, animals, barter, or sale. The most important new cash crop, by contrast, was suited only for market; the family could consume hogs, but not hops. The local economy also probably diminished, since the new crops were sold beyond local markets, rather than within them as had been the case formerly with such items as oats and hay.[71]

Moreover, because it was generated primarily by the need for cash, the new crop mix did not fit into an ecologically coherent overall production cycle; the farming system was beginning to dissolve into a set of unrelated fragments. Overreliance on nutrient-robbing timothy, with its accompanying deemphasis upon soil-sustaining clover, was only one sign of a less integrated system. Under the old ways hogs had consumed whey and produced manure, which was returned to the land. By contrast, under the new regime whey was often wasted at the factory; the farm supported fewer hogs and thus lost valuable fertilizer. Hops consumed manure and then were sold off the farm, contributing nothing to other farm operations. Corn, increasingly popular for fodder, also fed voraciously and returned little. The farm therefore became biologically less self-sustaining.

The centralization of cheese manufacture, then, brought profound changes for the farming system, primarily because cheese factories remunerated patrons exclusively according to volume. This emphasis set off responses that pervaded the farming system, reaching from the cows that made the milk to the foods they ate, the shelter they inhabited, and the four-legged company they kept. The pressures thus introduced fueled the new emphasis upon cash crops. The spiral escalated because the more farmers relied upon the market, the less they could choose their level of market participation.

The turn away from the farm for essentials coincided with the general cultural reorientation outward, away from the farm, for cultural ties, market opportunities, and employment. As the farm enterprise itself dissolved into loosely connected pieces, so did the household itself, with all the vulnerabilities and liberations that entailed.

9

Fragmentation and Reorientation
Dairying Households in the Factory Era

> And yet it seems hard, when you've worked from the dawn,
> Till the sun disappears from your sight,
> To think of the cows you have yet got to milk,
> Before you retire for the night.
>   Henry W. Herbert, 1872

Just as the factory-era farming system dismantled the older integrated system into individual, unrelated enterprises, the household also underwent fragmentation. Wives, daughters, husbands, sons, and hired help increasingly operated in different arenas and even pursued different occupations. As capitalist agriculture intensified, male workers especially were vulnerable to pressures for productivity. On the other hand, women enjoyed more opportunities for social and domestic activity than before. They did not reject farm work altogether, but they attempted to redefine "competency" in a way that would increase the emphasis upon comfort, consumerism, and culture.

Household Reorientation

As farm families delegated cheese processing to the factories, they also became more removed from marketing procedures. A factory committee took responsibility for finding buyers and arranging for shipping, and patrons received checks regularly. Factors still worked the cheese trade, but they now dealt with factories rather than with farmers.

At the household level other, fundamental, shifts accompanied these changes. Households altered their relationship to the neighborhood, appearing more as isolated units than as strands in a local economic web. Household account books from the factory era testify by their very format to the new orientation. No longer did farmers set out accounts by listing debts, in cash or in kind, for each individual partner in exchange. Now listings were organized not around people but around cash transactions, and around the total amounts "received" and "paid out." Farmers did not yet rigorously calculate farming profits, but household account books began to reflect some of the rationalized accounting procedures that agricultural reformers so cherished. The Jeremiah Sweet, Barnes family, and Nye account books show this. While the mix of income-generating items in these cases did not differ radically from that of the prefactory era, all are clearly exchanges for cash, and any noncash exchanges that might have taken place do not appear. Susan Nye's account book, from Vernon (Oneida County), suggests even more strongly that neighborhood transactions that formerly might have been simple barter were now matters for money exchange. She bought beefsteak, even cheese, and paid cash for pasturing a cow.[1] Peleg Babcock, by contrast, had traded for pasturing services and raised his own beef—and, of course, made cheese—in the 1850s.

These accounts point to the preeminence of cash exchange in the postwar period. Farmers had little choice but to meet the demand. The war itself played an important role in the shift, since taxation to finance the war intensified the demand for cash. In an escalating process, institutional and personal creditors—banks, corporations, manufactories, local merchants, and neighbors—dealt more exclusively in money. By 1877 many would have agreed with the correspondent in the *American Rural Home* who complained that "it is easy enough to sell anything if you will only exchange for goods or some other kind of mechanical labor. But a man must have some cash." The cash nexus helped to erode the old economic links between neighbors; as the account books graphically demonstrate, farmers now conceived their economic lives less in terms of multifaceted exchanges with neighbors than in terms of income and expenditure.[2]

The factory's action as social solvent extended to the customary work exchanges. The factory became an institutional focus in rural people's economic lives, and as a more capital-intensive agriculture demanded wage labor and investment in machines, "changing works" could not accommodate. Some welcomed this development, seeing an opportunity to disentangle themselves from the web of obligation that "changing works" demanded. In 1869 a commentator in the *Black River Herald* noted with satisfaction that the mower "gives a man independence." In the context of the times, his remark expressed implicit frustration with reliance upon neighbors and casual laborers.[3]

But others mourned a loss of sociability. An 1880 piece in the Rochester-based *American Rural Home* deplored the passing of "changing works." The author recalled the excitement of old-time haying, a sociable occasion when friendly but intense contests tested participants' manifold skills. He sadly contrasted this memory with contemporary "machine" farming, in his opinion tediously dull.[4]

Fragmentation occurred not only at the neighborhood level but also within households. A typical example illustrates the salient trends. In 1870, in the town of Steuben, 51-year-old Clinton Merrick lived with his wife, Abbey (49), sons John (25) and Frank (6), and daughters Marion (18) and Antoinette (15). The Merrick farm was worth $12,000 that year; the Merricks sent to the local cheese factory eleven thousand gallons of milk from thirty-five milch cows, while also raising a variety of crops and other livestock. No hired hands were present in the household at the time the census taker came around, but the family had paid out $875 in wages during the preceding year.[5] Merrick must have relied to a significant degree upon wage labor, because he apparently could not depend entirely upon his offspring. John was listed as a farm laborer but reported $7,000 worth of real property and $500 worth of personal property. Perhaps he had already claimed a share of his father's farm, but he might have farmed somewhere else while still living at home. Marion and Antoinette both claimed the occupation of teacher, so their contribution to the farm was likely also limited.

In hundreds of other households like the Merricks', similar trends appeared.[6] Though age structure had not altered substantially, striking changes marked other aspects of the household economy. Milk-producing households were markedly smaller (five persons per household) than their predecessors (six). One reason for this decline was that live-in hired hands were disappearing. In the seven Oneida County towns in 1870 only a third of milk-selling households reported live-in help, a decline from nearly half of cheesemaking farmers in 1850. Moreover, each household contained fewer live-in workers: 1.5 on average, as compared with 2.0 in 1850. Thus milk-selling households now more often consisted only of related people.

This did not mean that milk-selling farmers hired less labor, because two-thirds of them paid out wages for farm labor. In fact, milk-producing farms in the seven towns averaged about $300 per year on wage expenditures, nearly enough for one full-time hired man at prevailing wages. Despite the difficulties involved in comparing different censuses, it seems clear that there was a shift away from live-in help and toward wage labor; this reflected a national trend, realized within a specific local context. In Oneida County, milk-selling farmers could, and evidently did, use the money from milk, hops, and forest products to substitute cash payment for in-kind remuneration.[7]

The trend away from live-in labor was reinforced by changing views of class relationships. Throughout the North status-conscious farm people who could afford to pay wage workers increasingly erected social barriers to separate themselves from their "employees," as they now were called. Farm wives voiced their reluctance to "steam, and work, and drive, in order to try to cook enough to satisfy the gluttonous greed of some of the lower classes of farm help." One especially parsimonious farm wife became so exasperated that she expelled from her house the Irish laborer who had the audacity to declare, "The more I ates, the more I gits for me work." Others complained about workmen's table manners and worried about the "contagion" farmhands' behavior would spread among children. "We are not sufficiently dignified with our employees," one woman complained. This change in the relationship between hired hands and their employers had begun earlier; the Babcocks' trouble with their hired help is a case in point. But now the same shift was occurring all over the North, as wage labor replaced "changing works" and its associated patterns.[8]

A New World of Work for Men

The growing predominance of wage labor was but one aspect of the increasing influence of capitalistic forms and pressures on the milk-selling household.[9] Indeed, a new world of work was taking shape, and its greatest impact was upon male workers, whether farmers, their sons, or their hired men. In 1872 the *Utica Weekly Herald* published some verses by Henry W. Herbert:

> How pleasant it seems to live on a farm,
> Where nature's so gaudily dressed,
> And sit 'neath the shade of the old locust tree,
> As the sun is just sinking to rest;
> But not half so pleasant to hoe in the field,
> Where the witch grass is six inches high,
> With the hot, scorching sun pouring down on your back—
> Seems each moment as though you would die!
>
> 'Tis pleasant to sit in the cool porch door,
> While you smoke, half reclined at your ease,
> Looking out o'er your beautiful fields of grass,
> That sways to and fro in the breeze;
> But not quite so pleasant to start with your scythe
> Ere the morning sun smiles o'er the land,
> And work till your clothes are completely wet through,
> And blisters shall cover your hand.
>
> In keeping a dairy there's surely delight,
> And it speaks of contentment and plenty,
> To see a large stable well filled with choice cows
> Say, numbering from fifteen to twenty!
> And yet it seems hard, when you've worked from the dawn,
> Till the sun disappears from your sight,

> To think of the cows you have yet got to milk,
> Before you retire for the night.
>
> But the task fairly over, you cheer up once more,
> And joyfully seek your repose.
> To dream of the cream-pots with luxury filled,
> And milk pans in numberless rows;
> But the sweet dream is broken when, early next day,
> You're politely requested to churn.
> And for three weary hours, with strength ebbing fast,
> The victim despondingly turns. . . .
>
> But no one disputes that the farmer is blessed
> With true independence and labor—
> Whose food don't depend on the whims of mankind
> Like that of his mercantile neighbor;
> For God, in His mercy, looks down from above,
> And paternally gives him his bread,
> Provided—he works eighteen hours every day,
> And devotes only six to his bed![10]

Just as in home-dairying times farm women had satirized the supposedly bucolic life of the dairymaid, so Herbert bitterly critiqued factory-era farm life for dairymen. With clearly ironic intent, he titled his poem "The Independent Farmer." Similarly, by the latter part of the century writers like Hamlin Garland were comparing dairy farming in the Midwest, especially Wisconsin, with slavery.[11] Their imaginative renderings accurately identified an unmistakable trend in dairy work for men: it was getting longer, harder, more anxiety-ridden, and less independent. In this respect, even though farmers owned land and machinery, they exercised less and less control over their work under pressure of market competition and factory demands.

To be sure, the basic elements of men's work had not changed drastically; after all, in milk production as in cheesemaking, men took responsibility for the tasks involved: planting, tending, and harvesting field crops; selecting, feeding, and milking cows. But the way they performed their tasks was significantly altered, as were the hours they toiled and the attitude with which they approached their tasks.

Improved milking breeds required "high feeding" in order to thrive, and so even though farmers possessed fewer beef cattle, sheep, and hogs, they invested more effort than ever in caring for their stock. As we have seen, they achieved an increase in milk yields simply by feeding the cows more and sheltering them better and longer. Farm men found themselves expending more time and effort growing, storing, processing, and harvesting feed, and indeed in the feeding itself. When manure accumulated in the barn, men had to remove it and spread it on the fields, a task less necessary under the old grazing regimen.

Hay making—a critical task—was, of course, mechanized to a significant extent. But corn cultivation still demanded slow and arduous labor, and corn grown specially for fodder required meticulous attention—frequent fertilizing, hoeing, and later hand-harvesting, husking, and shelling—for the best quality and yield.[12] Root culture too was highly labor-intensive; most root crops grown for feed demanded careful planting, weeding, and thinning, and at harvesttime they had to be dug by hand.

Storing and processing feed also probably took up more time than previously. True, new barns were easier to work in, but the desire for high-quality feed prompted farmers to pay more careful attention to its condition when storing it. Too, they carefully chopped, mixed, measured, cut, and sometimes even cooked; to farm-grown items they added purchased feeds. The "mixing boxes" in the new-style farm barns attest to the importance attached to this process. Experimenters and practical farmers began to recognize how productive it was to work out precise rations and appropriate feed mixes for cows of different ages, sizes, and breeds.[13] This implied individualized portions, again involving more work.

Since cows were confined for a longer time each year, labor also increased because workers were essentially bringing food to the cows rather than letting them forage for themselves. Some commentators advocated six-times-a-day feeding; though not all went to such extremes, the literature makes clear that cows were beginning to receive the kind of attention usually reserved for human babies. In addition to being fed more frequently, cows were also cleaned more often. Twice-daily milking—also increasingly done by men—never varied, and as the season extended, it occupied more days in the year. In 1870 an *American Agriculturist* reporter watched in amazement as workers on a Long Island dairy farm went about their chores; he was impressed at the way "the whole labor force of the farm is kept on the jump."[14]

Fred Bateman's study of labor productivity in dairying uses econometric analysis to arrive at a similar conclusion: that dairy farmers devoted an increasing amount of time and labor to their herds. Indeed, by the 1930s dairying would be less seasonal than cotton, corn, or grain culture. Bateman further argues that the cows' productivity increase did not keep pace with the escalating investment of human labor. The evidence from individual farms and from descriptive accounts bears out Bateman's conclusion and further exposes its social ramifications.[15]

The dairy writer Henry Stewart's 1888 assessment of "What a Dairyman

Should Be" illustrates late-nineteenth-century ideal notions of the personality and skill requisites for dairying, and shows how the minutiae of dairying practice assumed new prominence. The dairyman should possess a "natural instinct to cleanliness," or "Purity." This the author regarded as an "exceedingly comprehensive virtue. It refers to the air breathed by his cows, to the condition of their skin, . . . their food, water, and to the person, clothing, and habits of the dairyman." Second, a dairyman should be patient and good-tempered. Third, he should follow strict regularity in his habits: "A man who can never do the same thing twice in the same way or at the same hour each day . . . will soon have to change his business." Fourth, a dairyman should be observant, and fifth, he needed to be studious in order to keep up with advances in his rapidly changing profession.[16] None of these recommendations was new; cleanliness, for example, had always been thought important for dairymaids. Rather, it is the codification and the degree of emphasis and detail evident in each point that carried a cumulative thrust, showing how dairying was growing more routinized, demanding, and masculinized.

When the mania for hops swept through central New York, another demanding work regimen evolved. While women were prominent in the harvest, men saw to most of the preliminaries, which were both time- and labor-intensive. The "hop yard" was plowed in the fall and again in the spring, before the hop roots were planted in hills about eight feet apart. As soon as the plants made their appearance, the grower placed poles next to the seedlings, to provide essential climbing places for the vines. These were no mere sticks; an "Old Hop Grower" described the poles as being fifteen feet to twenty feet in height, "sharpened like fence-posts," and set (by using crowbars) two feet deep and a foot apart.[17] The fields of poles must have had the menacing look of an army of medieval pikemen. Once the plants were well established, the grower watched them carefully, cutting back frequently to maintain the desired bushy growth, weeding frequently with hoe and cultivator, and destroying side shoots, which if allowed to grow would diminish the plant's vigor.

After harvesting (see below), the curing process also required round-the-clock attention. Farmers laid the delicate plants across specially prepared slats in the hop house, suspended above ground level. They then carefully stoked the stove in the hop houses to dry the hops, watching constantly to prevent excessive moisture loss or scorching on the one hand and underdrying on the other. After drying they pressed or baled the hops for market.[18]

The cumulative effect of the overall regimen upon dairymen is captured in the diary of B. F. Cloyes, who with his wife, Emma, farmed in Morrisville, Madison County—the heart of the cheesemaking region, along with Oneida and Herkimer Counties. In 1864 Cloyes's diary shows that he spent a lot of time on the road, carrying milk to the factory and "whay" back home, and "carrying" his wife to visit and to shop. He milked, worked in the fields, churned, and tied hops; he seldom recorded any work that his wife did, or whether or not she "helped" him, but he did mention when she went "visaton" and when she hired someone to do the housework. Cloyes, like Herbert's fictional farmer, frequently found himself tired and depressed. In 1878 he wrote that he "churned this morning & it tired me out." Later, when his wife "got on her dignity," Cloyes "got the blues." [19]

New Occupations for Wives

The Cloyes diary is revealing in the extent to which it shows how the couple's work worlds diverged. Mrs. Cloyes, whether by choice or by necessity, seems to have participated little in the farm work, except for poultry raising. Late-nineteenth-century dairying couples varied widely in their work allocation; the Cloyes family resembled those in which the woman distanced herself significantly from the main farm enterprise. While Mr. and Mrs. Cloyes represented an extreme, the trend among Oneida County dairying couples was indeed toward greater separation of roles. In this respect the Cloyes diary contrasts with those of prefactory dairy farming couples. Maria Whitford and her husband, Samuel, often worked together; so did Peleg and Cornelia Babcock, who even kept a diary jointly.[20]

Milk selling had quite different implications for women and men. Women could, or in some cases had to, strike out in different directions; the paths they took depended largely upon whether they were wives or daughters. The choices they made helped to accelerate and define rural change.

For farm wives the substitution of milk selling for cheesemaking brought about a vitally important change in their lives. At one stroke a time-consuming, physically taxing task was entirely eliminated. The effect was revolutionary. Local newspaper editors, farm journalists, and not least farmers themselves proclaimed farm women's new freedom; even though they occasionally indulged in rhetorical excesses, all agreed with the essential point expressed by Lewis Falley Allen, that the cheese factory had removed the "burden of cheesemaking labor from the hands of our over worked and tired down dairywomen, who, by a hard hearted and mistaken economy were the drudges of the curd tubs and cheese presses."[21]

Freedom from cheesemaking brought a reduction in related work as well. Boarding less hired help, for instance, often meant less cooking and housework for farmers' wives. More significantly, the evidence points to a diminished role in dairy work generally. Women in milk-selling families were not only making less cheese; they were also making less butter, because the factories took whole milk. Butter production on milk-selling farms fell off dramatically, to half the levels that had characterized cheesemaking farms (350 pounds versus 750 formerly). Most butter made in Oneida County was now made on farms that did not send milk to cheese factories.[22]

Indeed, some women even withdrew from buttermaking. Though this probably occurred in only a few households, it is nonetheless notable that some men were increasingly involved in butter production: Henry Herbert was not alone. In Madison County, for example, B. F. Cloyes not only milked, carried milk and whey to and from the factory, planted potatoes, spread manure, hayed, and tended hops, but also routinely churned.

The evidence on whether wives were also withdrawing from milking is scarce and ambiguous. The *Ohio Farmer* in 1869 estimated that women still did half the milking in the Western Reserve dairy region, and farm women in the Nanticoke Valley of New York continued to milk toward the close of the century. But the *American Agriculturist* maintained in 1878 that the cheese factory had "relieved the women not only from the care of the milk, but from the milking as well." Both *Harper's* and *Appletons'* sent reporters to investigate the cheese factory system in the 1870s, and both found that few women milked any longer. "The dairy-maid going singing to the pasture with milk-stool and pail," reported *Harper's,* "is either a myth or a tradition in the dairy regions. The milking is done chiefly by men, and in surroundings which suggest no poetry." "In Herkimer and Oneida counties," the *Appletons'* reporter said, "the milking is done in most instances by one of the farmer's boys. . . . He is far better suited for it than a masquerading maid would be." The masculinization of milking proceeded apace. The *Genesee Farmer* evaluated the virtues of a good milker in terms that were unmistakably gendered and, as in cheesemaking, tied these qualities to success in the capitalist system: "A boy who will always milk clean will have a good recommendation of being faithful wherever he goes, and such a recommendation always goes a great way among business men."[23]

Whatever the proportion of farm wives who still milked, the argument over its appropriateness for women continued into the factory era. An indignant contributor to *Moore's Rural New Yorker* in 1878 asked why a woman should feel out of place in a properly designed milking stable. Arguing that the presence of female milkers might even "purify" those surroundings, the author declared that it was "a gross charge against the business of dairying that it is managed in such a way that a woman is out of place in any part of it." Yet this correspondence was apparently prompted by a "charge" that some women did in fact feel "out of place" in the dairy barn; the choice of words is revealing, in that it implies that the barn was no longer perceived as a shared space. Some of the sentiments that inspired such feelings appeared in a piece of rustic poetry in the *Utica Weekly Herald* in May 1878. The poet remembered that the milkmaid of old pastoral odes had always been ruddy-cheeked and robust, but he wondered in retrospect if her hands had not in reality been rough, her feet blue with cold, her head like a scarecrow's. He preferred the soft, white hands of the "Modern Milk Maid." The first of these two pieces sought to make milking more attractive; the second assumed it could never be made so. But significantly, *both* implied that values of gentility were important to farm women and men.[24]

Women's decreasing involvement in milking and their total withdrawal from home cheesemaking had spatial implications. Even more than before, the barn became associated with men. Cheese houses decayed or were put to other uses. It is quite possible to postulate that in general, boundaries between "men's" and "women's" space were become more sharply defined, and that women's arena was more restricted than before.

But this pattern was not a clear and uniform one. Oneida County farm women expanded their activity in another area traditionally allocated to women: poultry raising. Perhaps cheese houses became henhouses, for the reported value of both poultry and eggs nearly doubled in Oneida County between 1865 and 1875. In his statewide survey completed in 1936 T. E. Lamont showed that numbers of chickens on New York State farms rose steeply between 1879—the first year for which he collected figures—and 1899, and the number of eggs increased even faster as productivity improved. In March 1878 the *Boonville Herald* exulted that eggs were cheap; put a new bottom in your frying pans, it told readers, and enjoy ham and eggs instead of buckwheat. The paper chose words with symbolic resonance to nineteenth-century readers; buckwheat was associated with poverty and scarcity, eggs and ham with times of plenty.[25]

While journal articles on poultry raising were usually addressed to men, other evidence indicates that women participated extensively in the hen and egg business. For example, male correspondents often used the plural instead of the singular possessive, suggesting a joint operation.[26] Manuscripts are still more definite. Mrs. Susan Nye of Vernon kept an account book from 1870 to 1873; she regularly bought hens, purchased feed for them, and sold their eggs.[27]

Butter and eggs were complementary products in the milk-selling farm wife's seasonal pattern. She could make butter during the fall and winter months, when the cheese factory was closed, and tend to the hens mainly during the summer months. The Barnes family account, kept in Boonville in the early 1880s, shows this pattern clearly: they sold butter from October to early May and collected milk money during the other months.[28]

Though they were an important source of cash,[29] butter and eggs probably did not occupy the kind of time and effort demanded by home cheesemaking. Other pursuits assumed greater prominence in farm wives' daily lives. Considered together, they form a picture of an unmistakable reorientation toward domestic work, with a new prominence for social pursuits.

One measure of this reorientation is the agricultural fair. In the 1860s and 1870s fairs in Oneida County proliferated and expanded in size.[30] The county fair attracted thousands every fall, and most towns also had their own well-attended fairs. Hundreds of categories for exhibits and competitions replaced the staid, conventional livestock and farm-tool shows of an earlier generation. Though all agricultural endeavors received more publicity in the newly expanded fairs, the increased participation of women is especially notable. Among the cheesemakers' generation, men had actively participated but their wives had been conspicuously absent from lists of prizewinners. Moreover, a noticeable disjunction had divided the women who did participate: those exhibiting flowers and decorative objects tended to reside in the cities, those showing utilitarian items such as linens, in rural areas.

By 1870 the pattern had changed. Among milk sellers women and men alike participated. In Trenton, for example, the increased interest on the part of women required the erection in 1869 of a spacious (84- by 25-foot) Ladies Hall as part of the fair's permanent building complex. Vegetables and dairy produce occupied the first story, floral and domestic exhibits the second.[31] Every autumn, women from Oneida County farm families displayed the results of their work in a staggering array, from unusual items such as a "horse breast protector" to more conventional entries of brown bread and canned fruit. Even if no official category existed for a particular item, the entrant could seek a "discretionary" premium. So many people entered so many items that the *Black River Herald* satirized the fair mania in 1869: a fictional "Miss Julia" entered a "bird-seed daguerreotype" and "one loaf of bread raised with yeast, one ditto raised with derrick."[32]

In the 1870s women from milk-selling families won prizes for (among other things) rag carpets, brown bread, woolen blankets, knitted stockings, canned fruit, pencil drawings, wax fruits and flowers, "worsted tidies," silk embroidery, sofa cushions, "embossed paper wreaths," and "hair wreaths." One Trenton woman won a premium for wax flowers in "a neat frame, surrounding some very natural autumn leaves in wax."[33] What all of these handcrafts had in common was their importance to domestic life, whether in feeding and clothing the family or in decorating the home. Of course, it is possible that Oneida County dairying women had been making these items all along without bothering to compete at the fair, but there are good reasons to believe that the new activity reflected a reallocation of labor in the home.

For some women the home itself was more elaborate and would require tasteful adornment. The *Black River Herald* frequently reported on local construction projects and took note whenever someone built a fine house. In 1880, for example, it carried a piece on milk-seller Walter Jackson's new house, which, it said, "a few years ago would have been considered magnificent for a governor's palace." Boonville farmer Walter Stickney retired in 1875 and built an elegant mansion in town. Real estate ads also give a sense of large and substantial houses, as in an 1875 ad for a "superior dairy farm" of 150 acres, with a two-story thirteen-room house and forty-six-cow dairy barn.[34]

These cases, however, were unusual. Available evidence indicates that most dairying families chose a more moderate course: they retained their older houses and elaborated their interiors. The *Appletons'* reporter, traveling through Herkimer and Oneida Counties in 1878, noted that most of the houses in the district were old-fashioned and unpainted. Pictorial evidence corroborates this information. At least four farmers who can be identified as milk-selling dairymen had their homesteads illustrated in the 1878 county atlas. None was up-to-date; in fact, the most modern of the four was Didymus Thomas's foursquare Italianate house, by 1878 fast passing out of fashion. Daniel M. Crowell, John C. Owens, and James Mitchell all resided in plain, rectilinear Greek Revival farmhouses characteristic of the mid-nineteenth century. Yet all four were successful dairymen and possessed modern outbuildings, an indication of the priority the farm operations had assumed.

The very disjunction between barn and house could be interpreted as a highly visible sign that men's power in the household still prevailed, perhaps even enhanced in an era when cash had taken on high importance. Indeed, both contemporaries and recent historians have made this argument. While these building patterns undoubtedly reflected household power dynamics with respect to gender, there are other interpretive factors to consider. One is that in the highly competitive climate of the later nineteenth century, dairying families may have had little choice about modernizing their production apparatus.

The other is that by preserving the old homestead with an altered interior environment, dairying families could develop an amalgam that fit more comfortably than would a brand-new structure with their background in the values of competency. Ties to the past were still important, and so an old house—often built by parents or grandparents—expressed values consistent with the evolving domestic realm. Dairying women's late-nineteenth-century activities are best interpreted as aiming toward an elaborated form of competency. For a new generation of dairying women, this trend toward comfort and decoration also represented a way, albeit a sometimes ambiguous one, to incorporate into farm life the "accomplishments" that had been so bitterly contested in earlier arguments over girls' educations. In this way we can make sense of why the same individuals might seek prizes for a utilitarian rag carpet and for a fashionable, decorative wreath of wax flowers, or why local businesses with large farm clienteles would carry mowers, reapers, fanning mills, washing machines, and pianos.[35] The elastic nature of competency allowed farm families to retain the ideal. Yet to stretch the boundaries of competency in this way was to jeopardize the entire endeavor, especially since the new "necessities" were often manufactured consumer goods acquired with hard cash.[36]

Oneida County manuscripts and inventories from the 1880s suggest how farm families elaborated the home environment with furnishings and decor ranging from the mass-produced to the hand-crafted. Susan Nye, for example, purchased a parlor sofa for $23 in 1870 and put up a wallpaper border in 1872.[37] Probate inventories from the early 1880s offer still more specific information. Both Nathan Moore of Marshall and Evan Hughes of Vernon had farmed fairly typical operations, as the numerous references to livestock, milk cans, and farm tools testified. Of interest here are their household furnishings. Unlike Nicholas Gardiner, who in 1820 had left chairs aplenty but few other articles of furniture, Moore had possessed two divans, a parlor carpet, window shades and curtains, two "covered chairs," three "shades and fixtures," a dining room carpet, a dining room table, two armchairs, a dozen wine glasses, a "dinner set," rocking chairs, and a piano. Hughes, who died in 1883, left two sewing machines, a "centre table," a sofa, twenty-four teacups and saucers, and forty yards of rag carpet, among other items.

While two inventories surely cannot represent all, they support evidence from contemporary observations and published sources: they show an increased emphasis upon comfort, on consumer goods, on ceremonial meals, on social amenities, and on what the farm writers liked to call "cultivation" in the farm home. In short, like their counterparts in the Midwest, these New York State farmers were helping to construct a "landscape of class" that expressed their identification with middle-class culture.[38] Women from milk-selling farms also expanded their participation in local social, cultural, and philanthropic activity. A fuller discussion of these activities and their implications for the community will follow in chapter 10.

New Directions for Daughters

In 1881 Elaine Goodale published a little book called *Journal of a Farmer's Daughter*. Goodale opened this semifictional portrait with a pointed disclaimer: "Of course, I do not mean by a farmer's daughter that sturdy, red-cheeked maid-of-all-work, who kneads bread and picks up potatoes with the same admirable vigor and elasticity." Rather, she refers to "one who stands a little aside from the engrossing toil and every-day interests of the farm, with a realizing sense of its value as art, as life." This romantic observer always describes farm life from the viewpoint of an onlooker. She consults with her father about the vegetable garden, but he cultivates it. She tends to the hens only when it suits her to indulge her romantic sensibility; she observes haying on a leisure-time "tryst" as she comes upon a mower in the field. Only when she spends "long mornings and long afternoons in the patient work" of putting up fruits and vegetables does this farmer's daughter carry on the work of her predecessor, yet even here a ceremonial as well as practical function comes into play, for the wares are partly intended for display at the county fair.[39]

Certainly Goodale's portrayal is fanciful, but it expresses much by its very exaggeration. The *Journal* gives strong expression, for example, to the farm daughter's growing detachment from the farm enterprise. Her involvement with the farm is now expressed in artistic evaluation or romantic reverie. Unlike her predecessor, who worked as part of a corporate body, Goodale's modern farm girl acts as an individual, participating on her own terms and in activities aimed as much at self-expression as at fulfilling farm goals. Here the household's increasing social and economic fragmentation appears clearly, if hyperbolically.

In ordinary dairying households, daughters' withdrawal from the farm enterprise took more prosaic forms. For daughters in the age group 20 to 29, the change to milk production was accompanied by fundamental personal choices. Most dramatically, the proportion of these young women present in milk-selling households fell dramatically to half of what it had been under the cheesemaking regime.

What had become of all the young women? Some had probably married, but more had migrated. All over the United States the surge of nineteenth-century rural-to-urban migration was led by young single women.[40] In 1830 there were roughly equal numbers of young men and women in Utica. But by 1865 women outnumbered men: in that year there were 131 women for every 100 men aged 20 to 29. By 1880 Utica's population consisted of 15,660 males and 18,248 females.[41] Outside Utica the sex ratio was a less extreme but still steep 113 women to 100 men. By contrast, on milk-selling farms men outnumbered women in this age group 109 to 100.

Wherever they went, young women were disappearing for a reason: except for a brief period during the Civil War, they participated less in farm work as men took over milking and even buttermaking. Daughters had a more limited role than wives in the milk-selling family economy. They might help with the milking, or help their mothers churn. In some families they drew the twice-daily task of driving the milk cart to and from the factory. The Camden (Oneida County) paper, for instance, had a young visitor to its offices one day who wanted to learn typesetting but had "to carry her ma's milk to the factory and wash dishes."[42] However, even this errand was mostly done by men, if period photos are any indication. In any case, none of these tasks approached the magnitude of work in home cheesemaking, so it would not have surprised readers of the *Utica Weekly Herald* to learn in 1874 that the Holland Patent town lecture series included a talk on "Our Girls, and What to Do for Them."[43]

The "girls" had notions of their own about what to do. In Utica textile and clothing factories supplied jobs to over twenty-five hundred women in 1880. The boot and shoe industry added several hundred more. Sex breakdowns for the city's commercial employments are unavailable, but women probably claimed some jobs there as well. These aggregate figures are not very satisfying, but we can perhaps discover some of the young women's new occupations by studying the pursuits of those who did not migrate but instead stayed at home. Among daughters who were living at home, far more declared some occupation in 1870 than in 1850. Among milk-selling households in the seven towns, the 1870 population census shows that daughters in the 20–29 age group declared a nonagricultural occupation more often than any other group of household members. At least one-third of these young women had found employment off the farm for all or part of the year; this number is probably a low estimate, because some census takers probably did not ask these women about their occupation. Their titles included teacher, music teacher, dressmaker, milliner, silk factory worker, and cook on a canal boat. As we have seen, some worked in cheese factories. These young women probably worked less and less on the farm.

The presence in the household of daughters with nonfarming occupations again points to a growing divergence between the economic activities of the farm's men and women. In many of these households there were no sons to supply labor; rather, the household head paid out wages, probably for hired men. In an earlier generation those daughters probably would have worked on the farm; now hired wage labor supplied their place. Thus the masculinization of dairying went hand in hand with daughters' "superfluity." In fact, daughters were to some extent becoming superfluous by evolving cultural definition and perhaps by common consent; they could have stepped in to milk as the sons left, but instead wage laborers were hired.[44]

Though these particular women did not migrate, their choices illuminate the motivation for migration among young single women, and therefore also for change in rural society. Their occupations suggest that migration occurred not only because women lacked land or work, but also because young women aspired to a degree of independence, the cultural advantages of urban life, and employment in feminizing occupations.[45]

During the postwar period the dynamic of household negotiation over daughters' careers culminated in the emergence of a sizable group of female schoolteachers. The most frequently mentioned occupation among farm women still living at home was teaching. Their presence in the teaching ranks probably reflected at least to some degree the influence of rural academies, as a new generation of academy-educated women obtained positions in the schools. Historians of education have observed that public-school teaching presented both restrictions and opportunities for women. Female teachers were often subjected to strict community control; their pay was half that of male teachers, and their working conditions were often difficult. But as teachers they also enjoyed a measure of autonomy from their families, they possessed status in their communities, and they received intellectual stimulation from their work.

The position of teachers in rural New York late in the century reflected this ambiguity. Mrs. Eloise Abbott remembered of Jefferson County that "if you come to reside among us [as a teacher] you must expect to be closely watched." The local papers corroborate her observation: teachers' effectiveness was a favorite subject for public comment. Each term the local papers extensively covered public closing ceremonies, and it is clear from the descriptions that the ceremonies were well attended, and that teachers as well as pupils were being evaluated.

In 1870 the *Rome Sentinel* published the proceedings of the Oneida County Teachers' Institute and printed the names of those attending; at least nine young women from milk-selling families in the seven towns were there.[46] Their participation is notable, for these institutes were not mandatory, and they manifested the professionalizing culture of education. Participants had a careerist outlook that stressed technical and professional matters involving pedagogy, theory, and curriculum.

The middle-class aspirations and educational background of farm daughters likely played an important role in their aspirations to a professional identity. This class dimension to their activities became apparent in an 1880 satire in the *Boonville Herald*. An indignant farmer comes to the school board to complain that his daughter had failed the teachers' examination, for the unfair reason that the history part asked her about people who had lived before she was even born. In a possibly related story, the paper also reported on the school board's teaching appointments. Apparently in this case the board had preferred a young woman, Fanny Kau, allegedly because she asked for less pay than her competitor, a Mrs. Buell who held a normal-school certificate, had six years' experience, and boasted "the most satisfactory references as to scholarship." Criticizing Miss Kau's father's claim that "she had taught fifteen terms in a country school, and had always given satisfaction," the editors dismissed parents' "satisfaction" as ignorance.[47] This incident illustrates the tensions between the professionalizing middle class and those having fewer resources and perhaps a more localistic outlook.

A source of seasonal employment for women was the hop harvest. The *American Rural Home* noted that "all classes engage in hop-picking during the season—farmers' daughters and sons, school-teachers, the children of professional gentlemen, in fact the *best society,* and they have social, merry times." Every fall masses of "girl" pickers flowed into the hop-growing townships from the vicinity of the hop-growing region; for example, Boonville sent large numbers of young women into the southern part of the county to pick hops. They worked for a daily wage, going from one hop yard to the next until the crop was in.[48]

Dora Walker's diary shows how one farm daughter treated employment in the hop harvest. She was from a well-to-do East Winfield (Herkimer County) dairying family. Dora picked hops between August 28 and September 30, 1860, and earned a total of $18.85. She was able to spend her hops money on jewelry, perfume, candy, yarn, a purse, a dress, and some cloth, among other items. It seems as though this young woman did not have to use her money for her education, which, to judge from the certificates from the East Springfield Academy, was obtained at that institution. Probably her parents paid for her education, while she was able to spend her money on herself. Even though her family was well-off, she was not idle at home; when not at school, visiting, or going to singing schools, she helped do the housework and churn.[49] The hop harvest allowed Dora to earn cash to spend and may have given her a little bit of independence from the family economy and budget.

A 1922 novel by Flavia Canfield called *The Hop Pickers: Girl Life in the Sixties* recreated a Wisconsin hop-picking season from the author's memories. The main characters, daughters of a respectable but poor missionary preacher, jump at the chance to pick hops because it represents the opportunity to earn money for a term at boarding school. These girls, unlike Dora Walker, had to earn their own tuition money. Even so, Canfield presents an unsympathetic portrait of their lower-class Irish and "Dutch" counterparts, assuming—and accepting—solid class barriers.[50]

Thus young farm women embarked on the uncertain path away from the farm household. These developments anticipated similar trends in Nebraska by a half-century. During the depression of the 1930s Nebraska farm women migrated more often, remained single more frequently, took advantage of nonfarm jobs, and reformulated the "agrarian message" to cast farming as a phase on the path to urbanism, to be abandoned in the course of social development.[51] While the contexts differed radically, of course, especially in the economic circumstances and the degree of government involvement, the salient trends are strikingly similar.

Capitalism in dairying had encouraged household fragmentation, as individual members ceased to be central to the farm enterprise. Neither were households any longer bound to each other by the economic ties forged through varied forms of exchanges involving work and goods. Under these circumstances it would be logical also to expect evidence of community disintegration. In some respects this is what happened. But the decline of community was not pervasive; it was offset by other trends. The demise of economic cohesion made it possible to strengthen other types of social bonds. The reorientation of dairy farming households in the factory era had significant implications for rural neighborhoods and communities. As the factory wielded increasing influence, farm families and households forged new relationships with one another. Both conflict and community assumed new forms.

⁂ 10

Rural Communities Transformed

> The moment you carry out the principle that no work of any kind shall be done on the Sabbath, you stop the wheels of the world's prosperity.—Hiram Walker, 1864

In the 1870s and 1880s residents of central New York dairy districts faced new challenges in sustaining their communities, just as they struggled to adjust household labor and farming practices. Rural communities faced the prospect of declining population, economic struggle, and limited land availability. Paul Gates has characterized the period 1870–90 as one of "relative decline, one might almost say depression," in New York State agriculture. Rural residents were migrating; by 1860 one in four had already left. Oneida County was not spared these losses. Of the seven Oneida County towns we have followed, most reached population peaks during the 1860s. Small manufacturing concerns struggled, unable to compete with large-scale operations. The number of farms began to decline after 1880, from about eighty-three hundred in that year to about seventy-five hundred in 1890 and seventy-two hundred in 1900. The total number of farm acres also fell, by about fifty thousand between 1880 and 1900.[1]

This pattern was characteristic of many regions in the North toward the end of the century. New England was especially affected, of course, but New York State and the Midwest followed. Historians of rural life have found varying responses to these new economic realities. A long tradition of scholarship represents rural responses to these changes as chiefly characterized by conflict.[2]

Recently, however, some historians have questioned the assumption that the capitalist transition always involved conflict. Life in the settled town of Chelsea, Vermont, actually became more stable, "tranquil and homogeneous." In the Nanticoke Valley the countryside "deindustrialized," but those who remained fashioned dense social networks based on kinship and proximity, and that many adjusted successfully to new markets. In local politics late nineteenth-century rural New Yorkers responded to the serious limitations upon their aspirations by attempting consensus as frequently as they conflicted.[3]

The course of community life[4] in rural central New York late in the century suggests that both conflict and community building occurred, but in different facets of social life. Conflict among farmers seems indeed to have intensified with the centralization of cheesemaking.[5] This conflict not only occurred along class lines but also developed from the peculiar nature of factory cheesemaking. At the same time, however, residents busied themselves with forging new social ties to supplement the weakening bonds of neighborhood and custom. This task fell largely, though not exclusively, to women. Freed from the heaviest work of home cheesemaking, farm women proceeded to orient themselves anew to the farming community. Through the agricultural fairs they expressed a new vision of agrarian life; in temperance and professional activity they aligned themselves with a national, conservative middle class; and in following new social activities they strengthened ties to the national culture. Thus, paradoxically, by the end of our period dairying communities in Oneida County were experiencing both a sense of crisis and one of unprecedented vitality.

The Social Basis for Conflict

Two issues in particular inflamed people's emotions across the dairy districts: milk adulteration and Sunday cheesemaking. A newly sharp class dimension exacerbated these divisions, but they also cut other ways, pitting patrons against managers, patrons against other patrons, and those who espoused religious values against those inclined toward secularization.

The factory system made class divisions more apparent than they had been under home cheesemaking. Factory cheesemaking required capital,[6] and it was a public endeavor in a way that home cheesemaking was not. Those who owned or organized cheese factories—and thus controlled them—came overwhelmingly from the ranks of wealthier farmers. David Brill was typical. Descended from early settlers, Brill established the second cheese dairy in the town of North Western, operated a store, and later opened the second cheese factory there. Merchants, bankers, and manufacturers augmented this local agrarian elite among cheese factory owners and shareholders; for example, J. A. Shearman was a New York Mills cotton mill owner, C. C. Weaver was involved in brick manufacturing, and Luther Leland was a merchant and bank director. The *Country Gentleman* in 1863 summarized trends in cheese manufacturing by stating that wealthy farmers formed "associated dairies," while the less wealthy became patrons.[7]

By the 1870s credit evaluation agents from R. Dun and Co. were screening potential cheese factory entrepreneurs with an eagle eye. Their observations confirm that these men came from the ranks of well-to-do farmers. In 1873, for example, the Dun analyst reported that the Marshall cheese factory was "composed of several farmers who are jointly worth $150,000; they pay cash for their wants." In 1879 fifteen "mostly Welsh" co-owners "watch[ed] the factory very closely." They were "considered perfectly good."[8]

Agricultural census manuscript data show that farmers who sent milk to the factories tended to occupy the middle social ranks in terms of wealth and farm size, and a sizable contingent of patrons owned or rented small farms. Thus the potential for conflict existed because of the way in which factories were organized, especially if they were joint-stock companies in which stockholders received earnings pro rata, or if the factory was owned by an individual. This potential was often translated into real conflict as everyday operations exposed the contradictory interests of poorer patrons and richer owners.

In 1868 an anonymous piece in the *New England Farmer*, entitled "Our Cheese," criticized cheese made from pooled milk. The terms employed revealed much about the social basis for factory-era conflict:

> *Our* cheese was the result of no such conglomerate, nor was it produced by any such multitudinous rabble. We know who strained the milk, who prepared the rennet and cleaned the utensils; whose fingers manipulated the curd, and whose hands smoothed the surface of this triumph of the dairy. These hands and these fingers belong to no factory superintendent or careless operative, but to a neat woman, the tidy, skillful wife of one of the best farmers. . . . [In factory cheese] the multitude, the mass, the mob had a hand; and who knows that all these hands were clean?[9]

The overt class references in this piece are striking. The references to a "multitudinous rabble," the "mass," the "mob," all contain undisguised contempt for the lower-class people the author assumes are now responsible for producing cheese. It is also notable that the woman possessing cheesemaking skill is portrayed as an upper-class agrarian, the "wife of one of the best farmers." These social differences informed conflicts over other issues as well.

Milk Adulteration

The factory system also possessed inherent potential for conflict because mixing milk from dozens of farms made patrons' fortunes literally far more intermingled than before; one person's negligence or unscrupulousness could reduce profits for everyone, because one bad batch of milk could ruin an entire day's worth of cheese. Yet the opportunities for neglect or outright deception were numerous and, for some, irresistible. Patrons were paid according to the number of pounds of milk they brought to the factory. Thus from the beginning they faced temptation to add water or to skim off the cream, since weight and not richness mattered. Another problem was cleanliness. As long as farmhouse cheesemakers had processed milk from their own cows, they could be fussy or complacent about the condition of the milk as they pleased. But when the milk from dozens of dairies and hundreds of cows was mixed together in the same vat, cleanliness in milk became crucial. These questions became bitterly contentious, because no reliable and accurate means of measuring butterfat content or of detecting contaminants existed. The Babcock test for butterfat content was not introduced until the 1890s, and bacteriology was barely understood. Conflict over milk watered, tainted, or skimmed divided patrons from managers and sometimes from one another. The fundamental conflicts were unresolved as long as testing was uncertain, but nonetheless they generated an uneven but unmistakable pressure for conformity in dairying practice.

In 1864, at the behest of the New York State Cheese Manufacturers' Association and others, the state legislature passed a law spelling out the obligations of factory patrons and managers:

> Whoever shall knowingly sell, supply or bring to be manufactured . . . any milk diluted with water, or from which any cream has been taken, or milk commonly known as skimmed milk, or whoever shall keep back any part of the milk known as "strippings," or whoever shall knowingly bring or supply milk . . . that is tainted or partly sour from want of care in keeping pails, strainers, or any vessels in which said milk is kept, clean and sweet, . . . or any butter or cheese manufacturer who shall refuse or neglect to keep . . . a correct account (open to the inspection of any one furnishing milk to such manufacturer) . . . shall, for each and every offence, forfeit and pay a sum of not less than twenty-five dollars nor more than one hundred dollars, with costs of suit.[10]

The law was a response to real and vexing problems in the cheesemaking business. *Harper's New Monthly Magazine* pointed to the potential for conflict: "At the receiving window of the cheese factory there arise questions which end sometimes in ill temper, sometimes in the courts of law. All is not milk which comes in cans, and all is not good milk."[11] As this comment indicated and as the law spelled out, problems with milk arose in several ways. Patrons could add water to the milk, in order to increase its weight. Or they could handle the milk carelessly—exposing it to high temperatures, say, or contaminating it with dirt—and thus "taint" it. Or they could skim it. The first two of these were the most common infractions.

Watering milk involved more deliberate deception than did careless handling, and so allegations of watering attracted wide public attention. The *Utica Weekly Herald* wryly noted in 1872 that "a farmer 'up north' has recently been detected watering the milk from his dairy to such an extent that his neighbors wonder whether he has put water in the milk, or milk in the water."[12] Factory owners and managers entertained perpetual suspicions of their patrons. The result was constant jockeying and persistent tension.

Nathan Huntington, an early factory manager in western New York, voiced an unequivocal mistrust of his patrons: "Nobody to help share the responsibility and every patron a critic. There was no legal test for milk then unless you saw them in the act. We kept our test bottles in sight and used the Lactometer and tried to convince them that we knew a good deal more than we could prove but we struggled through very satisfactorily. We only found one real fish three or four inches long and we did not advertise that."[13]

Like Huntington, factory managers tried to keep their patrons in line with various tests and sanctions. When a Boonville firm constructed equipment for use in several local factories to open in the spring of 1863, the owners pointedly mentioned that among their products was a "lactometer," for measuring the "exact quantity [of water] that *may* have *accidentally* got within the can." This claim for exact measurement was pure bluff, but it at least served warning of vigilance. Managers frequently used the local newspapers to expose offenders, attempting publicly to shame them into compliance and to deter others. Toward the end of the 1880 season a notice in the *Boonville Herald* brought attention to an offense: "John Owens promises to pay the patrons of a cheese factory in Floyd, the sum of $100 for the petty privilege and luxury of putting a little water with the milk he carried to the factory." Others resorted to more extreme tactics. An anonymous letter in the *Utica Weekly Herald* in 1875 reported that a factory manager in South Trenton (Oneida County) had hired a detective to spy on his suspect, so as to catch him in the act.[14]

Litigation under the New York State law offered another avenue for factory managers seeking to discipline recalcitrant patrons. "Milk suits" attracted wide interest throughout the dairy region; they indicate how intractable was the problem and how deep the feelings it inspired. In 1867 *Moore's*

Rural New Yorker reported on a two-day trial at the circuit court in Herkimer. A large crowd heard the testimony by the adversaries, the Frankfort Cheese Factory treasurer and one of its patrons. The factory official used a "hydrometer" and a "cream gauge" to argue that the patron's sample had from 12 to 15 percent less specific gravity than pure milk. The defense did not dispute the accuracy of the test but argued that there was a wide variance in milk, depending upon feed, weather, the cows' ages, and so on. As evidence the defense introduced ten unadulterated samples of milk that varied among themselves by up to 10 percent. The jury accepted the defense's argument and found the defendant not guilty.[15]

A local story from Oneida County oral tradition tells how the prominent cheese factory owner Gottlieb Merry also failed in his attempt at legal redress:

> When "Gud" Merry emptied the farmer's cans he would hold the empty can for a moment and watch to see if it kept dripping just a little. If all came out completely without a drop sticking inside it showed it had been watered some. A prominent family [whom] "Gud" suspected had the farm next to the Verona Presbyterian Church, so Gud got up in the bell-tower and saw them actually putting water in, but he lost out in court, the court would not admit evidence seen through field-glasses. The farmer's wife was jubilant that they got the best of Gud so the next time she saw him on the street, she told "Meestah Merry" just where he could kiss her![16]

Cleanliness was an even more ticklish problem than watering, because it was still harder to detect and remedy unsanitary conditions. So many conditions might cause a "taint" in milk that even conscientious makers and patrons were troubled with bad milk, especially in hot weather, which provided ideal conditions for bacteria to multiply rapidly. Harvey Day found this out, to his sorrow. In an era when some patrons objected to penalties for dipping drinks from their milk cans, it was a difficult thing to convince patrons of the necessity for sanitation.

The experience of John Chapman of Madison County shows how bitter were the feelings inspired by dirty milk. Chapman, a native of England's Lincolnshire, possessed extensive experience, first as a home cheesemaker and also as a factory patron, salesman, and clerk. He was involved with early factory production, and at one point, in 1866, he quit in frustration and returned to home cheesemaking, although he eventually returned to factory work. Chapman fumed about the "rascalities perpetrated by factory patrons," charging that they cared so little about cleanliness that they would even leave dead cows to rot in pastures; their surviving bovine sisters then consumed polluted food and so produced polluted milk. One "ignorant and pig-headed" patron even hired a lawyer to find out if he was legally obligated to bury his dead cows. Chapman further charged that carelessly cleaned cans resulted in tainted milk. He did not oppose the factory system in principle, but he criticized its "green, ignorant cheesemakers, low, rascally, thievish patrons, and incompetent committee men."[17]

Since legal and scientific remedies were lacking, factory men had to rely upon a combination of close oversight, sanctions, social pressure, and exhortations. X. A. Willard argued in the *Country Gentleman* for 1865 that managers must watch their patrons closely as to how they treated their cows and handled the milk; and again at the American Dairymen's Association meeting in 1870 he urged factory committees to "hunt up slack patrons" whose milk soured too quickly. The *Ohio Farmer* remarked in 1871 that patrons who delivered "tainted milk" incurred "sharp words." In 1873 William Gates, manager of the Wight factory in Oneida County, made a separate cheese from patrons' milk that had gone bad too soon: "There it stood, an eloquent witness of their delinquency, and a warning for them not to repeat the offense."[18]

Managers and owners were able to wield influence because the very conditions that generated conflict could also produce pressure for conformity. For example, factory owners could exploit a highly regarded factory's reputation to secure adherence to specified standards, since patrons wanted to take their milk to the best factories. L. L. Wight, for example, ran a well-esteemed factory; he required his patrons to give their cows access to running water and to observe "utmost cleanliness" in milking. "By no means wet the hands in the milk while milking," he instructed. He recommended careful straining and cooling of the milk and announced that "scalding all vessels used about milk at least once a day . . . is essential." Wight insisted that "keeping the strippings at home is morally and legally as bad as watering." Some factory managers were able to get patrons to use a certain type of can and to cool their milk using recommended techniques. Still others appealed to the dairyman's self-interest, pointing out that ultimately, because of lower prices, everyone lost out when pooled milk made inferior cheese. Management could also pit patrons against one another; in 1880 the *Boonville Herald* stressed that the patrons of the Sugar River factory were "quite excited," as two of their number were suspected of watering milk.[19]

Some members of the New York State Cheese Manufacturers' Association hoped to employ a carrot as well as a stick, by rewarding patrons who successfully produced clean milk. Indeed, patrons themselves sometimes argued that they deserved a premium for "good," clean, rich milk. Proponents of differential compensation argued that good patrons suffered from the negligence of bad ones. But several circumstances prevented any action. Managers' arguments for clean milk were compromised because edible cheese could be, and was, made with milk of widely varying quality. John Chapman implicitly conceded as much when he complained that the maker had to "hurry" the process when he suspected tainted milk. But more important, defining quality in the absence of reliable measures was a wholly subjective matter.[20]

Not surprisingly, suspicions were often mutual. Incompetent cheesemakers could ruin a cheese just as surely as could negligent farmers, and so patrons frequently threatened to take their milk elsewhere, as one group did in 1874, claiming disappointment with the manager. In the Oneida County town of Remsen the manager Didymus Thomas was even forced to quit cheesemaking by what he regarded as the "unreasonable exactions of patrons." In an acknowledgment that patron pressure was effective, the *American Agriculturist* advised managers in 1868 not to dip the curd too soft in order to make heavier cheeses just to please patrons. Especially where there were several factories in close proximity, patrons could play one off against the other, because the factories were competing for a finite number of patrons. The verdicts in the two court decisions mentioned above also suggest a strong community sentiment favoring patrons. In neither case could owners secure a legal victory. Not even an eyewitness account—albeit one depending on a telescope—could compel the jury to find in the owners' favor.[21]

Published accounts of patrons' meetings too suggest constant jockeying, as when the *Black River Herald* noted that patrons of the Willow Grove factory in Trenton had diplomatically but provocatively implied that there had been a conflict: "A series of resolutions were passed, not only making the patron interested for the welfare of the society, but making the manufacturer responsible for cleanliness and careful proceedings appertaining to cheesemaking in every respect." As if to underscore the point, the patrons added, "The manufacturer is not infallible." An irritated manager criticized the joint-stock form of organization in *Moore's* for 1870, complaining that it permitted patrons to "pester" managers with "dictatorial suggestions."[22]

Thus factory cheesemaking inherently possessed the potential to cause new tensions in the rural neighborhood. Factories contributed to creating an atmosphere in which the old "mutuality" could not function as well as it had before; neighborhoods were literally tied together, but not as voluntarily and not in as many varied ways. Indeed, the factory system brought a new kind of interdependence, one soured by competition; in this respect conflicts over milk were a metaphor for industrializing society everywhere. In turn, these tensions could further erode or change the character of older neighborhood ties. The new system, more so than the old, left little latitude for individual variation in farming practice; it pitted the "ambitious," systematic farmer against his reluctant neighbor.[23] The factory men's demands intruded in unprecedented ways upon the farmer's fundamental practices, and milk-producing farmers came under increasing pressure generally to conform to more exacting standards. Added to the stress that a quantity-based standard automatically created, this pushed farmers even harder to modernize.

In response, many farmers actively resisted cheese factory management's attempts to make them conform. Their activities present an interpretive challenge, for they directly address the question of resistance to capitalism in rural America. Before dairy centralization, home cheesemakers had fashioned a hybrid enterprise that blended precapitalist elements with more purely commercial production. The initial transition to centralization was accelerated by internal conflict, which was less over capitalism than over the gender division of labor. Once factories appeared, the locus of conflict shifted to the neighborhood. At the community level tension occurred partly along class lines, but it was also inherent in factory organization.

It is significant that these tensions occurred among men. As we have seen, most women did not resist but rather encouraged centralization. While most publicists for centralization were men, they tended to be men in a position to benefit—those with means to invest. But for ordinary patrons the benefits were more ambiguous; indeed, we have seen that on the farm, men were working harder.

But even disgruntled factory patrons did not reject capitalism's essentials any more than bickering household members had done. Their response differed substantially from that of upcountry Georgia's yeoman farmers, who invoked the rhetoric of self-sufficiency and communal values; in fact, in the North the notion of "competency," which arose from the mix of commercial and noncommercial, continued to serve large farmers better than small ones.[24] Small farmers were left with little choice but to try to maximize their position within an increasingly competitive system rather than attempt to dismantle the system altogether. We may explain some of this difference with reference to the hugely differential impact of the war upon the two sections, and especially to the subjection of southern cotton growers to outside control. But we must also note the essential continuity in the North between the pre- and postwar eras: the transitional system fed neatly into the more fully capitalist one developing after the war.

Sunday Cheesemaking

The limited nature of dairy patrons' resistance becomes clearer when we consider another debate that rocked dairying communities during the factory era. Despite loud opposition, patrons and managers alike chose to follow Sunday cheesemaking, and all invoked the profit motive.

The Sunday-cheesemaking question epitomized the values of leisure, secularism, materialism, and competition that apparently had come to be widely shared in dairy country. Home cheesemakers had generally not "run a curd" on Sundays, but most cheese factories operated on Sunday. Cows did not observe the Sabbath, and so milk must be processed. A small quantity of milk might be held over or allowed to stand overnight for cream, but the practical obstacles to suspending large-scale cheesemaking for a day during the busy season were very real. Moreover, powerful incentives to Sunday cheesemaking arose from the same motive that had impelled families to patronize the factory in the first place. One critic noted that patrons' families refused to withhold milk on a Sunday, preferring not to disturb their own day of rest. He characterized their reasoning this way: "We pay a large price for manufacturing, and the relief from Sunday work on the farm is one of the compensations."[25]

Nevertheless, it is not surprising that objections arose to Sunday cheesemaking. After all, the dairy region overlapped with the former "Burned-over District," noted for its evangelical fervor in the antebellum era.[26] What is surprising is the nature of the arguments both for and against Sunday cheesemaking. Only a minority argued exclusively on religious grounds. Even the opponents of Sunday cheesemaking framed their objections in terms of commercial, secular values.[27]

Transparently commercial motives particularly characterized arguments in support of Sunday factory cheesemaking. After having heard a sermon preached against the practice, a cheesemaker named Hiram Walker from Mexico, New York, detailed his own case in the *Country Gentleman* (1864). He opened with a bow to "observance of the Sabbath and its institutions." But immediately he began hedging: "I think no one—not even the Reverend Divins himself—lives up to the strict letter of the Jewish ritual." The "pastoral life" of the ancient Jews, he maintained, differed so materially from the "multitude of employments" practiced under the "cultivation of the arts and sciences" in modern civilization that the old rules were no longer practicable. Walker's central argument, though, was purely materialistic: "The moment you carry out the principle that no work of any kind shall be done on the Sabbath, you stop the wheels of the world's prosperity." He drew a motley assortment of analogies. The world of industry showed that "by a strict observance of the letter of the Jewish law, we could not have a burned brick, an earthen dish, quick-lime, a glass window, any kind of cast-iron, or charcoal with which to smelt the ores, because the articles could not be perfected in six consecutive days. We should have no commerce whereby an exchange of commodities could be made with distant nations." Bible stories furnished other proofs: "Christ said it was lawful to [do] good on the Sabbath." The argument from nature led Walker to assert, "If the laws of animal nature were such that milk would secrete only six days in the week, as the manna fell in the wilderness, I would go in for a truce on cheese-making on the Sabbath." Finally, utilitarian calculations of the greatest good for the greatest number prompted him to conclude that "more persons can attend church with the factory system than with single dairies." Walker finished on a note that combined sanctimoniousness and economic self-interest: "In these times of scarcity of labor, it behooves us all to save with a careful hand all the products of the farm. It is the more necessary at the present time to meet the calls made upon us for public and private charity."[28]

Walker's piece drew a cogent response from "R.N." in Randolph, Vermont. This correspondent addressed Walker's claims point by point, refuting each one. First he argued that, far from being irrelevant, the commandments were "given for all men and all *time*, as obligatory as ever. . . . God gave to us mankind, six days in which we shall do all our work, reserving the seventh for himself, and requiring of us . . . [that] we should keep that day holy, abstaining from all labor but that required by *charity, piety, and necessity*. All works prompted by *avarice, luxury, vanity, and self-indulgence*, are wholly forbidden." He proceeded to counter Walker's other arguments: "A burned brick, an earthen dish, quick lime, a glass window, cast iron and charcoal, can be all made in *six* days; not perhaps on so large a scale or so profitably." Further, "While living . . . in the dairy district [in Ohio], I was told by a dairyman that Sunday work in cheese-making was wholly unnecessary; he had formerly worked seven days per week at the business, but had experienced a change of views." The writer conceded that "commerce with distant nations . . . *cannot* be done in six days, . . . but the necessary Sunday ship labors do not prevent the observance of the day on those ships with pious captains." Then he took up the Bible stories Walker had cited. "The fact is, God has made no *exceptions* in his revealed will, in favor of making cheese in factories. . . . I think no man is justified in engaging in any occupation that requires the habitual disregard of his day, unless he can show an express permit from Jehovah. . . . The helping the ox from the pit, Christ alluded to, was an unforeseeable *necessity*, wholly different from Sunday cheesemaking." Finally, "R.N." pointed to the hypocrisy of justifying Sunday cheesemaking with references to charity: "Soon our Sabbaths will be among the things that *were*, with all their innumerable blessings! That more money can be made to enable us 'to meet the calls made upon us for public and private charity!' Alas for any remaining disposition to exercise that grace, when we shall become so debased as to rob the Lord of his holy day."[29]

"R.N." grounded his case in conservative religious sentiment, invoking a stricter interpretation of scriptural law than was evident in Walker's freer reading. Yet the later essay was not more notable for the quality of its biblical or religious thinking; where "R.N." scored his most telling point was in locating money at the root of the controversy, and in arguing that religious values were of higher importance than money.

"R.N." alone among opponents of Sunday cheesemaking stressed this point. The others shared the same ideological ground as Walker. Some appealed for the spiritual and physical well-being of five thousand factory operatives, arguing that lack of rest lowered productivity. But all sought to show that farmers and factory men would not really lose anything by suspending Sunday work. One even claimed that a Sabbath-keeping and church-going populace enhanced "the value of real estate in a community"! L. F. Mellen, writing in the *Ohio Farmer*, cited the usual reasons involving divine authority and health—and, imaginatively, the possibility that business failure could result from overtaxing employees. But his clinching argument was that the factory patron could make just as much money by saving Sunday's skimmed milk for Monday, making butter from the cream, as by taking his whole milk to the factory on Sunday.[30]

We may imagine the debate on Sunday cheesemaking taking place in local gatherings of people in the dairy regions, among church congregations, perhaps, or factory committees, or within families and neighborhoods. Sermons were preached and pamphlets distributed on the subject. The public debate shows commercial values significantly challenging traditional religious mores. Even opponents of Sunday cheesemaking could blithely invoke the profit motive in ways that would have appalled earlier Sabbatarians. Though their material stakes were puny by comparison with those of their counterparts in giant urban industry, these rural entrepreneurs shared in the spirit of the "Gilded Age" so aptly labeled by Mark Twain.

Thus again the factory brought unanticipated change. Proponents originally justified factories by making profit a secondary consideration to saving women's health and time, and by invoking values of leisure and personal cultivation. But once factories were up and running, factory routine and structural arrangements created their own dynamic. The argument over Sunday cheesemaking points up how centralized cheesemaking could create conditions that permitted farm families to stretch the meaning of competency. Sunday cheesemaking at the factory enabled patrons and their families to observe the Sabbath and to enjoy a few leisure hours. Deliberate suspension of work fit within the customary spirit of competency, in which nonpecuniary values held an important place. Yet in order to obtain these benefits, patrons willingly endorsed Sabbath-breaking for factory operatives. Moreover, even as they passed their Sundays they were making money, so they need not sacrifice profit. Because the milk check accounted for a significant portion of the family's income, the rhetoric of profit—however "profit" was perceived—took on heightened assertiveness. Though they still retained an attachment to some forms of competency, dairying families were in other respects moving decisively away from the old ideal.

The factory era thus brought with it new neighborhood tensions and helped to dissolve customary neighborhood ties. But at the same time it made possible new kinds of bonds. These new relationships were formed more in the context of sociability, and were more firmly anchored in class affiliation, than the old ones. Though they were actively pursued by both men and women, women were more visible in community building.

New Community Ties

Because dairy centralization allowed for a reorganization of household work patterns, it implied time for some household members, especially women, to devote to other activities. The activities they chose speak to their reorientation. Earlier critics had accused daughters of abandoning cheesemaking for "sewing, quilting, missionary and many other societies,"[31] and in the postwar period more and more women in rural Oneida County followed this pattern. Rural people helped to redefine the agrarian community through symbolic activity at agricultural fairs. By participating in village social occasions, they helped to cement a "mutuality" not dependent wholly on work relationships, but partaking more extensively of social pastimes. In actively sponsoring a cultural life that was highly varied by comparison with that of prior years, they added new perspectives and formed links with national trends and organizations. By joining temperance organizations, for example, they aligned themselves with national middle-class reform. In sum, their new activities helped them to refocus their lives toward a greater emphasis upon class identity, upon national as well as local culture, and upon social ties as well as work lives.

These activities helped women to revive their communities in the face of economic uncertainty, out-migration, and social conflict.[32] Dairying communities were certainly under stress, but at the same time they generated an unprecedented vibrancy. This was achieved partly through a reorientation on the part of dairying families away from the immediate neighborhood and toward national and regional culture.

For those who possessed the wherewithal and time, the prospect beckoned of becoming consumers in the emerging mass-production economy; inventories show how villagers' homes had filled with mass-produced furniture, mealtime accoutrements, and appliances. The *Appletons'* reporter visiting Oneida and Herkimer Counties in 1878 remarked that perhaps the milk collector "has for a companion as far as the railway depot some rosy girl who is going to Utica or Syracuse on a shopping expedition."[33]

Rural people's purchasing habits tied them more firmly to the national economy and culture, as they acquired products available to thousands of other middle-class families from Maine to Minnesota.

Participation in the consumer economy signaled both a reorientation toward domestic preoccupations and an alignment with national culture, and the agricultural fair accomplished the same. Though critics pointed to the popular element of spectacle and neglect of serious substance, late-nineteenth-century fairs in Oneida County appealed to a wider rural audience, and attracted a broader range of participants, than had earlier fairs. Through the county or town fair the community represented itself symbolically both to its members and to outsiders.

By participating in and attending the agricultural fairs in unprecedented numbers, dairymen and their families therefore not only gave evidence of new household activity but also took part in a redefinition of the agrarian community. By the later nineteenth century the local agricultural fairs in Oneida County unequivocally validated the new agriculture. The new-style fairs expressed a vision of agrarian life that emphasized farm families' membership in the wider society; by affirming values of competition, domestic comfort, and progress, they aligned themselves with the national culture, especially as it was shaped by the middle classes.

Indeed, class, rather than kinship or geographical proximity, increasingly shaped social life in the rural community. In 1906 James Mickel Williams undertook an extensive study—published as *An American Town*—of the Oneida County town of Waterville. Williams thought that when rural population densities dropped and economic functions moved to cities and towns, the village center assumed increasing importance and the rural neighborhood correspondingly less. As migration accelerated, kinship ties within the neighborhood decreased, and those between neighborhood and village increased. Young people and women in particular became more active in village life. Williams believed that class lines were hardening; for him the rise of the Grange, "*the* society of the pleasure-loving class," epitomized this process.[34]

The same process took place in other Oneida County towns. Where village social and cultural life was concerned, the variety of possible activities, as reported in the local paper, increased noticeably from the antebellum period. Earlier, local papers had devoted their coverage mostly to local and national politics. But by the 1870s they were overflowing with local news about social gatherings, various club meetings, entertainments, and travels.[35] Moreover, in a departure from earlier days when village residents had dominated social life, farm people, especially women, began to appear in the news as participants in village society and philanthropy.

As people from milk-selling households found time for more off-the-farm activities, they both patronized and generated a lively social scene.[36] While often the goal was pure entertainment, the work behind the scenes—women's work—should be kept in mind. Rural women were redirecting their energies and their labor toward creating new community ties. The cheese factory itself became the locus for socializing as the young people of North Gage held a supper right in the cheese factory in the winter of 1874. There they danced and ate oysters—a newly popular food imported on the extended transportation network. Villagers could also hear the itinerant lecturers, musicians, and dramatists who came by rail to perform in specially built halls such as Boonville's Opera House. Verona's young women pressed for, and obtained, admission to the Lyceum, where they debated important and trivial issues of the day.

An old-timer in Westmoreland in 1873 contrasted the social whirl with the previous generations' sober comportment: "We are deviating somewhat from the example set by our estimable ancestors." Before, he explained, they heard three sermons on a Sunday, met on two other evenings for religious purposes, and went to bed by nine each night. Now "almost every evening of the week is passed in some kind of amusement, and we are fast falling into the prevailing practice of frolicking during the night and sleeping during the day." He thought the "young ladies" were especially guilty.[37] Indeed, milk-selling farmers' daughters turned up often at events as varied as literary society debates, singing schools, Sunday school concerts, masked balls, and sleighing expeditions. Socializing had always been a favorite pastime for the young, but it seems that in the late nineteenth century, village life was richer and more varied than ever before.[38]

These events exposed villagers to the world beyond. Like the voluntary associations, lyceums, and public lecture systems of the antebellum era, they helped participants not only to locate themselves in relation to the national society and culture, but also to maintain the local community by helping people to form new ties with each other.[39]

Rural women pursued community building not only through their social activities but also through religion, philanthropy, and reform, as Pennsylvania farm women had exploited their command of dairying skills to develop new educational, religious, and reforming activity. In central New York, for example, teachers made important contributions to community.[40] Schoolteachers enhanced this function through the content of the education they offered, and through their work with students and their families.

Farm women assumed greater visibility in religious and philanthropic work. In this regard they continued involvement initiated during the Civil War, when they had organized relief operations for Union troops. Now the emphasis shifted to other causes. In 1871, for example, William S. Jackson's Boonville home was the site of a church "sociable" undoubtedly hosted by Mrs. Jackson. Another church "mite society"—these were invariably women's organizations—met at dairyman Utley's house. Mrs. William Wheeler, wife of a prominent Trenton farmer and cheese factory owner, served for a time as head of the Utica Orphan Asylum.[41]

The temperance cause attracted significant numbers of rural women in the postwar era. Women from milk dairying farm families appeared frequently as officers or members in chapters of the Independent Order of Good Templars. Mrs. Mariah Waterman, for example, served as an officer of the Taberg Good Templars (town of Annsville) in 1872. The village of Westmoreland also had a particularly active chapter of the International Order of Good Templars, with strong representation from farm households.[42] These activities contributed in a significant way to the re-formation of rural communities in the postfactory era.

Oneida County farm wives' temperance activity in the factory era seems to reinforce John Mack Faragher's observation:

> It was only after the beginning of the break-up of the older household order that [the] assertive side to women's character broke free from a culture of accommodation. It was in the postbellum years—after the process of full capitalization of midwestern agriculture had well begun, after the swelling of the ranks of rural wage laborers and tenants, after the destruction of women's domestic industries and the end of household production—that women's self-assertion found cultural vehicles of protest and resistance.[43]

While their prior adherence to a "culture of accommodation" is open to question, there seems to be little doubt that dairying wives aggressively pursued new avenues when conditions permitted.

Beyond its very general nature as protest, it is difficult to know just what to make of this temperance activity. We do not know enough about the history of temperance generally, and so the interpretive remarks made here are tentative at best. Rural women's temperance involvement seems, to some extent, consistent with Ruth Bordin's characterization of the movement as a search for "power and liberty," the "major vehicle through which women developed a changing role for themselves in American society." Bordin argues for a connection between temperance reform and the women's politicization, which in turn led to the suffrage movement. While it is not clear that Oneida County temperance activity led directly to suffrage activism, it is quite plausible to see temperance activism as the outgrowth of a generation's experience in questioning family authority and in pursuing alternative employments, whether on or off the farm. Bordin concentrates upon the movement in urban areas, and upon the all-female Women's Christian Temperance Union. But it seems that rural women tended to prefer the mixed-sex organization of the Good Templars, probably because it reflected longstanding rural customs of mixed-sex socializing and work.[44]

In some respects temperance activity in the postfactory period also follows the pattern outlined by Ian Tyrrell for the antebellum era. Tyrrell argues that antebellum temperance supporters came from the ranks of agrarian "improvers," businessmen, and industrialists who embraced the new industrial order and saw in temperance—and later in prohibition—a means of enforcing the self-discipline necessary for success.[45] To the extent that they were from well-to-do farm families with strong, positive ties to the factory system, Oneida County temperance reformers fit this description.

But it is not so clear that the label *conservative* should be discarded when the movement in rural Oneida County is assessed. We have already seen that rural residents received the industrialization of agriculture with ambivalence, and that poorer and richer farmers jockeyed for position within the new order. During the Civil War era women's work in benevolence, for example, became more narrowly defined, more conservative, and more devoted to institution building.[46] Women benevolent workers increasingly moved to protect and consolidate their class standing, and they supported the professionalization of benevolent work. In the shattered postwar nation pessimistic assessments of human nature and society also influenced the drift toward more conservative goals. Similar forces appear to have been at work in the postwar, industrializing dairy country. As class divisions hardened, temperance could become one instrument of middle-class identification and thereby take on conservative goals. Like the elite families in Sugar Creek, Illinois, Oneida County temperance advocates sought to consolidate their position.

The context must also be taken into account. In an atmosphere of declining expectations, perhaps temperance possessed attractions because fewer and fewer factors affecting the community's vitality were amenable to local control. Moreover, a factor peculiar to central New York must be considered: this temperance activity took place in the nation's leading hop-growing region. Hops, as a cash crop pure and simple, symbolized the new order perhaps better than any other crop. Were temperance advocates expressing ambivalence about the new agriculture as well as against the drink made from hops? Where the temperance issue is concerned, these speculation will have to suffice until more research clarifies the nature of temperance activity in rural areas.

The reorganization of production in dairy country had thus allowed women to develop and strengthen new community ties. Industrialization seems to have played a different role here than in other times and places. In Sugar Creek, community institutions—extended family, church, neighborhood—helped to facilitate the transition to modernity. In Oneida County it almost seems as if the reverse occurred: industrialization facilitated the building of communities. More accurately, it enabled some aspects of community to strengthen while others declined. Those that prospered tended to be activities that women sponsored, while men tended to experience a decline in communal forms such as the "changing works" system and the "cooperative" ethos.[47]

Whether through migration or other means, ties between country and city had intensified throughout the century. Some historians, indeed, have regarded change in rural America as primarily a process of urbanization.[48] Others contest this characterization, arguing that it presents an unrealistically one-sided model that leaves out rural people as actors.[49] In the case of rural Oneida County a very useful counterpoint exists in Mary Ryan's path-breaking *Cradle of the Middle Class: The Family in Oneida County, New York, 1790–1865*, which concerns Oneida County's major city, Utica. Ryan associates middle-class formation with urbanization; she follows Utica's transformation from agrarian frontier village to canal city, arguing that in the former an "old middle class" consisted of farmers, artisans, and shopkeepers, while the "new middle class" of managers, professionals, clerks, and office workers was essentially an urban phenomenon. The agrarian phase was characterized by the "corporate" family economy in which production took precedence, dominated by fathers, discouraging individualism, and lacking in distinctions between private and public. Non-kin often resided with the family, and face-to-face relationships reigned. By contrast, according to Ryan the "new" middle-class family was nuclear; it privileged mothers, encouraged individualism, emphasized consumption, and sharpened the distinctions between private and public. In keeping with this development, social relationships took on more impersonal qualities. Mothers enjoyed enhanced importance as the period of home residence for sons and daughters lengthened, and families put considerable resources into educating their offspring.[50]

The history of rural Oneida County in the nineteenth century helps to sort out the importance of "urbanization" relative to other factors. There were some ways in which rural families did follow a different path from their city cousins. In rural areas, for example, while impersonal relationships increased as they did in the city, a dense network of face-to-face relationships persisted. Motherhood does not seem to have played so strong a role in farm women's identities as it did among middle-class city dwellers, and neither was the public/private distinction so well drawn as it seems to have been in the city.

But the most fundamental transformations took place simultaneously in the countryside and in the city. The authority of men in households eroded, individualism strengthened, non-kin hired help disappeared from the farm household, and education became an important part of family strategies. Agrarian workers learned to deal with complex technology and with rigorous work discipline. In turn, this also helps to explain why the transition between country and city, and eventually to an industrialized society, was accomplished with relative ease by so many people.

As we survey the range of activities in postfactory dairy country, it becomes apparent that most of them involved modernization more than urbanization. Most scholars agree that the fundamental qualities of modernization include economic development—usually involving industrialization and urbanization—accompanied by what Richard Brown calls "social and political change in the direction of rational, complex, integrated structures."[51] According to Brown, members of "modern" societies tend to cherish "secular-rational values," to assign status on the basis of achievement rather than heredity, to possess a cosmopolitan rather than a local outlook, to value mobility rather than stability, to look for information sources beyond word of mouth, and to regard innovation favorably rather than holding fast to tradition. Daniel Walker Howe adds other traits to the description of the "modern" personality: an ethic of work and improvement; competitiveness; and a high value placed upon time.[52]

Historians' clashes over the issue of "modernization" occur less over definitions than over impact. Some celebrate modernization as a triumph for individualism, applauding the economic development that brought diversity of opportunity. But for others modernization means the destruction of traditional societies, a process accompanied by terrible social strains, displacement, and anomie.[53] Still others point out that the entire concept applies more to men than to women; Nancy Cott argues that "women and men experienced the impact of modernization in different ways."[54] Yet there is disagreement too among women's historians; Regina Morantz interprets antebellum health reform as a means for women to experience the "efficacy of individual action" and to develop values of self-improvement and personal freedom.[55]

Certainly the experience of rural Oneida County residents seems to have involved "modernization" to some extent; for example, "secular-rational values" seem to be at work in the conflict over Sunday cheesemaking. Yet for the most part rural people were coming to terms with modernity in ways that call into question the clarity with which "modern" and "traditional" are conceptually separated, and with which "modernization" is assessed. Throughout their experience, for example, Oneidans had combined longstanding "traditions" with innovations in a way that makes it extremely difficult to characterize the outcomes as one or the other. They slowly loosened some of their neighborhood ties, only to replace them with others. They expressed a degree of individualism but also felt enormous ambivalence about competition. Women pursued education and self-improvement but were instrumental in building communities and in reshaping the notion of competency. Perhaps the very lack of clear polarities made the transition less wrenching, easier to assimilate.

Conclusion

As the twentieth century draws to a close, American dairying is in crisis. Its astonishing productivity has been achieved at considerable cost to rural society and the environment. Heavy reliance upon purchased feed, fertilizer, pesticides, balers, silos, milking parlors, and computerized feeding systems results in an escalating cycle of indebtedness. Dairy farming families live under constant pressure. They work long hours, year round. Many struggle to turn a profit in a market flooded with milk from ever more productive cows, even as consumer demand has slowed as other beverages replace milk for popular consumption. Government programs help some dairying families, but each year more people leave farming, and rural communities wither. Dairy farming practices also often lead, directly or indirectly, to serious ecological degradation. Waste from large herds can create runoff that pollutes water. Heavy equipment can cause erosion. Petroleum-based pesticides and fertilizers result in pollution both during manufacturing and after application.

The seeds of these contemporary dilemmas were sown a century or more ago. Central New York farmers continued the New England "tradition" of ecological disruption by cutting down the forests,[1] introducing the plow, insisting upon an inflexible notion of property, and aggressively attempting to control animals. Later, as dairy farmers began to buy supplies, to raise high-yielding cattle, and to grow cash crops, they embarked on the path to modern ecological and social problems. As they extended the milking season and intensified feeding and care routines, they set the stage for chronic overproduction and competition, while they themselves faced unending toil.

Yet the path of historical dairying did not point inevitably to environmental and economic crisis. During the antebellum period the imperatives of the market actually encouraged farmers to circulate products, energy, and nutrients. They farmed with greater consideration for the land and its productivity and tried to accommodate to natural processes, rather than to overcome them as their careless wheat-farming forebears had done. Under the evolving dairying system, the careful combination of rotations, grassland care, and livestock management helped to maintain soil fertility and to foster a commitment to long-term stewardship. This episode reminds us that historically, dairying has pointed in multiple directions, to more than one kind of future.

Neither was the people's response to change automatic or predictable. With hindsight the often painful social costs of the rise of capitalism are apparent, but to participants in the process, choices carried more complexity. For women especially, the abandonment of home cheesemaking appeared to be not so much a loss of skill and economic responsibility as redress for an acutely unfair imbalance in household labor allocation. Dairy centralization also brought benefits for men and for entire communities. Women devoted their newfound time to increasing home comforts for all family members. At the same time, they expanded poultry raising, thereby maintaining their commitment to market activity, continuing strategies of mutuality in farm work, and contributing to farm diversification.

Women also engaged in new activities that were integral to reshaping rural society. In increased activity at agricultural fairs, they helped to redefine the agrarian community, pushing it toward modernity. Women took a central role in sponsoring, organizing, and running village social and cultural occasions, from charity fairs to theatrical performances to public debates to temperance activities. Through these activities they helped to cement social ties that did not depend wholly on increasingly tenuous work relationships. This village cultural life, when compared with that of the past, possessed vibrancy and excitement. It performed the important function of allowing rural people to orient themselves toward national culture and organizations. Temperance work, for example, fostered alignments with a prominent national reform. Through their participation in these activities, rural women forged community networks that would help to compensate for the erosion of older forms of community such as work exchanges, and would help to offset the conflict that the factory system brought to men's world of work.

As the process of change played itself out, members of Oneida County's dairying families and communities pondered its meaning. If they had lost a measure of economic security, they had also shed burdensome obligations. They had not yet come entirely under the sway of absentee capitalists; local entrepreneurs still controlled cheese manufacture. Nor had they fully felt the impact of late-nineteenth-century agricultural depression and out-migration. Indeed, rural and small-town life seemed more diverse and varied than ever. Conflict divided it, to be sure. But development had emancipated women from onerous burdens, ensured rural families longed-for material comfort and social enjoyment, and brought new opportunities for entertainment, social and political activism, and cultural exposure. While from a century's distance it is easy to see the transformation of dairying as a regrettable decline in farm family independence and women's craft skill, perhaps we ought to take seriously the strength of evidence that from the participants' perspective, life was wonderfully improved.

Notes

Abbreviations

AA	*American Agriculturist*
AC	*Albany Cultivator*
Ag History	*Agricultural History*
BRH	*Black River Herald*
CG	*Country Gentleman*
GF	*Genesee Farmer*
OC	*Ohio Cultivator*
NEF	*New England Farmer*
NYSAS	New York State Agricultural Society *Transactions*
UWH	*Utica Weekly Herald*

Introduction

1. H. E. Alvord, "Dairying of America," in *Dairy Farming*, ed. J. P. Sheldon (London, c. 1880), 366.

2. Mark Wilde, "The Industrialization of Food Processing in the United States, 1860–1960" (Ph.D. diss., Univ. of Delaware, 1988), chaps. 3 and 4.

3. For example, in Oneida County, New York, while butter production had stood at 3–4 million pounds in the 1870s and 1880s, by 1900 only 1.9 million pounds of butter was reported for Oneida County farms, yet the number of cows in the county was still growing at sixty-seven thousand. Countywide fluid-milk production jumped to twenty-one million gallons in 1900, from less than one million in 1870. U.S. census, published schedules, Agriculture (1850, 1860, 1870, 1900); New York State census, published schedules (1845, 1855, 1875); Eric Brunger, "Changes in the New York State Dairying Industry" (Ph.D. diss., Syracuse Univ., 1954), 271–72.

4. Marjorie Griffin Cohen, *Women's Work, Markets, and Economic Development in Nineteenth-Century Ontario* (Toronto: Univ. of Toronto Press, 1988), 109.

5. Joan Jensen, *Loosening the Bonds: Mid-Atlantic Farm Women, 1750–1850* (New Haven: Yale Univ. Press, 1986), 85–88.

6. Though everyone called them "factories," these plants did not fundamentally change cheesemaking in the way that, for example, textile factories revolutionized cloth production; methods remained the same, and tools were merely enlarged versions of those that had been used at home. Capital requirements were modest, and these were met by the pooling of small contributions made by individual farmers. Early

cheese "factories" resembled modern cooperatives more than large-scale industrial capitalist ventures. Nonetheless, because the term *factory* was universal, I use it here.

7. See notably Mary Ryan, *Cradle of the Middle Class: The Family in Oneida County, 1790–1865* (New York: Cambridge Univ. Press, 1981).

8. This use of the term *farming system* differs substantially from its meaning among contemporary social scientists, who use it to refer to practical, project-based extension work, usually taking place in Third World locations. Where the notion of *culture* is concerned, the parameters in use here are, of course, loose, but they fit within most generally accepted definitions.

9. A distinguished body of agricultural-history scholarship concentrated upon the economic, technological, and institutional forces impelling agricultural change. Paul Wallace Gates, *The Farmer's Age: Agriculture, 1815–1860* (New York: Holt, Rinehart & Winston, 1960), established how farmers acquired land, what they raised, what farming techniques they used, how they got access to markets, and how government policy affected agriculture. David Maldwyn Ellis, *Landlords and Farmers in the Hudson–Mohawk Valley, 1790–1850* (1946; rpt. New York: Octagon Books, 1967), described the conflict between landlords and farmers and showed how New York farmers switched from grain to dairying. Clarence Danhof, *Change in Agriculture* (Cambridge: Harvard Univ. Press, 1969), concentrated on the sources of innovation in agriculture, primarily technological and institutional, and celebrated the ingenuity and success of innovating farmers.

More recent historians have been preoccupied with unraveling the cultural and social dimensions of rural life. For a recent historiographical introduction, see Christopher Clark, "Economics and Culture: Opening Up the Rural History of the Early American Northeast," *American Quarterly* 43 (June 1991): 279–301. For a description of the "new rural history," see Hal S. Barron, "Rediscovering the Majority: The New Rural History of the Nineteenth-Century North," *Historical Methods* 19 (fall 1986): 142–52. See also Allan Kulikoff, "The Transition to Capitalism in Rural America," *William and Mary Quarterly* 3d ser. 46 (1989): 120–45, and *The Agrarian Origins of American Capitalism* (Charlottesville: University Press of Virginia, 1992); Robert Swierenga, "Theoretical Perspectives on the New Rural History: From Environmentalism to Modernism," *Agricultural History* 56 (July 1982): 495–503.

10. See, for example, Sean Wilentz, *Chants Democratic: New York City and the Rise of the American Working Class, 1788–1850* (New York: Oxford Univ. Press, 1984); Steven Hahn, *The Roots of Southern Populism: Yeoman Farmers and the Transformation of the Georgia Upcountry, 1850–1890* (New York: Oxford Univ. Press, 1983); Donald Worster, *Dust Bowl* (New York: Oxford Univ. Press, 1979).

11. Notable exceptions include John Mack Faragher, *Sugar Creek: Life on the Illinois Prairie* (New Haven: Yale Univ. Press, 1986); Hal S. Barron, *Those Who Stayed Behind: Rural Society in Nineteenth-Century New England* (Cambridge: Cambridge Univ. Press, 1985); Christopher Clark, *The Roots of Rural Capitalism: Western Massachusetts, 1780–1860* (Ithaca: Cornell Univ. Press, 1990); Jane Pederson, *Between Memory and Reality* (Madison: Univ. of Wisconsin Press, 1992); Jensen, *Loosening the Bonds;* and Nancy Grey Osterud, *Bonds of Community: The Lives of Farm Women in 19th-Century New York* (Ithaca: Cornell Univ. Press, 1991).

Notes to Pages 6–10

Chapter One: Agriculture in the Dairy Zone

1. Elkanah Watson, *History of the Rise, Progress, and Existing Condition of the Western Canals in the State of New York* (Albany, 1820) (excerpts from journals Watson kept beginning in 1788), 30; Pomroy Jones, *Annals of and Recollections of Oneida County* (Rome, N.Y., 1851), 36; New York State census (1865), xlii; Daniel Wager, *Our County and Its People* (Boston, 1896), 55; Timothy Dwight, *Travels in New-England and New-York*, vol. 3 (New Haven, 1822), 178–79; E. B. O'Callaghan, ed., *Documentary History of the State of New York* (Albany, 1849), 3:1135; Horatio Spafford, *Gazetteer of the State of New York* (Albany, 1824), 521.

2. Jack Campisi and Laurence M. Hauptman, eds., *The Oneida Indian Experience* (Syracuse: Syracuse Univ. Press, 1988), 5–16; Peter Pratt, *Archaeology of the Oneida Iroquois*, Occasional Publications in Northeastern Anthropology, no. 1 (n.p., n.d.), 13; Robert D. Kuhn and Dean R. Snow, eds., *The Mohawk Valley Project: 1983 Jackson-Everson Excavations* (Albany: Institute for Northeast Anthropology, State Univ. of New York at Albany, 1986), 1, 122.

3. Campisi and Hauptman, *Oneida Indian Experience*, 46; O'Callaghan, *Documentary History*, 2:1103.

4. David Maldwyn Ellis, "The Yankee Invasion of New York, 1783–1850," *New York History* 32 (Jan. 1951): 1–18; Jones, *Annals*, 14; Thomas F. Gordon, *Gazetteer of the State of New York* (Philadelphia, 1836), 564; T. Maxon and M. E. Carr, *Soil Survey of Oneida County, New York*, U.S. Department of Agriculture, Bureau of Soils, 15th report (Washington, D.C., 1915), 41; John H. Thompson, ed., *Geography of New York State* (Syracuse: Syracuse Univ. Press, 1966), 75.

5. William Cooper, *A Guide in the Wilderness* (1810; rpt. Rochester, 1897), 26; William Darby, *A Tour from the City of New York to Detroit* (1819; rpt. Chicago: Quadrangle Books, 1962), 54, 58, 60; Dwight, *Travels*, 3:188; John Fowler, *Journal of a Tour in the State of New York* (London, 1831), 79; Spafford, *Gazetteer*, 55; Gordon, *Gazetteer*, 567; O'Callaghan, *Documentary History*, 3:1108, 1134–34; Watson, *History of the Rise*, 36.

6. Cooper, *Guide*, 27; *Journals of John Lincklaen* (1791; rpt. New York, 1897), 70; James Stuart, *Three Years in North America* (New York, 1833), 1:58; Cooper, *Guide*, 24; O'Callaghan, *Documentary History*, 3:1143; Cooper, *Guide*, 41–42; Darby, *Tour*, 58–60; Ellis, *Landlords and Farmers*, 72.

7. Watson, *History of the Rise*, 12–13; letters, 1808, Mary Archbald papers, Sophia Smith Collection, Smith College, Northampton, Mass.

8. Cooper, *Guide*, 42; Spafford, *Gazetteer* (1813 ed.), 369.

9. Darby, *Tour*, 58.

10. Gordon, *Gazetteer*, 565; Stuart, *Three Years*, 1:61.

11. John Taylor, "Journal of a Missionary Tour through the Mohawk and Black River Countries in 1802," in O'Callaghan, *Documentary History*, 3:1136; Watson, *History of the Rise*, 12–13.

12. Darby, *Tour*, 59–60; Ebenezer Emmons, *Natural History of New-York*, 5 vols., vol. 1 (Albany, 1846), 24.

13. "Description of the Country between Albany and Niagara in 1792," in O'Cal-

Notes to Pages 11–14

laghan, *Documentary History*, 3:1103; Taylor, "Journal," 3:1137, 1148; Spafford, *Gazetteer* (1813 ed.), 20; Fowler, *Journal*, 77.

14. Ellis, *Landlords and Farmers*, 90, 105, 126; Taylor, "Journal," 1135, 1137; *AC* 1 (Oct. 1834): 97; *AC* 2 (Oct. 1836): 98; see also Ebenezer Emmons, *Natural History of New-York* (New York, 1846–53), 2:258; *NYSAS* 1844:367; Millard I. Roberts, *History of Remsen* (Remsen, N.Y.: privately printed, 1914), 54.

15. Quoted in Thomas Green Fessenden, *The Complete Farmer and Rural Economist* (Boston, 1835), 78–79.

16. Quoted in Ellis, *Landlords and Farmers*, 135; see also Percy Wells Bidwell and John Falconer, *History of Agriculture in the Northern States, 1620–1860* (1925; rpt. New York: Peter Smith, 1941), 21–25; see also William Cronon, *Changes in the Land: Indians, Colonists, and the Ecology of New England* (New York: Hill & Wang, 1983), 129, 138.

17. Ellis, *Landlords and Farmers*, 118–22.

18. *AC* 4 (Oct. 1837): 129; *AC* n.s. 6 (July 1849): 204–6; Whitney R. Cross, *The Burned-over District: The Social and Intellectual History of Enthusiastic Religion in Western New York, 1800–1850* (1950; rpt. New York: Harper Torchbooks, 1965), 59–67. Cross regards the Utica hinterland as a "mediocre" source of markets and produce, but perhaps this had as much to do with the ethos of "competency" as with potential.

19. "Norway History, No. 26," Norway (New York) *Tidings* 3 (Feb. 1889), Ferris family papers, Knox College, Galesburg, Ill.; *NYSAS* 1849:424; Bidwell and Falconer, *History of Agriculture*, 228.

20. Howard S. Russell, *A Long, Deep Furrow: Three Centuries of Farming in New England* (Hanover, N.H.: University Press of New England, 1976), 160, 37, 249, 271–72, 283–85; "Norway History, No. 26."

21. Cross, *Burned-over District*, 59–67, characterizes the central New York economy as having reached a "stable agrarian maturity" by about 1840. By then the best lands had been taken up, the rising city of Utica offered markets, and small-scale waterpower manufacturing abounded at the village level, complementing agricultural enterprises.

22. See William Cronon's comments from an Organization of American Historians roundtable session, transcribed in *Journal of American History* 76 (Mar. 1990): 1122–31.

23. William Townsend, *The Dairyman's Manual* (Vergennes, Vt., 1839), 5.

24. *NYSAS* 1849:427; Lewis Falley Allen, *American Cattle* (New York, 1868), 409.

25. Townsend, *Manual*, 5–6; see also *GF* 4 (Jan. 1844): 27.

26. Townsend, *Manual*, 7.

27. *AC* n.s. 2 (Jan. 1845): 15; *AC* 1 (July 1834): 30; *AC* n.s. 2 (Aug. 1845): 242.

28. *NYSAS* 1849:427; *NYSAS* 1854:216–17.

29. *NEF* 23 (Sept. 11, 1844): 81; Ohio State Board of Agriculture *Transactions* (1855), 256; *AA* 18 (July 1859): 201.

30. *AC* 3d ser. 5 (June 1848): 172; Townsend, *Manual*, 6–7.

31. *AC* 1 (July 1834): 61.

Notes to Pages 15–17

32. Townsend, *Manual*, 7.

33. *AC* 5 (May 1838): 55; see also *AC* 3d ser. 2 (Jan. 1845): 15; Joyce Appleby, "Commercial Farming and the Agrarian Myth," *Journal of American History* 68 (Mar. 1982): 833–49.

34. Nicholas Gardiner estate inventory, Nicholas Gardiner papers, Oneida Historical Society, Utica; Francis W. Squires diary, Apr. 7 and 20, June 28, 1842, Cornell Univ. Archives; *AC* n.s. 6 (Nov. 1849): 346.

35. *AC* 7 (Feb. 1840): 28; Francis Wiggins, *The American Farmer's Instructor* (Philadelphia, 1840), 294–95; *How to Make the Farm Pay* (Philadelphia, 1868), 340; John T. Schlebecker and Andrew W. Hopkins, *Dairy Journalism in the United States, 1810–1950* (Madison: Univ. of Wisconsin Press, 1957), 15, 41; Bidwell and Falconer, *History of Agriculture*, 23; James W. Thompson, *A History of Livestock Raising in the United States, 1607–1860*, U.S. Department of Agriculture Agricultural History Series, no. 5 (1942); G. A. Bowling, "The Introduction of Cattle into Colonial North America," *Journal of Dairy Science* 25 (Feb. 1942): 129–54; Peter Cook, "Domestic Livestock of Massachusetts Bay, 1620–1675," in *The Farm*, ed. Peter Benes (Boston: Boston Univ. Press, 1988), 113–18; Susan Geib, "Changing Works, Agriculture and Society in Brookfield, Massachusetts, 1785–1850" (Ph.D. diss., Boston Univ., 1981), 68.

36. *Rome Sentinel*, Mar. 12, 1851; see also *NEF* 12 (Oct. 30, 1833): 124.

37. *AC* 4 (July 1837): 81; *AC* 5 (1838), frontispiece advertisements; Schlebecker and Hopkins, *Dairy Journalism*, 11–23; *Moore's* 12 (Nov. 16, 1861): 365; *Moore's* 18 (Dec. 3, 1864): 389; *Moore's* 35 (Mar. 31, 1877): 199; George F. Lemmer, "The Spread of Improved Cattle through the Eastern United States to 1850," *Ag History* 21 (Apr. 1947): 79–93.

38. For an example, see Everett Edwards, "Europe's Contribution to the American Dairy Industry," *Journal of Economic History* 9 (supp., 1949): 72–84, and more recently, Fred Bateman and Jeremy Atack, *To Their Own Soil: Agriculture in the Antebellum North* (Ames: Iowa State Univ. Press, 1987), 147. Schlebecker and Hopkins, *Dairy Journalism*, document differences within the agricultural press.

39. *NYSAS* 1841:153. See also *NEF* 2 (1850): 51–52; *GF* 8 (Aug. 1847): 178–9; *NEF* 6 (June 1854): 259.

40. *NYSAS* 1847:247–59. The preeminent dairying authority of the nineteenth century, Xerxes Addison Willard, reported in 1861 that in his native Herkimer County —a leading cheese-producing county in central New York—a few local dairy farmers had experimented with pureblooded shorthorn Durhams, but most did not put "great effort" into acquiring purebred stock, preferring just a "dash" if indeed they took the trouble to do any special breeding at all (*NYSAS* 1861:499). See also "Dairy Business in New York," *GF* 16 (Aug. 1855): 245; *Rome Sentinel*, July 1, 1845, Sept. 25, 1850. More evidence for farmers' lack of interest comes in *GF* 7 (Oct. 1846): 232–33. In a notice about the "Importation of Cattle," the magazine noted that with prices as high as $100–300 per head, sellers of pedigreed animals found their stock "difficult to dispose of." The article chastised the Massachusetts Agricultural Society for spending large sums of money to "buy five Ayrshires, in no way superior to what they could obtain at home for one-third the money."

41. *GF* 21 (May 1860): 149–50; biographical information on Talcott from S. W. Durant, *History of Oneida County* (Philadelphia, 1877), 374.

42. See John Paton, "The Jersey Dairy Cow in Maine" (Ph.D. diss., Univ. of Maine, 1983), 110.

43. *AC* n.s. 3 (May 1842): 97; Allen, *American Cattle*, 125; see also *AC* n.s. 2 (June 1841): 98–99.

44. *AC* 3d ser. 5 (Aug. 1848): 238.

45. Fowler, *Journal*, 78; *How to Make the Farm Pay*, 343; Schlebecker and Hopkins, *Dairy Journalism*, 22ff.; *AA* 10 (Feb. 1851): 44; *NEF* 2 (Mar. 2, 1850): 85. However, Benjamin Hudson, "A Fine and Noble Beast" (unpublished paper, 1990), argues that the Milking Shorthorns were very good dairy animals.

46. *NEF* 2 (Jan. 19, 1850): 36.

47. Robert Trow-Smith, *A History of British Livestock Husbandry, 1700–1900* (London: Routledge & Kegan Paul, 1959); John R. Walton, "The Diffusion of the Improved Shorthorn Breed of Cattle in Britain during the Eighteenth and Nineteenth Centuries," Institute of British Geographers *Transactions* n.s. 9 (1984): 22–37; Walton, "Pedigree and the National Cattle Herd, 1750–1950," *Agricultural History Review* 34 (1986): 149–70; *AC* n.s. 1 (Feb. 1840): 29.

48. Charles Flint, quoted in *How to Make the Farm Pay*, 372–75, 343; see also Lauren B. Arnold, *American Dairying: A Manual for Butter and Cheese Makers* (Rochester, 1876), 43.

49. *NEF* 2 (Mar. 2, 1850): 85. For more on disagreements, see *AC* n.s. 2 (June 1841): 98–99; *AC* n.s. 3 (Dec. 1842): 191; *AC* n.s. 4 (May 1843): 73; see also Nicholas Russell, *Like Engend'ring Like: Heredity and Animal Breeding in Early Modern England* (Cambridge: Cambridge Univ. Press, 1986), for a very clear explanation of why this is still true to some extent.

50. Not until the Babcock test, artificial insemination, and modern genetics could breeders more confidently select for milk production. See John R. Paton and Barbara A. Barton, "The Development of the Ability to Select for Increased Milk Production: The Jersey Dairy Cow in Maine, 1900–1984," Maine Experiment Station *Bulletin*, no. 792 (Aug. 1984). See also E. P. Prentice, *American Dairy Cattle: Their Past and Future* (New York: Harper & Brothers, 1942), and Russell, *Like Engend'ring Like*.

51. Arnold, *American Dairying*, 56; John Shattuck, "Errors and Requisites in Making Butter," *AC* 3d ser. 9 (May 1861): 154–55; Allen, *American Cattle*, 56, 368; Asa Sheldon, *Yankee Drover* (1862; rpt. Hanover, N.H.: University Press of New England, 1988), 141; *AC* 4 (May 1837): 58; *How to Make the Farm Pay*, 350, 357, 358, 373; Wiggins, *American Farmer's Instructor*, 294–95; *NYSAS* 1854:207.

52. *NEF* 1 (May 10, 1823): 324; verse quoted in Gurdon Evans, *The Dairyman's Manual* (Utica, 1851), 86; see also *GF* 17 (Feb. 1856): 49.

53. *Working Farmer* 3 (1852): 137; Paton, "Jersey Dairy Cow," 121.

54. *AC* n.s. 4 (May 1843): 85. See also B. A. L. Cranstone, "Animal Husbandry: The Evidence from Ethnography," in *The Domestication and Exploitation of Plants and Animals*, ed. Peter Ucko and G. W. Dimbleby (London: Gerald Duckworth, 1969), 247–65; Keith Thomas, *Man and the Natural World* (London: Allen Lane, 1983), 16, 61; H. Epstein, "Domestication Features in Animals as Functions of Human Society,"

Ag History 29 (Oct. 1955): 137–47; Harriet Ritvo, "Race, Breed, and Myths of Origin: Chillingham Cattle as Ancient Britons," *Representations* 39 (summer 1992): 3–22. Faragher, *Sugar Creek*, 132, has found that in Sugar Creek, Ill., opposition to stock improvement arose out of class differences; poorer farmers feared that improved stock would simply allow the rich to get richer. The antiaristocratic rhetoric of New York State farmers suggests similar sentiments, but in the case of dairy cows at least, as we have seen, the expensive cows were not necessarily "improved" over the ordinary ones, and so their potential to give the richer farmers a competitive advantage was questionable.

55. Cynthia Eagle Russett, *Sexual Science: The Victorian Construction of Womanhood* (Cambridge: Harvard Univ. Press, 1989); Susan Conrad, *Perish the Thought* (New York: Oxford Univ. Press, 1976); Ann Douglas Wood, "The Fashionable Diseases: Women's Complaints and Their Treatment in Nineteenth-Century America," *Journal of Interdisciplinary History* 4 (summer 1973): 25–52.

56. For an overview, see Kulikoff, "Transition to Capitalism"; Clark, *Roots*; Faragher, *Sugar Creek*; Barron, *Those Who Stayed Behind*.

57. Sheldon, *Yankee Drover*, 146; *GF* 14 (May 1853): 105–7; *AC* n.s. 6 (Mar. 1849): 91. See also Allen, *American Cattle*, 309.

58. *GF* 14 (May 1853): 137; *Rural American* 10 (Apr. 15, 1866): 117; *GF* 8 (Apr. 1847): 82; *GF* 13 (Apr. 1852): 105–8. The implicit association of older ways with witchcraft—and therefore with women—also signified the shift of milking and cattle care to men's domain.

59. *Lowitt's Farmer's Almanac* (Concord, N.H.), Jan. 1848.

60. Fred Bateman, "Improvement in American Dairy Farming, 1850–1910: A Quantitative Analysis," *Journal of Economic History* 28 (1968): 255–73; Bateman, "Labor Inputs and Productivity in American Dairying, 1850–1910," *Journal of Economic History* 29 (1969): 206–29.

61. Jones, *Annals*, 75; Alfred Crosby, *Ecological Imperialism* (Cambridge: Cambridge Univ. Press, 1986), 151–57; Cronon, *Changes*, 41, 129, 138, 151; Bidwell and Falconer, *History of Agriculture*, 21, 14, 224, 105, 234.

62. On New England, see Danhof, *Change*.

63. New York State census, manuscript schedules, Oneida County (1855); see also Fowler, *Tour*, 77; *NYSAS* 1844:229; *NYSAS* 1851:238.

64. X. A. Willard, *Practical Dairy Husbandry* (New York, 1877), 255; *NYSAS* 1852:409; *NYSAS* 1844:229; *NYSAS* 1846:133.

65. For examples of rotation schemes, see Danhof, *Change*, 153ff.

66. *NYSAS* 1851:237; *AC* n.s. 4 (Aug. 1843): 130; Horatio Seymour, in an address to the New York State Agricultural Society, remarked that the "necessity for rotation of crops and improvement of the soil sometimes prevents [the farmer] from carrying this principle [of cash trade] out to its fullest extent" (quoted in ibid., 19). Seymour's statement reflects the compatibility of rotations with a system in which maximum gain was not the major consideration. When cash became prevalent, rotations declined.

67. *Rome Sentinel*, Sept. 3, 1851. For further discussion, see *AC* n.s. 4 (Mar. 1843): 44; *NYSAS* 1859:81; *NYSAS* 1843:130; *AC* n.s. 4 (Mar. 1843): 44. See also Robert L. Jones, *History of Agriculture in Ohio* (Kent, Ohio: Kent State Univ. Press, 1983), 65–

66; James Lemon, *The Best Poor Man's Country: A Geographical Study of Early Southeastern Pennsylvania* (1972; rpt. New York: W. W. Norton, 1976), 169–73.

68. *NEF* 6 (Jan. 1828): 212; *AC* n.s. 4 (Aug. 1843): 130; *NYSAS* 1854:204; *AC* 5 (May 1838): 56; *GF* 21 (Sept. 1860): 265–66; Charles Flint, *Milch Cows and Dairy Farming,* new ed. (Boston, 1858), 184.

69. Daniel Fink, *Barns of the Genesee Country, 1790–1915* (Geneseo, N.Y.: James Brunner, 1987), 86.

70. See Charles V. Piper and Katherine S. Bort, "The Early Agricultural History of Timothy," *Journal of the American Society of Agronomy* 7 (Jan.–Feb. 1915): 1–14; Ellis, *Landlords and Farmers,* 190, attributes increasing productivity to English hay. For more recent evaluations of these plants, see Robert Schery, *Plants for Man* (Englewood Cliffs, N.J.: Prentice-Hall, 1972).

71. *AC* n.s. 5 (Apr. 1848): 121; *AC* n.s. 6 (July 1849): 204–6; *AC* n.s. 7 (Feb. 1850): 79; *NYSAS* 1858:139; Samuel Hinckley general store account books, 1843–57 (see, for example, Mar. 4, 1856, 96–97), New York State Historical Association, Cooperstown; Stephen M. Babcock papers, State Historical Society of Wisconsin, Madison, box 1; Peleg and Cornelia Babcock diary, Oct. 20, 1852, S. M. Babcock papers, box 11; David Hughes diary, New York State Historical Association.

72. *GF* 8 (Aug. 1847): 178–79; *NYSAS* 1851:239. *NEF* 1 (May 10, 1823): 324 argued that clover was best for cheesemaking.

73. New York State census, manuscript schedules (1855).

74. *AC* 1 (June 1834): 48; *AC* n.s. 5 (Aug. 1844): 263; G. P. H. Chorley, "The Agricultural Revolution in Northern Europe, 1750–1880: Nitrogen, Legumes, and Crop Productivity," *Economic History Review* 34 (Feb. 1981): 71–94. Thanks to Steven Fales, agronomy department, Pennsylvania State Univ., for contemporary information. See Roy L. Donohue, *Soils: An Introduction to Soils and Plant Growth,* 2d ed. (Englewood Cliffs, N.J.: Prentice-Hall, 1965), 110.

75. *NYSAS* 1844:367; David Hughes diary, Feb. 1, Mar. 27, 1849. Yasuo Okada, "Squires' Diary: New York State Agriculture in Transition, 1840–1860," *Keio Economic Studies* 7 (1970): 195–213; Francis W. Squires diary, 1840, 1842.

76. *NYSAS* 1852:408; see also *NYSAS* 1844:369; *NYSAS* 1858:138; *NYSAS* 1851:238; *NYSAS* 1859:82; *NYSAS* 1853:424; *NYSAS* 1844:367; Richard Wines, *Fertilizer in America: From Waste Recycling to Resource Exploitation* (Philadelphia: Temple Univ. Press, 1985). Lemon, *Best Poor Man's Country,* maintains that rigorous rotations were possible only with livestock, so that adequate manure could be generated on the farm.

77. It is not clear just what advertisers meant when they mentioned "lime" or "plaster." Most likely both terms referred to ground limestone that had been processed through the application of heat. It is possible that making "plaster" involved processing the mineral gypsum rather than the composite rock of limestone.

78. New York State census, manuscript schedules, Oneida County (1855); *Rome Sentinel,* Apr. 30 and 16, 1851; *Atlas of Herkimer County* (Philadelphia, 1868); *BRH* (Boonville, Oneida County), May 12, 1857; *Rome Sentinel,* Apr. 22, 1845; see also *NYSAS* 1859:81; *NYSAS* 1851:238.

79. Thanks to Joseph Walsh, Cornell Cooperative Extension, Oriskany, N.Y., for this and other relevant information.

Notes to Pages 29–31

80. Assuming average yields for tillage crops, and using a ratio of pasture acreage to animals that is commonly mentioned (2:1), it is possible to extrapolate the acreages Oneida County cheese dairy farms had in tillage (seventeen), pasture (forty-two), and meadow (thirty-three). Since the average tonnage of hay on these farms was forty-three, the hypothetical yield of hay acreage works out to about one and a third—well above the county average of one. The manuscript state census for 1855 also can furnish evidence on hay yields, because it records for individual farms the acreage in hay and the total hay harvest. Of forty cheesemaking farms in the towns of Westmoreland and Boonville, thirty-two had average or above-average yields for this dry year. Among the farmers who reported using plaster, yields were even higher, sometimes—as in the case of Boonville's Spencer Pitcher—well over two tons per acre. There is one other way to estimate hay yields. From the 1855 census we can determine that on average, the forty cheesemaking farms examined in detail averaged 30 percent meadowland. Assuming that the percentage of meadow had not changed significantly since 1850, this means that 1850 cheesemakers averaged twenty-eight acres in meadow, and they harvested forty-three tons of hay, a yield of one and a half tons per acre. Thus it seems quite likely that cheesemaking farms had higher-than-average yields of hay. Of course, there were variations in yields; Boonville, more recently settled, had higher yields than Westmoreland. It is also possible that differences can be accounted for in part by the fact that cheesemaking farm families in some towns possessed better land than did those in others. However, this is not likely to have been the case everywhere, because dairying was not the first consideration during the time of settlement, when lands were being taken up. The yields extrapolated here are close to those Lemon, *Best Poor Man's Country*, found for southeastern Pennsylvania (one and a half tons per acre), where rotation and manuring improved output earlier in the century. The regimens described here had much in common with those prescribed by the agricultural reformers of the day. Carville Earle, "Myth of the Southern Soil Miner," in *The Ends of the Earth*, ed. Donald Worster (Cambridge: Cambridge Univ. Press, 1988), 175–210, finds that in the South, agricultural reform actually had a disastrous impact upon the ecological (and economic) viability of farming there, but the effects of reforming practices in the North are more ambiguous. Some practices—especially those that demanded draining—involved greater exploitation, but not all did.

81. GF 21 (Mar. 1860): 73; "Cows and Carrots," GF 13 (Apr. 1852): 105–7; NYSAS 1844:368; AC n.s. 5 (Feb. 1848): 69, 95; AC n.s. 3 (Nov. 1842): 181; AC n.s. 3 (Jan. 1842): 17; AC n.s. 2 (Nov. 1842): 176; AC n.s. 4 (July 1843): 114; Helene Phelan, ed., *"And a White Vest for Sam'l": The Diaries of Maria Whitford, 1857–1871* (Alfred, N.Y.: Sun, 1976), 35; AA 8 (Mar. 1849): 94.

82. NEF 5 (Feb. 1853): 81; see also GF 22 (Nov. 1861): 302; Rome city directory (1859–60).

83. NEF 6 (June 1854): 259–60; Fessenden, *Complete Farmer*, 76; see also AC n.s. 4 (July 1843): 114; NEF 9 (Dec. 1857): 548–49; NYSAS 1853; AA 9 (Mar. 1850): 88; NYSAS 1847:623–26; GF 2 (Feb. 1841): 18; AA 17 (Jan. 1858): 27.

84. X. A. Willard, "The Rise and Progress of American Dairying," *Moore's* 36 (Jan. 6, 1877): 6.

85. AA 8 (Nov. 1849): 331–32; AC n.s. 1 (Sept. 1840): 137; AC n.s. 3 (Nov. 1842): 176; AC n.s. 4 (July 1843): 114; NYSAS 1861:405; "Another Premium Plan of Barn,"

Moore's (July 6, 1865); "How a Jerseyman Keeps His Cows," *NEF* 12 (June 1860): 251–52. Stanchions were not universally popular; some farmers regarded them as being too confining and used them either selectively or not at all. In 1869 Larkin Safford of Maine, N.Y., offered a variation on a stanchion that he claimed would allow greater mobility than "ordinary stanchions." See *NEF* n.s. 3 (Feb. 1869): 69.

86. "About Barns," *NEF* 10 (Mar. 1858): 132; L. L. Pierce, "What Buildings Are Necessary for a Farm of One Hundred Acres?" *NEF* 12 (Jan. 1860): 53.

87. *NYSAS* 1844:224; *NEF* 16 (Feb. 1864): 50; Arnold, *American Dairying*, 160; *Moore's* 11 (Nov. 3, 1860): 350; *GF* 16 (May 1855): 145; *AC* n.s. 4 (Aug. 1843): 130; *NYSAS* 1852:409.

88. *GF* 16 (Aug. 1855): 246; *Moore's* 11 (Apr. 1860): 110. See also *NEF* 8 (Aug. 1856): 356–57.

89. *Rome Sentinel*, Mar. 18, 1845.

90. *NYSAS* 1853:429; *NYSAS* 1859:86; *NEF* 7 (Dec. 1855): 541.

91. *GF* 14 (Jan. 1853): 36. For similar versions, see also *NEF* 4 (June 1852): 272–73 and *AA* 12 (Aug. 23, 1854): 369.

92. "Ground Plan of a Convenient Barn," *NEF* 8 (Mar. 1856): 136; William D. Brown, "An Hour in a Great Barn," *NEF* 7 (Jan. 1855): 46; Thomas Hubka, *Big House, Little House, Back House, Barn* (Hanover, N.H.: University Press of New England, 1984).

93. Alonzo Fish, *AC* n.s. 4 (Aug. 1843): 130; B. Andrews, *OC* 15 (1859): 340–41.

94. *NYSAS* 1851:263, 268. See also *NYSAS* 1858:599; *GF* 14 (Apr. 1853): 128.

95. See, for example, *AA* 16 (Nov. 1857): 242.

96. For a superb introduction to the Pennsylvania barn, see Robert Ensminger, *The Pennsylvania Barn: Its Origin, Evolution, and Distribution in North America* (Baltimore: Johns Hopkins Univ. Press, 1992).

97. *OC* 6 (Apr. 1, 1850): 101. For another example, see an 1841 plan drawn up by Horace Humphrey of Winchester, Conn., for a 33- by 44-foot side-hill barn specifically designed for milking and stabling his dairy cows. Rather than arranging his levels along the lengthwise axis, Humphrey essentially redistributed the old bay/floor/stable arrangement into vertical layers, with the floor and two bays on the upper level and stables on the lower. This gave more space for cows and for hay to feed them. On the upper level he made chutes extending down to the mangers. "This is the most convenient way of feeding cattle that I know of," he wrote. "The hay is shoved off one side of the floor without the trouble of lifting. It falls into a double manger and fodders twenty head in the time it would take to feed ten in any other way." The lower level was similarly efficient; in addition to its by-now familiar pattern of central feed alley/rear troughs, it contained space for young cattle, oxen, horses, and a root cellar. *GF* 14 (Mar. 1853): 96. See also *AC* n.s. 6 (June 1849): 193.

98. Evans, *Dairyman's Manual*, 71–75.

99. Wager, *Our County*, 372–75, 380–88, 396–402, 544–51; Gordon, *Gazetteer*, 571; Wager, *Our County*, 553–64, 579–83, 604–12.

100. U.S. census, manuscript schedules, Agriculture (1850). See also Bateman and Atack, *To Their Own Soil*, 153, for northeastern figures. Where necessary, I have calculated milk yields based on equivalencies given in T. E. Lamont, "Agricultural

Production in New York, 1866 to 1937," New York State Agricultural Experiment Station *Bulletin,* no. 693 (1938).

101. U.S. census, manuscript schedules, Agriculture (1850); New York State census, manuscript schedules (1855); "Farming in Oneida County," *AC* n.s. 6 (Dec. 1849): 364.

102. Ellis, *Landlords and Farmers,* 188; Nicholas Gardiner inventory; X. A. Willard, "Prices of 54 Years," *CG* 44 (Apr. 10, 1879): 235–36; Carolyn Merchant, *Ecological Revolutions: Nature, Gender, and Science in New England* (Chapel Hill: Univ. of North Carolina Press, 1989), 175–80; Bettye Pruitt, "Self-Sufficiency and the Agricultural Economy of Eighteenth-Century Massachusetts," *William and Mary Quarterly* 3d ser. 41 (1984): 333–64; Peleg and Cornelia Babcock diary, Jan. 14, 1852; *BRH,* May 30, 1856; New York State census, manuscript schedules (1855). Ryan, *Cradle,* 11, argues that by 1820 "diversified agricultural production" had "lost hegemony" to small-scale commerce. The diversification in commercial cheese dairies calls this into question.

103. Clark, *Roots,* 85.

104. Karl Finison, "Energy Flow on a Nineteenth-Century Farm," Univ. of Massachusetts Department of Anthropology Research Report, no. 18 (Amherst, 1979), and John and Carol Steinhart, "Energy Use in the U.S. Food System," *Science* 184 (Apr. 19, 1974): 307–16.

105. Cronon, *Changes;* Worster, *Dust Bowl.* Carolyn Merchant, *Ecological Revolutions,* places more emphasis than does Cronon upon the subsistence element of early New England agriculture, which she characterizes as "ecologically similar" (p. 157) to the Indians' agricultural system. The "animate cosmos" (chap. 4) that she finds among the settlers differs significantly from the rationalist, profit-oriented mentality Cronon describes. She places the "capitalist ecological revolution" (p. 197) in the late colonial and early national period. Though they disagree on the timing of its impact, both Cronon and Merchant portray capitalist agriculture as environmentally destructive, because the push to raise commodities for profit stimulated activity beyond the land's capacity to renew itself, and because exchange with distant markets resulted in the permanent exit of nutrients from the farm. Brian Donahue, "Dammed at Both Ends and Cursed in the Middle: The 'Flowage' of the Concord River Meadows, 1798–1862," *Environmental Review* 13 (fall/winter 1989): 47–67, argues that dairying in eastern Massachusetts was not sustainable, but the farmers he studied apparently overgrazed their grasslands and did not use manure or practice rotation. Appleby, "Commercial Farming," 84, also argues that market opportunities at the turn of the nineteenth century favored diversification and, in turn, promoted maintenance of soil fertility.

Chapter Two: Sharp Dealings

1. James Henretta, "Families and Farms: Mentalitè in Pre-industrial America," *William and Mary Quarterly* 3d ser. 35 (Jan. 1978): 3–33; Michael Merrill, "Cash Is Good to Eat: Self-Sufficiency and Exchange in the Economy of the Rural United States," *Radical History Review* 4 (Sept. 1976): 42–71.

2. Percy Wells Bidwell, "The Agricultural Revolution in New England," *Ameri-*

can *Historical Review* 26 (July 1921): 683–703; Gates, *The Farmer's Age;* James Lemon, "Household Consumption in Early America," *Ag History* 41 (Jan. 1967): 59–71; Cronon, *Changes;* Winifred Rothenberg, "The Market and Massachusetts Farmers, 1750–1850," *Journal of Economic History* 41 (June 1981): 283–313; Appleby, "Commercial Farming," esp. 845.

3. Pruitt, "Self-Sufficiency"; Hahn, *Roots of Southern Populism,* 196; Michael Bellesiles, "The World of the Account Book" (paper presented at the annual meeting of the Organization of American Historians, New York, 1986); Clark, *Roots;* Kulikoff, "Transition to Capitalism."

4. Common estimates of U.S. per capita consumption for the late nineteenth century mention four pounds per year as an average figure. If we assume that cheesemaking households consumed double that amount, the total for a household of six people comes out to just forty-eight pounds.

5. Hahn, *Roots of Southern Populism,* 6; Pruitt, "Self-Sufficiency," has found that the same was true of Massachusetts farms in the 1770s.

6. Bellesiles, "Account Book," argues that "communal capitalism" paved the way for a smooth transition, because all transactions were conceived in cash equivalents, and because exchange partners clearly established a credit-and-debit system.

7. As Kulikoff, "Transition to Capitalism," has observed, "The hybrid nature of the economic order of the early American North provided a porous defense at best against capitalist advance" (p. 134). This is not to say that social conflict was absent in the antebellum phase of the transition. However, conflicts took place mostly out of the public arena, within households.

8. John Gould, "Pioneer Dairying in Ohio, #1," *Ohio Farmer* 91 (Mar. 11, 1897): 196; "To the Dairy Women of Our Country," *NEF* 16 (June 20, 1838): 397; *NYSAS* 1851:92. In *The Fruits of Merchant Capital* (New York: Oxford Univ. Press, 1988), Eugene Genovese and Elizabeth Fox-Genovese point out that "in the free states, cheap transportation could rapidly transform the family farm into a paying enterprise.... In the Lower South, however, cheap transportation would draw the yeomen into the cotton economy at high economic and social risk" (p. 256). This helps to explain why southern yeomen resisted capitalism so much more than their northern counterparts; in addition, they accepted the slave society as long as slaveholders helped to maintain their isolation and to provide services that protected them from the demands of merchant capital.

9. *AC* 3 (May 1836): 29; *NYSAS* 1851:92. See also *Farmer's Almanac* (Boston, 1849), 6; *AC* 4 (July 1837): 85.

10. Fred M. Jones, *Middlemen in the Domestic Trade of the United States, 1800–1860,* Illinois Studies in the Social Sciences 21, no. 3 (Urbana, 1937; rpt. New York and London: Johnson Reprint Corp., 1968).

11. Ephraim Perkins, "On the Dairy Farming of Herkimer," *AC* 1 (Aug. 1834): 74; Morton Rothstein, "The International Market for Agricultural Commodities, 1850–1873," in *Economic Change in the Civil War Era,* ed. David Gilchrist and W. David Lewis (Greenville, Del.: Eleutherian Mills–Hagley Foundation, 1964), 62–73.

12. *NYSAS* 1871:435–36; *GF* 4 (Aug. 16, 1834): 262; Brunger, "Changes," 43; *NEF* 12 (Oct. 30, 1833): 124; *AA* 18 (Sept. 1859): 266; *BRH,* Nov. 2, 1855; "American

Cheese Trade," *Farmer's Cabinet* 11 (May 15, 1848): 322; Dorothea Ives, *Herkimer County Cheese: The History of a Famous Industry* (Mohawk, N.Y.: privately printed, 1986), 23–24; see also *Ohio Farmer* 8 (Oct. 8, 1859): 323. On McHenry, see OC 7 (1851): 201, and see also his pamphlet, *Instructions for Curing and Packing Beef, Bacon, and Hams . . . for the English Market* (Liverpool, 1855), Hagley Library Pamphlet collection, Wilmington, Del.

13. Letter dated 12th mo 21st 1822, Ferris family papers, folder 5; Brunger, "Changes," 43; GF 4 (Aug. 16, 1834): 263; John Gage, "Matters in Central New York—Cheesemaking," GF 8 (Dec. 1847): 284; "Cheese Dairies," GF 8 (Aug. 1847): 179; OC 2 (July 1846): 61; NYSAS 1851: 89–94.

14. Peleg and Cornelia Babcock diary, Feb. 10, Mar. 6, Apr. 23, May 19 and 20, Sept. 20 and 24, Oct. 29, Nov. 20, 1852.

15. C. C. Wick, "Good Cheese Pays Best," OC 9 (Aug. 1853): 243; letters dated 5th of 4th month 1821 and Jan. 19, 1828, Ferris family papers, folder 5.

16. "To the Dairywomen," NEF 16 (June 20, 1838): 397.

17. Ives, *Herkimer County Cheese*, 18, 20, 29–30, 44; letter dated 12th mo 21st 1822, Ferris family papers, folder 5.

18. Letter dated Dec. 21, 1820, Ferris family papers, folder 5; Ives, *Herkimer County Cheese*, 43.

19. Letter dated Jan. 19, 1828, Ferris family papers, folder 5; Ives, *Herkimer County Cheese*, 43.

20. Quoted in Ives, *Herkimer County Cheese*, 19.

21. Peleg and Cornelia Babcock diary, Mar. 28, 1853; James Kilham account book, Dec. 1 and 2, 1852, New York State Historical Association; Gould, "Pioneer Dairying in Ohio," 196.

22. Kulikoff, "Transition to Capitalism," 128–29; Clark, *Roots*; see also Robert Mutch, "Yeoman and Merchant in Pre-industrial America," *Societas* 7 (autumn 1977): 279–302.

23. NYSAS 1851: 89–94; Bellesiles, "Account Book," n. 38.

24. *An American Dictionary of the English Language*, rev. and enl. Chauncey A. Goodrich (New York, 1852), 201; *Oxford English Dictionary* (1989); NYSAS 1858: 103. For an interpretation of early American "competency," see Daniel Vickers, "Competency and Competition: Economic Culture in Early America," *William and Mary Quarterly* 3d ser. 47 (Jan. 1990): 3–29. He argues that farmers wanted to avoid working for others, but the pervasiveness of labor exchange calls this into question.

25. Townsend, *Manual*, 7.

26. Barry Levy has argued that in Pennsylvania the "decently lucrative but communally safe" crop was wheat. Levy quoted in Clark, "Economics and Culture," 288–89. But Joan Jensen, whose study concerns the very same region, sees dairying, especially butter production, as a stable hedge in an era when rising land values made ownership more difficult (*Loosening the Bonds*, 91).

27. AC 6 (July 1839): 83; "Cattle Husbandry," AC 5 (May 1838): 55; NYSAS 1861: 543; see also AA 18 (Oct. 1859): 293–94. See also Appleby, "Commercial Farming," 841, for an argument that farmers of the early Republic could participate in market activity "without taking the risks associated with cash crops."

28. Clarence Danhof, by contrast, has argued that "a market orientation of farm management necessarily separated the farm from the family. The farm became a profit-making enterprise to be managed with a calculus independent of the family, and the family became a resource-supplying unit that received a money return to be used independently of the farm." But as we have seen, family and market orientation were made compatible through the notion of competency. Danhof, *Change*, 135–38.

29. Mutch, "Yeoman and Merchant," esp. 280.

30. For examples of profit-and-loss statements, see Norman Gowdy's, published in *NYSAS* 1858:135–37; "The Dairy—Its Profits," *NEF* 15 (Aug. 10, 1836): 34–35; "Profitable Dairying," *NEF* 6 (Feb. 15, 1828): 239; Albert Ford's farm accounts, *NYSAS* 1851:243–44; "Dairy Husbandry," *Farmer's Cabinet* 3 (May 15, 1839): 300–302; "Profitable Dairy," *Farmer's Cabinet* 2 (Sept. 1, 1837): 47; *CG* 2 (Mar. 6, 1856): 157; *NYSAS* 1844:221. See also Merrill, "Cash Is Good to Eat"; Bellesiles, "Account Book," describes this economy well.

31. E. H. Arr [Ellen Chapman Hobbs Rollins], *New England Bygones* (Philadelphia, 1880), 146.

32. Robert Gross, "The Great Bean-Field Hoax: Thoreau and the Agricultural Reformers," *Virginia Quarterly Review* 61 (summer 1985): 483–97, esp. 490. See also Tamara Thornton, "Between Generations: Boston Agricultural Reform and the Aging of New England, 1815–1830," *New England Quarterly* 59 (June 1986): 189–212.

33. "Dairy Business in New York," *GF* 16 (Aug. 1855): 246; *NYSAS* 1851:89–94; Ohio State Board of Agriculture *Annual Report* (1851), 221; C. C. Wick, "Good Cheese Pays Best," *OC* 9 (Aug. 1853): 243–44; see also U.S. Patent Office *Annual Report* (1842), 2:999.

34. A. L. Fish, "Management of Cheese Dairies," *AA* 7 (July 1848): 201–5.

35. U.S. Treasury Department, McLane Report, *Documents Relative to the Manufactures in the United States* (Washington, D.C., 1832), 21, 44.

36. *NYSAS* 1841:135. For further descriptions of this system, see Clark, *Roots;* Merrill, "Cash Is Good to Eat;" Benes, *The Farm;* Geib, "Changing Works."

37. Peleg and Cornelia Babcock diary, July 14, Aug. 30 and 22, Sept. 25–26, Dec. 18 and 31, 1852. The Ezra Birdseye papers, Oneida Historical Society, show that Ephraim Perkins, one of the region's leading commercial farmers, traded with storekeeper Birdseye candles, logs, cheese, oats, seed, hay, his team, and barley for sawing and for work; David Hughes's diary mentions in an 1849 entry that there was no cash in the house at all.

38. U.S. census, manuscript schedules, Agriculture and Population (1850).

39. Ives, *Herkimer County Cheese*.

40. Charles S. Brown to Cornelia Babcock, Dec. 19, 1859, S. M. Babcock papers, box 1.

41. Evidence for competition among factors is cited in Danhof, *Change*, 44–45.

42. Ives, *Herkimer County Cheese*, 44.

43. Samuel Hinckley account books; X. A. Willard, "Cheese Dairying in Herkimer County," *NYSAS* 1860:189; Charles S. Brown to Cornelia Babcock, Dec. 19, 1859, S. M. Babcock papers, box 1. Jones, *Agriculture in Ohio*, 66, 182–92, associates cash payment with the rise of canal and commission agents. Clark, *Roots,* 200–202,

Notes to Pages 58–63

suggests that panics forced businesses to adopt cash policies. The panic of 1857 probably contributed to the rapid shift in the cheese business during the late 1850s and early 1860s.

44. Wager, *Our County*, 225; Ellis, *Landlords and Farmers*, 173–80; *CG* 17 (Jan. 31, 1861): 77; *BRH*, May 25, 1855. Okada, "Squires' Diary," shows that in 1850 farmer Squires made only 28 percent of his transactions in cash, but by 1860 that proportion was up to 64 percent.

45. Clark, *Roots*, 195–220; *Rome Sentinel*, Dec. 25, 1850; *BRH*, May 30, 1856. For contemporary testimony about the importance of cash, see Horatio Seymour's remarks in *NYSAS* 1852 (quoted in Danhof, *Change*, 19).

46. *Rome Sentinel*, Sept. 10, 1851, Apr. 1, 1845; *BRH*, May 30, 1856; Clark, *Roots*, 197.

47. Nelson Greene, *Mohawk Valley, Gateway to the West* (Chicago: S. J. Clarke, 1925), 3:48; U.S. Patent Office *Annual Report* (1858), vol. 2, Agriculture, p. 221; U.S. Patent Office *Annual Report* (1852), 4; *NYSAS* 1849:429. American cheese was exported to over fifty countries, but Great Britain formed by far the biggest single market.

48. Ives, *Herkimer County Cheese*, 32–40; "Rise and Progress of Dairying in America," *Moore's* 35 (Jan. 6, 1877): 6.

49. Jones, *Agriculture in Ohio*, 191–92, argues that differential pricing was slow to appear.

50. Inventory, Mar. 5, 1857, S. M. Babcock papers, box 1.

51. *OC* 15 (Aug. 1, 1859): 233; *Rome Sentinel*, Aug. 5, 1845; Gould, "Pioneer Dairying"; Thomas Dublin, "Rural Putting-out Work in Early Nineteenth-Century New England: Women and the Transition to Capitalism in the Countryside," *New England Quarterly* 64 (Dec. 1991): 531–73, esp. 571. In the South, by contrast, yeomen believed that their interest was served by periodic, irregular forays into the market for specific purposes; northerners' contact with the market was more constant, even institutionalized. Eugene Genovese and Elizabeth Fox-Genovese argue that southern yeomen's estimate of their situation was accurate, because in a slave society, yeomen could not hope to gain a competitive position in the market economy. See "Yeomen Farmers in a Slaveholders' Democracy," in *Fruits*.

52. *NYSAS* 1849:428. See also *AA* 12 (June 1854): 195. See also *NYSAS* 1842:169; *NYSAS* 1854:215 ("Dairymen in Herkimer County, as a body, are in most comfortable circumstances as to money"); *BRH*, Apr. 16, 1863.

Chapter Three: The Social Organization of Cheesemaking Households

1. This may also explain the relatively large size and wealth of cheesemaking farms. As Jeremy Atack and Fred Bateman have demonstrated, older farmers had accumulated more property and stock throughout their careers in agriculture and so tended to be wealthier than younger farmers. Atack and Bateman, "Egalitarianism, Inequality, and Age: The Rural North in 1860," *Journal of Economic History* 41 (Mar. 1981): 85–94.

2. "Neatness in a Dairy," *NEF* 8 (Sept. 11, 1829): 58–59.

Notes to Pages 63–72

3. Of course, these families may have hired day labor from time to time, but it is reasonable to assume that wage labor hired on a temporary basis did not supply a significant portion of their labor needs, since most had several adult workers. Moreover, it is unlikely in this period that they would have hired long-term wage laborers who did not board with the family. The fact that those households with live-in labor seem to have been using hired people to substitute for family members further reinforces this assumption.

4. Interestingly, the general pattern here diverges from Clark's finding (*Roots*, 107) that in the Connecticut Valley, families with live-in laborers tended to be either young or old families. This difference highlights the fact that cheesemaking households were more geared toward capitalism than Clark's families, since they were reinforcing a situation in which they were already provided with enough adult laborers.

5. Forty percent of cheesemaking households produced less than 500 pounds; 8 percent, from 501 to 1,000 pounds; 22 percent, from 1,001 to 5,000 pounds; 19 percent, from 5,001 to 10,000 pounds; and 11 percent, more than 10,000 pounds. U.S. census, manuscript schedules for the seven Oneida County towns described in chap. 2 (1850).

6. *NYSAS* 1858:126–37; Peleg and Cornelia Babcock diary, Apr. 1, 1853.

7. Lavinia Mary Johnson memoir, Oneida Historical Society, 52, 56.

8. Faye Dudden, *Serving Women: Household Service in Nineteenth Century America* (Middletown, Conn.: Wesleyan Univ. Press, 1983).

9. Clark, *Roots*, 87; see also *Moore's* 17 (Aug. 4, 1866): 246.

10. Peleg and Cornelia Babcock diary, Apr. 28, 1852; Andrew Hurlburt diary, Jan. 26, Apr. 10, Oct. 11 and 24, 1865, Rome (New York) Historical Society; Rosetta Hammond Bushnell diary, Sept. 8, 1857, Feb. 26, Aug. 3, 1860, New York State Historical Association.

11. *NYSAS* 1847:653; *NYSAS* 1844:221; Ives, *Herkimer County Cheese*, 62–63; see also *Moore's* 12 (Mar. 2, 1861): 70.

12. Andrew Hurlburt diary, Apr. 4 and 10, 1865; Ives, *Herkimer County Cheese*, 63–64; see also *Moore's* 11 (Nov. 10, 1860): 358, for an instance in which the male farm owner clearly controlled "his" dairymaid's work.

13. *NYSAS* 1862:238.

14. Correspondence in the possession of Mrs. Dorothea Ives, Dolgeville, N.Y. Thanks to Mrs. Ives for permission to quote these letters.

15. *AA* 12 (May 17, 1854): 133–34; *OC* 10 (Aug. 11, 1854): 228–29.

16. Cornelia and Peleg Babcock diary, Jan. 15, 1852; letter of C. H. Babcock, May 18, year illegible, S. M. Babcock papers, box 1.

17. Cornelia and Peleg Babcock diary, Dec. 18, 1852.

18. Ibid., July 15, 1852.

19. *AA* 12 (May 10, 1854): 133–34.

Chapter Four: "Intense Interest and Anxiety"

1. Deborah Fink, *Open Country Iowa: Rural Women, Tradition, and Change* (Albany: State Univ. of New York Press, 1986); Faragher, *Sugar Creek;* Merchant, *Ecological Revolutions*. Linda Kerber, "Separate Spheres, Female Worlds, Women's Place:

The Rhetoric of Women's History," *Journal of American History* 75 (June 1988): 9–39, calls for historians to examine how women's and men's activities have intersected.

2. Osterud, *Bonds of Community*; Jensen, *Loosening the Bonds*. Carolyn Sachs, "The Participation of Women and Girls in Market and Non-market Activities on Pennsylvania Farms," in *Women and Farming: Changing Roles, Changing Structures*, ed. Wava Haney (Boulder: Westview, 1988), 123–34, criticizes all of the conventional dichotomies—the Marxist, neoclassical, and structural-functional.

3. M. M. Bagg, "The Earliest Factories of Oneida," Oneida Historical Society *Transactions* (1879), 112–24; Ellis, *Landlords and Farmers*, 128–29; BRH, May 30, 1856. See also Ryan, *Cradle*, 44–45.

4. Thomas Dublin, "Women and Outwork in a Nineteenth-Century New England Town: Fitzwilliam, New Hampshire, 1830–1850," in *The Countryside in the Age of Capitalist Transformation*, ed. Steven Hahn and Jonathan Prude (Chapel Hill: Univ. of North Carolina Press, 1985), 51–71.

5. *Rome Sentinel*, Sept. 3, 1851.

6. Geib, "Changing Works," 81.

7. Andrew Hurlburt diary, Jan. 7, 1865; *Moore's* 11 (Sept. 8, 1860): 286.

8. Nancy Grey Osterud, "'She Helped Me Hay It as Good as a Man': Relations among Women and Men in an Agricultural Community," in *"To Toil the Livelong Day": America's Women at Work, 1780–1980*, ed. Carol Groneman and Mary Beth Norton (Ithaca: Cornell Univ. Press, 1987), 87–98; Clarence Danhof, "Gathering the Grass," *Ag History* 30 (Jan. 1956): 169–73; William Baron and Anne Bridges, "Making Hay in New England: Maine as a Case Study," *Ag History* 57 (Apr. 1983): 165–81; R. Douglas Hurt, *American Farm Tools, from Hand-Power to Steam-Power* (Manhattan, Kans.: Sunflower Univ. Press, 1983; from the Jan. 1982 *Journal of the West*), 85.

9. *GF* 3 (Oct. 1842): 151.

10. *UWH*, Sept. 9, 1873.

11. Osterud, *Bonds of Community*; Sarah Burgess diary, May 13, 1860, New York State Historical Association; Peleg and Cornelia Babcock diary, Jan. 10, 1852.

12. New York State census, manuscript schedules (1855). The relative absence of these supposedly ubiquitous items may be explained by the fact that men reporting to the census undervalued this women's work, but it is more likely, given that other products of women's work *were* reported, that poultry was unimportant in the context of other demands on women's time.

13. *GF* 12 (May 1851): 114; *GF* 8 (Aug. 1847): 178; *Rome Sentinel*, July 1, Sept. 23, 1845, Feb. 13, June 26, Sept. 25, 1850. Women were invited to participate and to judge in some fairs; see *NEF* 2 (Jan. 19, 1850): 36; *NEF* 13 (June 1861): 286; *CG* 14 (Nov. 24, 1859): 337.

14. Joel Hatch, Jr., *Reminiscences, Anecdotes, and Statistics of the Early Settlers and the "Olden Times" in the Town of Sherburne, Chenango County, New York* (Utica, 1862), 80; *NEF* 18 (Oct. 2, 1839): 112; *Moore's* 12 (Aug. 17, 1861): 261; Geib, "Changing Works," 125.

15. Receipt, 1860, S. M. Babcock papers, box 1; anonymous diary, Apr. 3, 1868, J. G. Parkhurst Collection, Bentley Library, Univ. of Michigan.

16. Phelan, *White Vest*, entry dated Apr. 28, 1859; Sheldon, *Yankee Drover*, 141.

Notes to Pages 76–81

17. Peleg and Cornelia Babcock diary, Jan. 9, June 4, Sept. 6, 1852; Andrew Hurlburt diary; Solon Robinson, *Facts for Farmers* (Cleveland, 1866), 451; Phelan, *White Vest*, 72.

18. GF 7 (July 1846): 161. According to Debra Reid of the Farmers' Museum, Cooperstown, N.Y., sometimes dairywomen could buy rennets from local butchers.

19. K. D. M. Snell, *Annals of the Labouring Poor* (Cambridge: Cambridge Univ. Press, 1985).

20. "Milk-Maids Turned Pianists," GF 9 (Apr. 1848): 115; GF 18 (Mar. 1859): 90; "Farmers' Daughters," NEF 2 (Aug. 31, 1850): 284.

21. "Women Milking," GF 1 (May 1840): 77. See also NEF 19 (June 30, 1841): 414.

22. GF 13 (Apr. 1852): 111.

23. Flint, *Milch Cows* (1858 ed.), 149. See also GF 9 (Feb. 1848): 55.

24. Laurel Ulrich, *Goodwives* (New York: Oxford Univ. Press, 1980), 16; Jensen, *Loosening the Bonds*, 93; NEF 1 (May 10, 1823): 324.

25. NEF 6 (Feb. 15, 1828): 239; see also AC 7 (Sept. 1840): 139.

26. GF 9 (Feb. 1848): 55. Deborah Fink, "Farming in Open Country, Iowa: Women and the Changing Farm Economy," in *Farm Work and Field Work*, ed. Michael Chibnik (Ithaca: Cornell Univ. Press, 1987), 121–45, notes that Iowa farm women are assumed to possess nurturing qualities and are therefore assigned to work with animals.

27. Lavinia Mary Johnson memoir, 56; GF 1 (May 1840): 77. Maria Whitford (see Phelan, *White Vest*) and her husband shared the milking; Osterud, *Bonds of Community*, finds that late-nineteenth-century farm women in the Nanticoke Valley also shared in milking chores with men.

28. *Moore's* 10 (June 1859): 197; "What a Woman Saw and Thought," *Moore's* 11 (Oct. 20, 1860): 334; NEF 24 (Dec. 10, 1845): 187; *Moore's* 12 (Apr. 27, 1861): 133; *Farmer's Cabinet* 1 (Aug. 1, 1836): 23.

29. "Management of Milch Cows," AC 7 (June 1840): 96; *Moore's* 11 (Sept. 8, 1860): 230; see also *Moore's* 11 (Apr. 19, 1860): 158 and NEF 23 (May 28, 1845): 377.

30. *Moore's* 11 (July 14, 1860): 222.

31. NEF 8 (Jan. 1856): 39.

32. *Moore's* 12 (Mar. 9, 1861): 78–79; NEF 9 (Sept. 1857): 411; "Farmers' Wives and Daughters," *Michigan Farmer*, Nov. 1, 1844, p. 126; *Moore's* 11 (Sept. 8, 1860): 286; *Moore's* 11 (Oct. 20, 1860): 334; *Moore's* 11 (Nov. 17, 1860): 366; NEF 15 (July 13, 1836): 4; NEF 22 (July 26, 1843): 26; NEF 18 (July 31, 1839): 29–31; NEF 2 (Feb. 2, 1850): 51–52; NEF 2 (July 20, 1850): 242; *Moore's* 12 (Mar. 9, 1861): 78; *Moore's* 13 (May 17, 1862): 159; CG 7 (Mar. 6, 1856): 157; CG 14 (July 14, 1859): 28; CG 17 (Mar. 21, 1861): 194; GF 7 (Feb. 1846): 27; AA 22 (June 1863): 165; AA 9 (Nov. 1849): 331.

33. BRH, May 21, 1863; see also references in AC 4 (May 1837): 54; AC n.s. 2 (July 1845): 225.

34. UWH, Aug. 26, 1873.

35. NYSAS 1851: 262–63.

36. Sally McMurry, "Women's Work in Agriculture: Divergent Trends in England and America, 1800–1930," *Comparative Studies in Society and History* 34 (Apr. 1992): 248–70.

37. Frank Kosikowski, "Cheese," *Scientific American* 252 (May 1985): 99; Joseph Harris, "Butter and Cheese Making," *Rural Affairs* 1856:242–55; "Rules for Making Cheese," *Rural Affairs* 1863:262; "A Day at the Dairy," *CG* 18 (Sept. 26, 1861): 207.

38. *Rural Affairs* 1863:262; *CG* 17 (Jan. 31, 1861): 79.

39. An example of milk quality varying with diet is that the milk's Vitamin A content is higher when cows are eating fresh grass than when they are eating hay or silage. Thanks to Dorothy Blair, Pennsylvania State Univ. nutrition department, for this information.

40. *GF* 22 (Dec. 1861): 364; *GF* 12 (Mar. 1851): 62; *GF* 22 (Feb. 1861): 55; *Rural Affairs* 1863:262; *Rural Affairs* 1856:246.

41. *Rural Affairs* 1863:263; *GF* 22 (Feb. 1861): 56; *GF* 22 (Dec. 1861): 364; *UWH*, Aug. 26, 1873; Fish, "Management of Cheese Dairies."

42. *CG* 20 (Nov. 27, 1862): 351; *Rural Affairs* 1863:263–64; "Cheesemaking in Herkimer County," *CG* 18 (Dec. 12, 1861): 383; *GF* 13 (May 1852): 145. Some of these explanations were probably incorrect. "Heaving," for example, probably occurred not because too much rennet was added, but because carbon dioxide–producing bacteria somehow made their way into the cheese; sometimes these bacteria can come from the cow's feed. Hard, dry, or cracked cheese resulted from insufficient humidity during curing. Leaky cheese usually had its origin early in the cheesemaking process, often occurring because the curd was insufficiently drained. Thanks to Manfred Kroger, Department of Food Science, Pennsylvania State Univ., for help with this information.

43. Arr, *New England Bygones*, 126; *CG* 14 (July 7, 1859): 19; *GF* 15 (Aug. 1854): 277–80.

44. Arr, *New England Bygones*, 126; "Cheese and Cheese Making," *CG* 17 (May 30, 1861): 54; *CG* 18 (Nov. 7, 1861): 301.

45. Mrs. M. E. Stephenson, "Experience in Cheesemaking," *AA* 20 (July 1861): 206; see also *GF* 4 (Jan. 1844): 27.

46. "Jefferson County Dairy Farming," *AA* 9 (Nov. 1849): 331–32; see also "A Farmer's Wife on Cheesemaking," *AA* 15 (June 1856): 204–5.

47. Nicholas Gardiner inventory.

48. Arr, *New England Bygones*, 124.

49. Nicholas Gardiner inventory. For descriptions of old and new tools, see *AC* 4 (July 1837): 87; *CG* 18 (Aug. 15, 1861): 111; *CG* 18 (Sept. 26, 1861): 207; *CG* 18 (Nov. 28, 1861): 354; *GF* 22 (Feb. 1861): 55; *NYSAS* 1845:125; *NEF* 9 (Aug. 1857): 367; *GF* 13 (June 1852): 177; *NYSAS* 1855:45.

50. *NEF* 14 (July 14, 1833): 89; *NEF* 15 (Mar. 22, 1837): 290–91; *NEF* 6 (May 30, 1828): 356, 362–63; *AA* 7 (July 1848): 201; *CG* 14 (July 14, 1859): 28; *GF* 17 (May 1857): 151; *NYSAS* 1858:135; "A Farmer's Wife on Cheesemaking," *AA* 15 (June 1856): 204.

51. "Verses on Maria Hocrij," rpt. in *Notions of the Americans*, ed. David Grimsted (New York: George Braziller, 1970), 155–56. The original broadside is in the Houghton Library, Harvard Univ.

52. Phelan, *White Vest*, entry dated Sept. 29, 1860; *BRH*, Mar. 12, 1856; see also *AC* 3d ser. 6 (Dec. 1849): 366; *GF* 12 (Mar. 1851): 62.

53. *Moore's* 13 (Jan. 4, 1862): 6; *AC* n.s. 5 (Mar. 1848): 96; *AC* n.s. 7 (Mar. 1850): 104. For patent designs, see U.S. Patent Office *Annual Report* (1861), vol. 1, pt. 1, pp.

316 (patent no. 32,198) and 267 (patent no. 31,825); (1862), vol. 1, pp. 204–5 (patent no. 34,511) and 365 (patent no. 35,396).

54. *Moore's* 13 (Oct. 11, 1862): 326.

55. *CG* 7 (May 1, 1856): 284.

56. *AA* 20 (July 1861): 206.

57. *Moore's* 12 (Apr. 27, 1861): 133; *CG* 18 (Sept. 26, 1861): 207; *Moore's* 12 (June 22, 1861): 197; *GF* 22 (Feb. 1861): 55; *GF* 25 (Apr. 1864): 111; *CG* 18 (Aug. 15, 1861): 111; *Moore's* 10 (Mar. 19, 1859): 94.

58. Lauren S. Poese, "The History and Design of Specialized Architecture: Cheese Factories in Central New York" (M.A. thesis, Cornell Univ., 1985), 70.

59. *AC* 2 (Nov. 1835): 118.

60. *UWH*, Aug. 26, 1873.

61. An observer writing in *OC* 14 (1858): 215 noted that women used the vats regularly. Judith McGaw, "Women and the History of American Technology," *Signs*, 7 (summer 1982): 798–829; Jeanne Boydston, *Home and Work* (New York: Oxford Univ. Press, 1990); Corlann Bush, "The Barn Is His, the House Is Mine: Agricultural Technology and Sex Roles," in *Energy and Transport*, ed. George Daniels and Mark Rose (Beverly Hills: Sage, 1982), 235–59; Susan Armitage, "Farm Women and Technological Change" (paper presented at the Fifth Berkshire Conference, Vassar College, Poughkeepsie, N.Y., June 16–18, 1981). See also Ruth Schwartz Cowan, *More Work for Mother: The Ironies of Household Technology from the Open Hearth to the Microwave* (New York: Basic Books, 1983).

62. Stephen Innes, ed., *Work and Labor in Early America* (Chapel Hill: Univ. of North Carolina Press, 1988), suggests that there were qualitative differences between men's and women's work in the nineteenth century. Men's work, he argues, was commercial, solitary, separated from leisure, and rationalized in terms of costs and benefits, while women's work was noncommercial, communal, often interrupted, and lacking in cost-benefit calculations. The allocation and nature of labor in antebellum home cheesemaking certainly does not fit this pattern, at least to the extent that both men's and women's work was commercial and alternatively done alone or in groups.

63. "Butter and Cheese," *Harper's New Monthly Magazine* (Nov. 1875): 814; *Prairie Farmer* n.s. 19 (Apr. 20, 1869): 258.

64. Lavinia Mary Johnson memoir, 50.

65. *OC* 7 (1851): 151. As we have seen, Maria Whitford too had a "cheese room" in her house in Almond, N.Y. See also *NYSAS* 1854:473, for a plan in which manufacturing and curing were separated; *Michigan Farmer*, June 1847, p. 46, for a plank cheese room, clapboarded, with no windows; *NYSAS* 1858:102, for a cheese room measuring 44 by 24 feet, with two fifteen-light windows in each side and ends.

66. Thanks to Mrs. Hopson, Salisbury Historical Society, Salisbury Center, N.Y., to Mrs. Dorothea Ives, Dolgeville, N.Y., and to James Huxtable, West Winfield, N.Y., for sharing information about these buildings.

67. Evans, *Dairyman's Manual*, 75. See also the description and plan of Moses Wheeler's cheese house, *NYSAS* 1844:227, and the "Cheese Dairy House" in Lewis Falley Allen's *Rural Architecture* (New York, 1852), 331.

68. *Michigan Farmer*, June 30, 1860, p. 206.

69. "Butter and Cheese," *Harper's*, 814.

Notes to Pages 96–106

70. *Michigan Farmer,* Oct. 1846, p. 111.
71. Allen, *Rural Architecture,* 330.
72. John Mack Faragher, "History from the Inside-Out: Writing the History of Women in Rural America," *American Quarterly* 33 (winter 1981): 555.
73. Osterud, *Bonds of Community,* 13.
74. Mary Blewett, *Men, Women, and Work: Class, Gender, and Protest in the New England Shoe Industry, 1780–1910* (Urbana: Univ. of Illinois Press, 1988).
75. Dublin, "Women and Outwork," 65.
76. Wilentz, *Chants Democratic;* Hahn, *Roots of Southern Populism;* Pete Daniel, *Breaking the Land: The Transformation of Cotton, Tobacco, and Rice Cultures since 1880* (Urbana: Univ. of Illinois, 1985).

Chapter Five: The Social Dynamics of Household Dairying

1. Bidwell, "Agricultural Revolution," 695.
2. Clark, *Roots,* 38, 55. Vickers, "Competency and Competition," states that "family members shared a commonly recognized interest in the achievement of household competency, and so the frank confrontations of power that were not usually acceptable between households were quite appropriate within them" (p. 23). While there is ample evidence that intrafamily conflicts were frequent and important stimuli to change, Vickers's explanation for them errs in assuming a common goal; if indeed this were the case, one might expect to find less conflict rather than more.
3. Arr, *New England Bygones,* 135; "The Farmer's Daughter," *NEF* 16 (Feb. 21, 1838): 264. For a suggestion that farm wives too were expected to defer to their husbands, see *NEF* 6 (Oct. 1854): 456–57.
4. Deborah Fink, *Agrarian Women: Wives and Mothers in Rural Nebraska, 1880–1940* (Chapel Hill: Univ. of North Carolina Press, 1992); Nancy Folbre, "Patriarchy in Colonial New England," *Review of Radical Political Economics* 12 (summer 1980): 4–13; see also articles by Sonya Salamon and others in *Women and Farming,* ed. Haney. Heidi Hartmann, "The Family as the Locus of Gender, Class, and Political Struggle: The Example of Housework," *Signs* 6 (1981): 366–94, explicates the general argument well. See also Osterud, *Bonds of Community,* 8.
5. Carl Degler, *At Odds: Women and the Family in America from the Revolution to the Present* (New York: Oxford Univ. Press, 1980), 144, argues that women challenged the family more in the nineteenth century than they did earlier.
6. *GF* 18 (Mar. 1857): 90.
7. Ibid.
8. *GF* 18 (July 1857): 208–9.
9. Ibid., 209; *GF* 18 (May 1857): 164.
10. For a thoughtful analysis of how men and women assessed their work within this dual system, see Nancy Grey Osterud, "The Valuation of Women's Work: Gender and the Market in a Dairy Farming Community during the Late Nineteenth Century," *Frontiers* 10 (1988): 18–24.
11. *GF* 18 (July 1857): 209.
12. *GF* 18 (May 1857): 164; *GF* 18 (July 1857): 208; Lavinia Mary Johnson memoir, 9.

13. *GF* 18 (July 1857): 210; *OC* 16 (Apr. 1, 1860): 111.
14. *GF* 18 (July 1857): 209–10; *GF* 18 (Mar. 1857): 90.
15. *NEF* 20 (Aug. 25, 1841): 61.
16. Ibid.
17. *AC* 6 (Dec. 1839): 203.
18. *Prairie Farmer* 10 (June 1850): 199.
19. *Rome Sentinel*, Sept. 3, 1851.
20. *Moore's* 12 (Apr. 27, 1861): 133; *NYSAS* 1847:721; *Ohio Farmer* 18 (Jan. 9, 1869): 18; Mrs. Eloise Abbott, *Personal Sketches and Recollections* (Boston, 1861), 31. See also "Farmers' Wives and Daughters," *Prairie Farmer* 9 (Dec. 1849): 366; "To Housekeepers," *Michigan Farmer*, July 1856, p. 214; "From a Farmer's Daughter," *Michigan Farmer*, June 1854, pp. 180–81. For a statement that pastoral farming involved less labor than arable agriculture, see Clark, *Roots*, 78, 99.
21. Mrs. M. B. Bateham, "At Home Again," *OC* 4 (July 15, 1848): 111; anonymous diary, Jan. 19, 1868, J. G. Parkhurst Collection.
22. Phelan, *White Vest*, 152–53.
23. *AA* 18 (July 1859): 201.
24. *AA* 11 (Oct. 12, 1853): 54.
25. *OC* 6 (Jan. 1, 1850): 29.
26. *NEF* 24 (Dec. 10, 1845): 187.
27. Blewett, *Men, Women, and Work*, 21–22, interprets this same story from her perspective as a historian of the New England shoe industry. She stresses the reality of women's comparative idleness and argues that outwork (among other occupations) substituted for farm work. In central New York similar substitution took place, but the timing was different: dairying substituted for home textile manufacture, then later, teaching and other employments substituted for home employment.
28. *GF* 9 (Apr. 1848): 115; *NEF* 2 (Aug. 31, 1850): 284; see also *AC* n.s. 3 (Apr. 1842): 65; *AC* n.s. 6 (Feb. 1849): 45.
29. "Farmers among Fashionable Young Ladies," *Michigan Farmer*, July 1853, p. 216; *Prairie Farmer* 1 (Jan. 1841): 8; *AC* n.s. 3 (Apr. 1842): 65.
30. Phillip Greven, *Four Generations: Population, Land, and Family in Colonial Massachusetts* (Ithaca: Cornell Univ. Press, 1970), 95, 133; Robert A. Gross, *The Minutemen and Their World* (New York: Hill & Wang, 1976), chap. 4; Clark, *Roots*, 129. For sons, diminished inheritance prospects had long acted as a solvent of family and social ties, motivating young men to migrate westward or cityward. Daughters, because of legal and cultural restrictions placed upon women, had not been able to seek independence as their brothers had. For the complexity of women's relationship to the land, see Nancy Grey Osterud, "Land, Identity, and Agency in the Oral Autobiographies of Farm Women," in *Women and Farming*, ed. Haney, 73–88.
31. For an example, see "Farmers' Daughters," *Michigan Farmer*, July 1854, pp. 211–13.
32. *GF* 1 (Dec. 1840): 185.
33. *GF* 9 (Nov. 1848): 283; see also *Michigan Farmer*, Jan. 19, 1859, p. 38.
34. *GF* 9 (Nov. 1848): 283.
35. Faragher, "History from the Inside-Out," 552.

Notes to Pages 113–15

36. Phelan, *White Vest*, 141, 12, 33, 39, 72.
37. *GF* 2 (Feb. 1841): 31.
38. See Dublin, "Rural Putting-out Work," 571.
39. Jensen, *Loosening the Bonds;* See also Nancy Folbre, "The Wealth of Patriarchs: Deerfield, Massachusetts, 1760–1840," *Journal of Interdisciplinary History* 16 (autumn 1985): 199–220, and "Patriarchy in Colonial New England." Folbre argues that economic development was instrumental in loosening fathers' control over sons. The same appears to be occurring for family control over women as new opportunities develop for them. Laurel Ulrich, *A Midwife's Tale* (New York: Knopf, 1990), writes that similar forces appear to have been at work during the home textile-producing era. Martha Ballard's "girls" exploited their weaving skills to leave home and marry early. Marriage opportunities for nineteenth-century dairying women were probably not as great as they had been late in the eighteenth century, but this may have been offset by the emergence of other employment possibilities. Ulrich also notes that this intrafamily dynamic could contribute to perceptions of drudgery; Martha Ballard found her work more taxing when her supply of "help" became undependable. Also, Jonathan Prude, *The Coming of Industrial Order: Town and Factory Life in Rural Massachusetts, 1826–1860* (Ithaca: Cornell Univ. Press, 1983), 101, notes that off-farm opportunities helped farm sons and daughters seek independence: they "may well have experienced an unprecedentedly vivid sense of autonomy." Dublin, "Putting-out Work," 567–71, finds the same for hatmaking.

40. Arr, *New England Bygones*, 138; *GF* 1 (July 1840): 109. Simon Brown, editor of the *New England Farmer,* agreed. In his travels through the "farming districts of New England" he found young women who were "well educated, and read papers and current books, and are often acquainted with classical literature; they are quick to catch the manners of living as they rise. . . . Honitons . . . and crinolines rustle on the hills or by their crystal streams with more unaffected grace than on the dusty paves of Boston." Despite their attainments, these young women were justifiably dissatisfied, because as one of them put it, the farm "lack[s] society; it is a life of confinement; the dairy demands constant attention and hard work, and the hired men are sometimes difficult to please, after we have labored carefully to provide for them." *NEF* 10 (Sept. 1858): 321. For an example of a moderate view, see "To the Young Ladies of Michigan," *Michigan Farmer,* Nov. 15, 1843, p. 147.

41. Thomas Woody, *A History of Women's Education in the United States,* 2 vols. (New York: Science Press, 1929); Maris Vinovskis, "Have We Underestimated the Extent of Antebellum High School Attendance?" *History of Education Quarterly* 28 (winter 1988): 551–67; Geraldine Jonçich Clifford, "Home and School in Nineteenth Century America: Some Personal-History Reports from the United States," *History of Education Quarterly* 18 (spring 1978): 3–35; Sally Schwager, "Educating Women in America," *Signs* 12 (winter 1987): 333–72, esp. 341.

42. Moses M. Bagg, *Memorial History of Utica* (Syracuse, 1892), 99; Wager, *Our County,* 276.

43. School catalogs for the Clinton Liberal Institute, Home Cottage Seminary, Houghton Seminary, Domestic Seminary, and Delancey Institute are among the holdings of the Clinton Historical Society and the Hamilton College Library, Clinton, N.Y.,

Notes to Pages 115–19

and the Oneida Historical Society, Utica. Notices also appeared frequently in the local newspapers; see, for example, *Rome Sentinel*, Apr. 24, 1850; *BRH*, Mar. 26, Dec. 31, 1857, Apr. 13 and 20, 1855.

44. School catalogs, Clinton Historical Society, Hamilton College Library, and Oneida Historical Society; Oneida County directory (1859–60); *BRH*, Mar. 26, 1857; U.S. census, manuscript schedules, Agriculture and Population (1850, 1860). On coeducation in the Rome school, see Martha Maclear, *The History of the Education of Girls in New York and in New England, 1800–1870*, Howard Univ. Studies in History, no. 7 (Washington, D.C.: Howard Univ. Press, 1926), 39.

45. Maclear, *Education of Girls*, 42.

46. U.S. census, manuscript schedules, Agriculture and Population, seven towns (1850); *GF* 1 (July 1840): 109.

47. Harriet A. Jacobs, *Incidents in the Life of a Slave Girl*, ed. Jean Yellin (Cambridge: Harvard Univ. Press, 1987), 189, 288 n. 2; school catalogs, Clinton Historical Society, Hamilton College Library, Oneida Historical Society.

48. Mary Root diary, Mar. 1 and 6, Apr. 13, May 2 and 7, Sept. 25, 1860, Utica College.

49. Sarah Burgess diary.

50. Unsigned, undated letter, about 1860, Barnes family papers, Oneida Historical Society.

51. Anne Firor Scott, "The Ever Widening Circle: The Diffusion of Feminist Values from the Troy Female Seminary, 1822–1872," *History of Education Quarterly* 19 (spring 1979): 3–27; David Allmendinger, "Mount Holyoke Students Encounter the Need for Life-Planning, 1837–1850," *History of Education Quarterly* 19 (spring 1979): 27–46.

52. Nancy F. Cott, "Young Women in the Second Great Awakening in New England," *Feminist Studies* 3 (1975): 15–29; Linda Kerber, "Women and Individualism in American History," *Massachusetts Review* 30 (winter 1989): 589–609.

53. School catalogs, Clinton Historical Society; Sophia Holmes to Cornelia Babcock, Aug. 24, 1859, S. M. Babcock papers, box 1; Arthur C. Hackley diary, May 25, 1861, New York State Historical Association; "Education of Farmers' Children," *GF* 2 (May 1841): 76–77.

54. David P. Davenport, "Migration to Albany, New York, 1850–1855," *Social Science History* 13 (summer 1989): 159–85; John Modell, "Mobility and Industrialization: Countryside and City in Nineteenth-Century Rhode Island," in *Essays from the Lowell Conference*, ed. Robert Weible, Oliver Ford, and Paul Marion (Lowell, Mass.: Lowell Conference on Industrial History, 1981), 96–97.

55. David Jaffee, "One of the Primitive Sort: Portrait Makers of the Rural North, 1760–1860," in *Countryside*, ed. Hahn and Prude, 130; Jaffee, "The Village Enlightenment in New England, 1760–1820," *William and Mary Quarterly* 3d ser. 37 (July 1990): 327–46. Jack Larkin and William Gilmore have added to Jaffee's portrait of a dynamic village society. Larkin, "The Merrims of Brookfield: Printing in the Economy and Culture of Rural Massachusetts in the Early Nineteenth Century," American Antiquarian Society *Proceedings* 96 (Apr. 1986): 39–73, shows that small-town printers operated a flourishing trade within customary exchange networks, facilitating a trade

Notes to Pages 119–23

in ideas as well. Gilmore, *Reading Becomes a Necessity of Life: Material and Cultural Life in Rural New England, 1780–1835* (Knoxville: Univ. of Tennessee Press, 1989), argues that the print media were crucial in the commercialization of New England culture, and finds a measure of "intellectual emancipation" (p. 47) in the outcome. In New York State too the proliferation of schools is evidence of a similar process at work.

56. Phelan, *White Vest,* 19–23, 127.

57. GF 21 (July 1861): 216.

58. Houghton Seminary alumni newsletters, 1870–90, Clinton Historical Society; Lavinia Mary Johnson memoir, 142, 144. Thomas Dublin found that Lowell factory workers followed the same pattern. Dublin, *Women at Work: The Transformation of Work and Community in Lowell, Massachusetts, 1826–1860* (New York: Columbia Univ. Press, 1979).

59. Joan Jacobs Brumberg, *Mission for Life* (New York: Free Press, 1980). Cross, *Burned-over District,* suggests that women participated in evangelical revivals at a moment when they had new, but limited, life-choices. He associates participation with a settled agrarian economy, arguing that frontier women were too busy and urban women too distracted with secular concerns. He also associates revivalism with the seminary movement and with a rising age at marriage. I have found very little evidence of intense involvement in evangelical revivals among dairying families in Oneida County. It is quite possible that people simply did not record their conversion experiences, but in view of the extraordinary work load of cheesemaking, it is more likely that dairying women did not have time for extensive involvement in religious activity.

60. Houghton Seminary alumni newsletters, 1870–90.

61. Richard Bernard and Maris Vinovskis, "The Female School Teacher in Antebellum Massachusetts," *Journal of Social History* 10 (Mar. 1977): 332–45; Clifford, "Home and School"; Allmendinger, "Mount Holyoke Students," 40.

62. Rosetta Hammond Bushnell diary, Aug. 3 and 9, 1857, Sept. 6–26, Feb. 26, 1858.

63. Bernard and Vinovskis, "Female School Teacher"; Allmendinger, "Mount Holyoke Students."

64. Kathryn M. Kerns, "Farmers' Daughters: The Education of Women at Alfred Academy and University before the Civil War," *History of Higher Education Annual* 1986:20, 21, 24.

65. Osterud, *Bonds of Community;* Jensen, *Loosening the Bonds;* Arr, *New England Bygones,* pp. 136–38. Fink, *Agrarian Women,* 10, 129, suggests that Nebraska farm women encouraged their daughters to get out of farming if they could.

66. Henretta, "Families and Farms."

Chapter Six: The Rise of the Factory

1. The folklore surrounding this incident has been repeated but not verified in primary documents; it is an oral tradition, and the versions vary. Nonetheless, all of the printed versions pinpoint a family event as the trigger for innovation. See Frederick Rahmer, *Jesse Williams, Cheesemaker* (Holland Patent, N.Y.: Steffen, 1971), 8;

Notes to Pages 124–27

"Butter and Cheese," *Harper's*, 814; Rome Cheese Manufacturing Association papers, Cornell Univ.; J. P. Sheldon, *Dairy Farming*, 466.

2. Hahn, *Roots of Southern Populism*; Daniel, *Breaking the Land*; David Danbom, *The Resisted Revolution: Urban America and the Industrialization of Agriculture* (Ames: Iowa State Univ. Press, 1979); Danbom's portrait was more sympathetic than that of Earl Hayter, in *The Troubled Farmer, 1850–1900* (DeKalb: Northern Illinois Univ. Press, 1968), who maintained that the farmers' resistance to industrialization arose from their narrow, anti-intellectual, traditional outlook.

3. Faragher, *Sugar Creek*; Barron, *Those Who Stayed Behind*; Clark, *Roots*; Pederson, *Between Memory and Reality*; Paula Baker, *The Moral Frameworks of Public Life: Gender, Politics, and the State in Rural New York, 1870–1930* (New York: Oxford Univ. Press, 1991).

4. Fred Shannon, *Farmer's Last Frontier* (1945; rpt. White Plains, N.Y.: M. E. Sharpe, n.d.), 245–55; Ellis, *Landlords and Farmers*; Eric Lampard, *The Rise of the Dairy Industry in Wisconsin: A Study in Agricultural Change, 1820–1920* (Madison: Univ. of Wisconsin Press, 1963), 109; Paul Wallace Gates, *Agriculture and the Civil War* (New York: Knopf, 1965); Cohen, *Women's Work*, 91.

5. Clark, *Roots*.

6. Ryan, *Cradle*, 10; M. M. Bagg, "The Earliest Factories of Oneida," Oneida Historical Society *Transactions* (1879).

7. "How Swiss Cheeses Are Made," *Moore's* 12 (July 13, 1861): 222; see also Arnold, *American Dairying*, 14. Rome Cheese Manufacturing Association papers, Cornell Univ.; *CG* 16 (Nov. 29, 1860): 351; *Moore's* 11 (Nov. 17, 1860): 366; see also *Farmer's Cabinet* 7 (July 15 1843): 378; "Claims of American Dairy Inventors Vindicated," *Moore's* 26 (Nov. 23, 1872): 331.

8. Joan Thirsk, ed., *The Agrarian History of England and Wales*, vol. 5, pt. 1, *1640–1750: Regional Farming Systems* (Cambridge: Cambridge Univ. Press, 1984), 153; *Moore's* 17 (July 21, 1866): 230.

9. Poese, "History and Design," 9.

10. *Moore's* 17 (Aug. 4, 1866): 246; Sarah McMahon, "A Comfortable Subsistence: A History of Diet in New England, 1630–1850" (Ph.D. diss., Brandeis Univ., 1982), 99–100. See also *AA* 21 (Sept. 1862): 262; *AA* 21 (Nov. 1862): 367; Clark, *Roots*, 87.

11. Sheldon, *Dairy Farming*, 465; J. L. Sammis, *Cheese Making* (Madison: Cheese Maker Book Co., 1930), 110; J. G. Pickett, "Pioneer Dairying," Wisconsin Dairymen's Association *Annual Report* (1878), 95–99.

12. *Moore's* 36 (May 19, 1877): 311–12; examples of detailed contracts under the three-fifths system can be found in the contract between the Babcock brothers and Henry Vosburgh dated Apr. 1, 1868, S. M. Babcock papers, box 1. See also *CG* 21 (Apr. 30, 1863): 285; *NYSAS* 1858:135–37; *Moore's* 20 (Nov. 29, 1869): 746.

13. *NYSAS* 1847:264–68.

14. *Moore's* 11 (Feb. 18, 1860): 54. See also *CG* 16 (Dec. 20, 1860): 396; *AA* 23 (Jan. 1864): 15; *NEF* 12 (Nov. 1860): 509; Jones, *Agriculture in Ohio*, 192–94.

15. Rahmer, *Jesse Williams*; T. C. Peters, "Cheese Factories," *Moore's* 13 (Nov. 8, 1862): 358.

Notes to Pages 128–32

16. *Moore's* 12 (Aug. 3, 1861): 246; New York State Cheese Manufacturers Association Annual Meeting, 1864, reports of individual cheese factories; *NYSAS* 1866:452; *NEF* n.s. 4 (Sept. 1870): 405, 424; *NEF* n.s. 4 (June 1870): 291; *NEF* n.s. 4 (Oct. 1870): 482–83. Lampard, *Rise,* 88, 92–93, and H. E. Erdman, "The Associated Dairies of New York as Precursors of American Agricultural Cooperatives," *Ag History* 36 (Apr. 1962): 82–91, also recognize long-term precedents to factories.

17. Willard, *Practical Dairy Husbandry,* 362–67; *BRH,* Feb. 24 and May 19, 1870.

18. For a summary of advantages of the factory system, see Lampard, *Rise,* 97.

19. McLane Report, *Documents Relative to the Manufactures in the United States,* pp. 38–39.

20. Poese, "History and Design," 24.

21. Brunger, "Changes"; R. Andreano, ed., *The Economic Impact of the American Civil War,* 2d ed. (Cambridge, Mass.: Shenkman, 1967), 4–22; Walter T. K. Nugent, *Money and American Society, 1865–1880* (New York: Free Press, 1968), 9; Rothstein, "International Market."

22. Phillip Paludan, *"A People's Contest": The Union and Civil War, 1861–1865* (New York: Harper & Row, 1988), 159.

23. Margery Davies, *Woman's Work Is at the Typewriter* (Philadelphia: Temple Univ. Press, 1982); Emerson D. Fite, *Social and Industrial Conditions in the North during the Civil War* (New York: Peter Smith, 1930), 187, 246.

24. Abijah Barnes to his parents, Feb. 18, Mar. 6, 1864, and Mary to Abijah, May 19, 1864, Barnes family papers.

25. New York State census, published summary (1865). For the seven towns, about 835 served and fully a quarter of them died in the war. For the county as a whole, over two hundred returned permanently disabled, probably an underestimate. For a sensitive portrayal of several rural women's experiences on the home front, see Nancy Grey Osterud, "Rural Women during the Civil War: New York's Nanticoke Valley, 1861–1865," *New York History* 71 (Oct. 1990): 357–85.

26. Gates, *Agriculture and the Civil War,* 233; Gates, "Agricultural Change in New York State, 1850–1890," *New York History* 50 (Apr. 1969), 117, 126.

27. Wayne Rasmussen, "The Civil War: A Catalyst of Agricultural Revolution," *Ag History* 39 (Oct. 1965): 187–94; *BRH,* Aug. 6, 1863; *Oneida Weekly Herald,* Mar. 31, 1863. In some cases women operated the mowers; see George Winston Smith and Charles Judah, *Life in the North during the Civil War* (Albuquerque: Univ. of New Mexico Press, 1966), 166–67. By 1870 the mower was mentioned as an item "usually found" in barns. *Roman Sentinel,* Feb. 1, 1870. Extra help was still required for gathering and storing the cut hay, and for scythe work on hilly, steep, or irregular patches, but labor requirements were drastically reduced.

28. *NEF* 15 (May 1863): 161; *AA* 23 (July 1864): 206; *AA* 21 (Oct. 1862): 301; *AA* 22 (June 1863): 165; *AA* 24 (Feb. 1865): 46; *AA* 21 (Apr. 1862): 119; Osterud, "Rural Women during the Civil War." Indeed, the "helping" that women did in the fields during wartime was different in degree, perhaps, but not in kind from the work they ordinarily did; farm wives had always helped the men when they were needed, though the volume of such work was not always accurately perceived. See Osterud, "She Helped Me Hay It as Good as a Man."

29. "War, Work, and Woman," *Moore's* 13 (Oct. 4, 1862): 317; "Women Farmers," *Moore's* 13 (Nov. 15, 1862): 366; *Moore's* 14 (May 23, 1863): 165; *Moore's* 14 (July 11, 1863): 222.

30. *Moore's* 14 (Mar. 7, 1863): 77.

31. *BRH*, Feb. 3, 1863; see also *BRH*, Feb. 5, 1863. Lori Ginzberg, *Women and the Work of Benevolence* (New Haven: Yale Univ. Press, 1990), sets out some implications of war work for women involved in reform. It is not clear how, at the local level in rural areas, the newly rationalized organization of benevolent work affected women's outlook. It seems entirely possible that participants acquired new perspectives from their contacts with national networks; the new activities and preoccupations of Oneida County rural women in the postwar decades suggest such a possibility.

32. New York State census, published summary (1865); Brunger, "Changes," 55.

33. Frank Metcalf to his parents, Oct. 15, 1863, Metcalf family papers, Oneida Historical Society. Abijah Barnes, in a letter to his parents dated Apr. 1, 1864, Barnes family papers, also urged his mother to see to her own health first.

34. Charles L. Flint, *Milch Cows and Dairy Farming*, rev. ed. (Boston, 1889), 369; Willard, *Practical Dairy Husbandry*, 214–27; Brunger, "Changes," 60; *CG* 27 (Dec. 1865): 401. The figures for output and for the numbers of cheese factories are not consistent; change occurred so rapidly, and reporting to the state Cheese Manufacturers' Association and to the census was so incomplete, that a fully accurate account is impossible. The federal census figures for 1870, however, show that a decisive change had occurred in the previous decade.

35. New York State census, published summaries (1865, 1875).

36. Brunger, "Changes," 88.

37. *Moore's* 23 (Feb. 25, 1871): 126; *Moore's* 32 (Aug. 21, 1875): 123; *Moore's* 20 (Nov. 20, 1869): 746; *Moore's* 20 (Nov. 27, 1869): 762; *NEF* n.s. 5 (Feb. 1871): 133; X. A. Willard, "English and American Dairying," U.S. Department of Agriculture *Annual Report* (1866), 367–72; *NYSAS* 1864:232–38; *NEF* n.s. 5 (Jan. 1871): 43; *CG* 34 (Aug. 12, 1869): 121; *CG* 42 (Nov. 22, 1877): 751; *CG* 38 (July 17, 1873): 459; New York State Cheesemakers Association *Annual Report* (1864), 34–37. *CG* 23 (June 2, 1864): 353 noted that better storage in steamships reduced spoilage.

38. *AA* 37 (July 1878): 258; T. R. Pirtle, *History of the Dairy Industry* (1926; rpt. Chicago: Mojonnier, 1973), 139–40; the census of Great Britain for 1925 estimated that in Britain that year 2.84 million hundredweight of cheese had been imported to the country, and only 7,500 hundredweight exported. On the relative quality of American and British cheeses, see *Moore's* 15 (Oct. 1, 1864): 318. On the decline of English cheesemaking, see David Taylor, "English Dairy Farming" (Ph.D. diss., Oxford Univ., 1971); C. S. Orwin and E. Whetham, *History of British Agriculture, 1846–1914*, 2d ed. (Newton Abbot: David and Charles, 1971); P. J. Perry, *British Farming in the Great Depression, 1870–1914* (Newton Abbott: David & Charles, 1974).

39. *Moore's* 29 (May 23, 1874): 331; *Moore's* 38 (Sept. 27, 1879): 623; "Over Production of Cheese," *CG* 40 (Nov. 11, 1875): 712–13; "Our Surplus Cheese," *CG* 43 (Nov. 28, 1878): 763–64; "Over Production of Cheese," *Ohio Farmer* 16 (June 6, 1868): 358; "Is the Factory System a Failure?" *Moore's* 38 (Sept. 28, 1878): 623; H. Stewart, column in *Moore's* 38 (Oct. 4, 1878): 643.

Notes to Pages 135–42

40. *Moore's* 17 (Dec. 15, 1866): 398; *NEF* n.s. 1 (Sept. 1867), 419; Willard, *Practical Dairy Husbandry*, 233; *Moore's* 23 (Apr. 22, 1871): 254. Lampard, *Rise*, 106, suggests that Wisconsin farmers might have found it more profitable simply to send milk to the factory, since the capital outlay for setting up domestic cheese production was substantial. This might not have obtained to the same degree in New York State, where domestic production was already well established.

41. *Ohio Farmer* 15 (Mar. 3, 1866): 67. On the value of time lost in hauling, see also *GF* 26 (Feb. 1865): 44; *Moore's* 21 (June 14, 1870): 366; *CG* 39 (Apr. 2, 1874): 217; *AA* 27 (Feb. 1868): 58; *CG* 43 (Mar. 28, 1878): 203; *CG* 37 (July 18, 1872): 459; *Moore's* 26 (Aug. 10, 1872): 91; "Is the Factory System a Failure?" *Moore's* 38 (Sept. 27, 1878): 623. T. C. Peters, "Cheese Factories," *Moore's* 13 (Nov. 8, 1862): 358, reckoned that all things considered, the costs were the same to the farmer either way. Another disadvantage of factory production was the short season; farmers often would have to produce at home when the factories were closed. See Lampard, *Rise*, 98.

42. *Moore's* 21 (Mar. 12, 1870): 157.

43. *NYSAS* 1864:242; "Cost of Cheese," *NEF* n.s. 2 (Apr. 1868): 168.

44. Factory list, Gang Mills cheese factory, New York State Historical Association. Mary Neth, "Preserving the Family Farm: Farm Families and Communities in the Midwest, 1900–1940" (Ph.D. diss., Univ. of Wisconsin, 1987), 186, also notes that cooperatives in the Midwest grew out of neighborhood networks. Dun and Co. credit records for a cheese factory in Marshall, Oneida County (1870), 473:58, noted that most of the patrons were Welsh.

45. Faragher, *Sugar Creek*, 118; Faragher, "History from the Inside-Out"; Fink, *Agrarian Women;* Neth, "Preserving the Family Farm"; Osterud, "Valuation of Women's Work." For a more general survey of historians' evaluations of the impact on women of capitalism and industrialism, see Janet Thomas, "Women and Capitalism: Oppression or Emancipation?" *Comparative Studies in Society and History* 30 (1988): 534–48.

46. *CG* 26 (Aug. 31, 1865): 144–45; *Moore's* 20 (Feb. 20, 1869): 123.

47. *AA* 27 (Feb. 1868): 58; *NYSAS* 1862:238–41; "Spinning Wheel and Cheese Press," *AA* 23 (Jan. 1864): 5; *NYSAS* 1864:227; see also *Moore's* 23 (Apr. 15, 1871): 238; *NEF* n.s. 5 (Mar. 1871): 114. The *Oneida Weekly Herald* editorial appeared Mar. 17, 1863, p. 5. For other examples of the rhetoric of how the factories would relieve drudgery, see *AA* 24 (Oct. 1865): 340; *CG* 36 (Sept. 1871): 615; *CG* 38 (June 19, 1873): 395; *GF* 24 (1863): 246; *GF* 22 (Feb. 1861): 55; *Moore's* 14 (Dec. 5, 1863): 398.

48. *CG* 31 (Apr. 9, 1868): 263; *NEF* n.s. 5 (Mar. 1871): 114.

49. Wisconsin Dairymen's Association *Annual Report* 3 (1875):17–20.

50. *NEF* 15 (May 1863): 141; *NEF* 16 (June 1864): 182–83. See also Sally McMurry, *Families and Farmhouses in Nineteenth-Century America* (New York: Oxford Univ. Press, 1988), chap. 4.

51. *OC* 18 (Nov. 1862): 343; "Make the Farm Beautiful," *Prairie Farmer* 20 (Aug. 24, 1867): 115. (It should be noted that this article was critical of women who chose sophistication over values of work, but the choice of an occupation to contrast with "worldly" pursuits may reveal something about how cheesemaking was regarded.) Hopkins's speech was reprinted in the *Rome Sentinel*, Sept. 27, 1870.

52. Donald Marti, *Women of the Grange: Mutuality and Sisterhood in Rural America, 1866–1920* (New York: Greenwood, 1991), notes that Grange women tended to express values very similar to those described here.

53. Fink, *Agrarian Women*, 94, argues that for isolated Nebraska farm women, companionate marriage was not a possibility.

54. *AA* 27 (June 1868): 212.

55. *AA* 37 (Feb. 1878): 53; see also *NYSAS* 1863:431. For the observation that farmers understood their wives' work, I am indebted to Osterud, "Strategies."

56. Osterud, "Strategies," 19–21; Boydston, *Home and Work*. Boydston has perceptively traced the shift in cultural and rhetorical perceptions of housework in the antebellum period. She notes that as the money economy grew in importance, men associated themselves with it and came to regard only paid work as "real" work. Similar forces appear to be operating here, though in a different context.

57. Martha Verbrugge, *Able-Bodied Womanhood: Personal Health and Social Change in Nineteenth-Century Boston* (New York: Oxford Univ. Press, 1988). Russett, *Sexual Science*, 126, discusses how late-nineteenth-century scientific thinkers associated women's physiology with vulnerability and decline.

58. Jensen, *Loosening the Bonds*.

59. Bengt Angkarloo, "Agriculture and Women's Work: Directions of Change in the West, 1700–1900," *Journal of Family History* 5 (summer 1979): 111–20; Lena Sommestad, "From Dairymaids to Dairymen" (paper presented at the Berkshire Conference, Douglass College, New Brunswick, N.J., June 9, 1990); Marjorie Griffin Cohen, "The Decline of Women in Canadian Dairying," *Social History* 17 (Nov. 1984): 307–34. Keith Snell, "Agricultural Seasonal Unemployment, the Standard of Living, and Women's Work in the South and East, 1690–1860," *Economic History Review* 34 (1981), 421–23, and Leonore Davidoff, "The Role of Gender in the 'First Industrial Nation': Agriculture in England, 1780–1850," in *Gender and Stratification*, ed. Rosemary Crompton and Michael Mann (Cambridge: Polity Press, 1986), 90–114, convincingly document a decrease in women's employment in arable farming in southern and eastern England early in the nineteenth century. Women were gradually excluded from the field work of harvesting, weeding, stone-picking, planting, and the like. Snell and Davidoff attribute the decline (as does Angkarloo) to the rise of large-scale specialized capitalist grain farming, which in the specific case of England was encouraged by enclosure and reinforced by the New Poor Law. For the poultry industry in more recent times (the 1940s), Fink argues in *Open Country Iowa* that male-dominated capitalist enterprises shut women out.

60. Clark, *Roots*, 134, has argued that women responded to their intensified work burden by reducing fertility and reorganizing household production. Lee Craig, *To Sow One Acre More: Childbearing and Farm Productivity in the Antebellum North* (Baltimore: Johns Hopkins Univ. Press, 1993), makes a similar argument. There is no way to test whether dairying women's fertility dropped, though neither is there any reason to suppose it did not. But clearly we can see a reorganization of household work taking place.

61. Kulikoff, "Transition to Capitalism," 139.

Notes to Pages 146–56

62. Nancy Grey Osterud, "Gender and the Transition to Capitalism in Rural America," *Ag History* 67 (spring 1993): 14–30.

63. Fink, *Agrarian Women*, 10.

Chapter Seven: The Social Organization of Factory Work

1. Hannah Day to Harvey Day, June 8, 1870, Harvey Day papers, Bentley Historical Collections, Univ. of Michigan.
2. Ibid., July 6, 1870.
3. Ibid., Sept. 6, 1870.
4. Ibid., July 6, 1870, July 25, 1871.
5. Ibid.
6. H. W. Smith to Harvey Day, May 10, June 12, 1871, Harvey Day papers.
7. See *UWH*, Jan. 24, 1865; H. W. Smith to Harvey Day, June 12, 1871, Harvey Day papers.
8. H. W. Smith to Harvey Day, Sept. 5, 1871, July 14, 1872, Harvey Day papers.
9. Ibid., Oct. 2, 1872, July 10, 1871.
10. Ibid., June 23, 1871, June 10, 1872.
11. Ella Comstock to Harvey Day, July 24, 1871, Sept. 3, 1872, Harvey Day papers.
12. H. W. Smith and Ella Comstock to Harvey Day, Oct. 2, Aug. 4, 1872, Aug. 1, June 29, 1871, Harvey Day papers.
13. Typescript poem, Harvey Day papers; Washtenaw County, Mich., census (1894). Thanks to Karen Jania, Bentley Historical Collections, for information on Day's wife.
14. Poese, "History and Design"; Willard, *Practical Dairy Husbandry*; *AA* 22 (Feb. 1863): 47.
15. "The American at Work: II, The Cheese-maker," *Appletons' Journal* 4 (Apr. 1878): 300; Beach Nichols, *Atlas of Herkimer County New York* (New York, 1868).
16. *Twelfth Annual Report of the American Dairymen's Association* (Ingersol, Ont., 1877), 154, quoted in Poese, "History and Design," 64; see also 63.
17. *NYSAS* 1862:243.
18. *Moore's* 12 (Aug. 3, 1861): 246.
19. *NYSAS* 1861:243; James Gibb, "Making Cheese: Organizational and Technological Strategies in a Nineteenth-Century Industry" (unpublished paper, n.d.), 16–17.
20. *Moore's* 23 (Feb. 18, 1871): 110; Poese, "History and Design," 88, notes that Gottlieb Merry's otherwise conventional timber-frame factory in Verona, N.Y., also had iron trusswork.
21. *Moore's* 36 (Feb. 10, 1877): 85–86.
22. *Moore's* 13 (Nov. 8, 1862): 358. Patent grants, from the U.S. Patent Office, show that inventors concentrated on the problem of heating the milk evenly, coming up with a variety of solutions, from pipes pierced with holes to fabric pieces intended to spread the heat evenly. Some also made an effort to design vats that could be easily

Notes to Pages 157–66

operated, with easily manipulable and accessible valves. The 1870 U.S. census of Manufactures, vol. 3, Wealth and Industries, p. 594, lists 818 cheese factories in which 160 steam engines and 1,966 waterwheels were in use.

23. *Moore's* 12 (Aug. 3, 1861): 246; *Moore's* 13 (Nov. 8, 1862): 358; *Moore's* 15 (July 23, 1864): 238; "American at Work," *Appletons'*, 301; Arnold, *American Dairying*, 352–53; Sheldon, *Dairy Farming*, 460, 475; Poese, "History and Design," 70; "American at Work," *Appletons'*, 303.

24. *BRH*, Feb. 5, 1863.

25. *Moore's* 12 (Aug. 3, 1861): 246 (steam engines); *NEF* n.s. 5 (Jan. 1871): 4 (steam-powered curd mill); *NYSAS* 1871:452 (hoister); *CG* 24 (July 7, 1864): 13 (pulleys); Willard, *Practical Dairy Husbandry*, 440, 453, 456; *Rural Affairs* 1864:88; *CG* 22 (July 20, 1863): 204; *CG* 25 (May 18, 1865): 319. Poese, "History and Design," 70, notes eighty-three cheesemaking patents from 1790 to 1860, three-quarters of which were for presses; three hundred from 1860 to 1900, sixty-five of which were for presses and fifty-one for vats.

26. Poese, "History and Design," 59.

27. Ibid., 72–150; *NEF* n.s. 4 (Oct. 1870): 453. Gary Kulik notes similar continuities in "A Factory System of Wood: Cultural and Technological Change in the Building of the First Cotton Mills," in *Material Culture of the Wooden Age*, ed. Brooke Hindle (Tarrytown, N.Y.: Sleepy Hollow Press, 1981), 300–337.

28. Willard, *Practical Dairy Husbandry*, 369–70, describes the Sanborn factory; the Willow Grove factory appears in Arnold, *American Dairying*, 308; useful general descriptions appear in *Moore's* 13 (Nov. 8, 1862): 358; *Moore's* 16 (July 23, 1864): 238; *Moore's* 23 (Feb. 18, 1871): 110; *CG* 21 (Jan. 1, 1863): 16–18; *CG* 24 (July 7, 1864): 13; *CG* 26 (July 20, 1865): 45; *AA* 22 (Feb. 1863): 47.

29. Daniel Haviland to Harvey Day, 2nd mo. 21st, 1871, Harvey Day papers; see also "Butter and Cheese," *Harper's*, 820; Sheldon, *Dairy Farming*, 477. Lynn Larkin's "West Lima's Blue Ribbon Cheesemaker," *Wisconsin Magazine of History*, autumn 1984, 43–53, is an oral history showing that family cooperation in cheesemaking continued into the 1930s; because factory work was a ten-hour, seven-day job, the maker's wife and children were trained to help.

30. U.S. census, manuscript schedules, Population and Agriculture (1870). See also U.S. census, manuscript schedules, Population, Boonville, p. 19, l. 29; p. 21, l. 35; p. 1, l. 38.

31. *AA* 27 (June 1868): 212.

32. *UWH*, June 24 and 17, 1873, Sept. 20, 1864; *Prairie Farmer* n.s. 18 (May 19, 1855): 340; Oneida County atlas (1878), town of Ava; *BRH*, Aug. 22, 1867; Utica city directory (1873), Whitesboro; Utica city directory (1869), Deerfield; *Rural American* 10 (Mar. 1, 1866): 67 (this couple was paid $897 for the year); *Prairie Farmer* n.s. 13 (Jan. 16, 1864): 35; *NEF* n.s. 4 (Sept. 1870): 405; *Moore's* 21 (Apr. 2, 1870): 222; "Butter and Cheese," *Harper's*, 823; *CG* 21 (Apr. 30, 1863): 285.

33. *BRH*, May 19, 1870; *Moore's* 12 (Aug. 2, 1861): 246; *Rome Sentinel*, Apr. 12, 1870; *UWH*, Jan. 4, 1876; "Butter and Cheese," *Harper's*, 820, 823; Nathan Huntington letter, Mar. 25, 1914, Mann Library, Cornell Univ.

34. "American at Work," *Appletons'*, 300.

Notes to Pages 166–72

35. "Butter and Cheese," *Harper's*, 816; *Rome Sentinel*, Apr. 12, 1870; *UWH*, Mar. 24, 1874.

36. *AA*, 33 (Apr. 1874): 137–38; *NEF* n.s. 4 (Jan. 1870): 453; *Moore's* 31 (Mar. 13, 1875): 171; *NEF* n.s. 4 (Jan. 1870): 30; see also *CG* 21 (Jan. 1, 1863): 18, which argued that larger farmers associated, and small ones supplied milk.

37. *Seventh Annual Report of the American Dairymen's Association* (Syracuse, 1872), 181, quoted in Poese, "History and Design," 63. See also *Moore's* 41 (Aug. 1882): 561.

38. *AA* 18 (July 1859): 201; *Utica Morning Herald*, quoted in *AA* 22 (Feb. 1863): 47; "On Cheesemaking," *GF* 22 (Feb. 1861): 55; "Butter and Cheese," *Harper's*, 813; *Oxford American Dictionary*; *CG* 36 (Oct. 5, 1871): 631; *NEF* n.s. 5 (Mar. 1871): 133; *Prairie Farmer* 43 (Aug. 1873): 253; *AA* 23 (Oct. 1864): 295. See also "American at Work," *Appletons'*, 298.

39. New York State Cheese Manufacturers' Association *Annual Report* (1864), 96; *Moore's* 15 (Oct. 1, 1864): 318; *Moore's* 23 (Mar. 4, 1871): 142; *CG* 31 (June 18, 1868): 439; see also Steven Hamp, "From Farm to Factory: The Development of Equipment and Process in the American System of Cheese Manufacture" (M.A. thesis, Univ. of Michigan, 1978).

40. This significant change in method was introduced by X. A. Willard. In 1866 he traveled to Somerset to learn from Joseph Harding, renowned in England for having developed an "improved" way to make cheddar cheese. Willard returned home to publicize Harding's methods in speeches at dairy conventions, in visits to factories, and in articles. As Lauren B. Arnold reported in the mid-1870s, "The English method is not strictly followed by American manufacturers. It is varied in different ways, but the underlying principle is not lost sight of" (*American Dairying*, 322). The flavor of American cheese, however, still did not match that of the best English cheddar.

41. "Butter and Cheese," *Harper's*, 817.

42. Hannah Day to Harvey Day, Sept. 6, 1870, Harvey Day papers.

43. Twelfth Census (1900), vol. 7, Manufactures, pt. 1, U.S. by Industries, table 1, p. 5, table 4, pp. 138–39.

44. Osterud, *Bonds;* Angkarloo, "Agriculture and Women's Work"; Cohen, *Women's Work;* Sommestad, "From Dairywomen to Dairymen"; Joanna Burke, "Dairywomen and Affectionate Wives: Women in the Irish Dairy Industry, 1890–1914," *Agricultural History Review* 38, no. 2 (1990): 149–65; see also Patrick Nunnally, "From Churns to 'Butter-Factories': The Industrialization of Iowa's Dairying, 1860–1900," *Annals of Iowa* 49 (winter 1989): 555–69. Ginzberg, *Women and the Work of Benevolence*, has argued that these trends occurred nationally in Civil War–era benevolent work, as masculinization, system, efficiency, and organizational consolidation took place.

45. Kulikoff, "Transition to Capitalism," 139.

Chapter Eight: The Dairy Zone Embattled

1. Willard, *Practical Dairy Husbandry*, 1. See also "American at Work," *Appletons'*, 298; Arnold, *American Dairying*, 27; "American Dairying, Its Rise, Progress,

Notes to Pages 173–77

and National Importance," U.S. Department of Agriculture *Annual Report* (1865), 431.

2. Lampard, *Rise;* Willard, in New York State Cheese Manufacturers' Association *Annual Report* (1864), 24; Brunger, "Changes," 60.

3. *Moore's* 28 (Sept. 13, 1873): 171.

4. "Among the Farmers," *AA* 38 (May 1879): 179–80.

5. *Rome Sentinel,* Apr. 12, 1870.

6. Figures for factory cheese are based on the New York State census for 1875. No comparable published data exist for the earlier period, so those estimates were calculated by comparing the number of cattle milked on cheesemaking farms of the seven towns in 1850 with the total number of cows milked in each town in 1845, the closest year for which this information is available.

7. These calculations are based upon manuscript and published data from the U.S. census of Agriculture and Population (1850, 1870), and from the New York State census (1845, 1855, 1865, 1875).

8. The term *milk-selling* is not technically accurate, since most farmers were actually sending milk to the factory rather than selling it outright. I use it for convenience here.

9. Again, data are calculated from the U.S. and New York State census manuscript schedules for the seven Oneida County towns. Yields for dairy cows are consistent with 1880 estimates by J. P. Sheldon, who thought that average cows yielded from thirty-six hundred to five thousand pounds per year, and the best up to eleven thousand. Sheldon, *Dairy Farming,* 411.

10. "American at Work," *Appletons',* 298.

11. New York State Cheese Manufacturers' Association *Annual Report* (1864), 56–57; see also *NEF* n.s. 5 (July 1871): 344.

12. *Boonville Herald,* Feb. 4, 1875.

13. *UWH,* Sept. 16, 1873. According to R. John Dawes, *The Art of American Livestock Breeding* (exhibit catalog, 1991), 44, this auction was the culmination of competition between American and British Shorthorn breeders.

14. *UWH,* quoted in New York State Cheesemakers' Association *Annual Report* (1865), 128.

15. Ibid., 148; see also 159, 161, 162, 128, 143.

16. *UWH,* Dec. 30, 1873; see also Nov. 4, Dec. 2, 1873; Sheldon, *Dairy Farming,* 396.

17. *Rome Sentinel,* Apr. 12, 1870.

18. *UWH,* Oct. 11, 1864; see also Charles L. Flint, *Milch Cows* (1889 ed.), 55, where he says that in the last twenty-five years purebred bulls have improved the dairy herd.

19. *Rome Sentinel,* Sept. 27, 1870; *UWH,* July 11, 1865, Oct. 11, 1864.

20. *Boonville Herald,* Feb. 13, 1873, Sept. 16, 1875, Aug. 26, 1880; *UWH,* Mar. 18, Sept. 23, 1873; *Rome Sentinel,* June 14, 1870.

21. *AA* 30 (Nov. 1871): front page; *Boonville Herald,* Sept. 18, 1873; *Michigan Farmer,* June 2, 1885, cover; *UWH,* Jan. 25, 1876. The milk may indeed have traveled well, as we now know that the Holstein's milk has a substantially lower butterfat content than that of other breeds, especially the Channel Islands cattle hitherto favored

Notes to Pages 177–83

for the dairy. Gates, *Agriculture and the Civil War*, 218, notes that pleuropneumonia slowed the importation of Holsteins during the war. On hardiness, see *Moore's* 32 (Sept. 4, 1875): 155; *Moore's* 38 (Apr. 19, 1879): 246; *Moore's* 37 (Dec. 7, 1878): 778.

22. Allen, *American Cattle*, 167; *American Rural Home* 10 (July 10, 1880): 224; Arnold, *American Dairying*, 29–35; *History of Oneida County* (Philadelphia, 1878), 623; *American Rural Home* 7 (Aug. 10, 1877): 253–54; *Moore's* 24 (May 6, 1871): 286; *Moore's* 32 (Sept. 4, 1875): 155; "Dairy Stock of Herkimer County," *Moore's* 35 (Mar. 31, 1877): 199; C. H. Eckles and G. F. Warren, *Dairy Farming* (New York: Macmillan, 1916), 17.

23. John Borneman, "Race, Ethnicity, Species, Breed: Totemism and Horse-Breed Classification in America," *Comparative Studies in Society and History* 30 (1988): 25–51, presents an intriguing comparison of the American and European systems for classifying and evaluating horses. He maintains that European systems were performance-based, while Americans emphasized appearance, strict hierarchy, and purity of bloodlines. While his explanation for this difference—that in America "nation building preceded state building"—is not fully persuasive and his chronology unclear, the parallel to cattle breeding is suggestive. It seems as though in the evolving system of dairy cattle classification, Americans valued performance as well as appearance, hierarchy, and breed.

24. Also, the fluid-milk industry was expanding, and this also would contribute to an interest in high-producing stock.

25. Bateman, "Improvement in American Dairy Farming"; Bateman, "Labor Inputs."

26. *NYSAS* 1866:17–20.

27. *UWH*, Nov. 21, 1865; *Rome Sentinel*, July 26, 1870; "Butter and Cheese," *Harper's*, 817.

28. *BRH*, Aug. 12, 1869.

29. *Boonville Herald*, July 27, 1871.

30. *NEF* n.s. 4 (Feb. 1870): 98–101; see also *Boonville Herald*, Apr. 20, 1871, ad; *UWH*, Nov. 19, 1872; *CG* 24 (July 7, 1864): 13. "Butter and Cheese," *Harper's*, 817–18, noted that the best barns were large, oblong, and two-storied, with a cupola and a whole basement where cows stayed "nearly all the time in winter."

31. Henry Glassie, "The Variation of Concepts within Tradition: Barn Building in Otsego County New York," *Geoscience and Man* 5 (June 10, 1974): 182, 229–30, 231. For an example of how one farmer literally raised an old barn onto a new basement, see "Improving Old Barns," *AA* 25 (June 1866): 215.

32. For more discussion of dairy barns, see I. F. Hall, "An Economic Study of Farm Buildings in New York" (Ph.D. diss., Cornell Univ., 1926); published in part in Cornell Univ. Agricultural Extension Service *Bulletin*, no. 478 (May 1929).

33. *Moore's* 23 (June 10, 1871): 361; *Moore's* 29 (June 6, 1874): 361, 364. See also Willard, *Practical Dairy Husbandry*, 36.

34. *AA* 28 (Oct. 1869): 373–74. See also *American Rural Home* 9 (Sept. 6, 1879): 297. Similarly, in William Wheeler's 1870 barn in Trenton, Oneida County (cited by X. A. Willard as typical of the newer barns then being erected), "the cows are fed from a central alley, which is fourteen feet wide." On either side was a range of stables

eleven feet wide, behind which manure chutes led to a collection area below. *NEF* n.s. 4 (Feb. 1870): 98–99. See also a full description in New York State Cheesemakers' Association *Annual Report* (1864), 136–37, rpt. from *UWH*, July 1, 1864. See also a description of another Oneida County dairy barn in *UWH*, June 21, 1864.

35. Loyal Durand, "Dairy Barns of Southeastern Wisconsin," *Economic Geography* 19 (1943): 37–44.

36. See Fink, *Barns of the Genesee Country*, 95; Hurt, *American Farm Tools*, 94–96.

37. Oneida County directory (1869); *UWH*, Jan. 21, 1873. For illustrations of choppers and feed mills, see Hurt, *American Farm Tools*, 96.

38. *Boonville Herald*, Apr. 20, 1871; *UWH*, Feb. 4, Aug. 12, 1873.

39. *UWH*, Jan. 9, 1866.

40. *AA* 36 (Nov. 1877): 425; see also *NEF* n.s. 3 (Aug. 1869): 358–59; *AA* 28 (Oct. 1869): 373.

41. *Moore's* 23 (Apr. 15, 1871): 233.

42. See also *AA* 32 (Jan. 1873): 46.

43. For more descriptions and plans, see *NYSAS* 1858:599; *NYSAS* 1863:587; *AA* 26 (June 1867): 214.

44. *Moore's* 38 (May 3, 1879): 277.

45. Ibid. See also *Boonville Herald*, Apr. 20, 1871; *UWH*, Sept. 23, 1873.

46. *Ohio Farmer* 14 (Nov. 11, 1865): 353.

47. *CG* 43 (Jan. 24, 1878): 59–60; *NEF* n.s. 2 (Apr. 1868): 169.

48. *AA* 42 (Apr. 1883): 173.

49. Willard, *Practical Dairy Husbandry*, 108–9; *Moore's* 28 (Dec. 28, 1873): 379. Brucellosis is also dangerous for humans, since it causes a serious and sometimes fatal undulant fever. Thanks to Dr. Larry Hutchinson, extension veterinarian at Pennsylvania State Univ., for information on these diseases. On the history of veterinary medicine, see B. Bierer, *A Short History of Veterinary Medicine in America* (East Lansing: Michigan State Univ. Press, 1955); J. F. Smithcors, *The American Veterinary Profession: Its Background and Development* (Ames: Iowa State Univ. Press, 1963).

50. *NYSAS* 1863:582.

51. Lamont, "Agricultural Production in New York," 1.

52. *AA* 33 (1874): 180–81.

53. Flint, *Milch Cows* (1858 ed.), 185; Willard, *Practical Dairy Husbandry*, 52; *Moore's* 34 (Aug. 12, 1876): 103; *UWH*, Aug. 29, 1865.

54. "Butter and Cheese," *Harper's*, 816; *NEF* n.s. 5 (Mar. 1871): 114. See also H. T. Stewart, "Does Dairying Exhaust the Soil?" *American Rural Home* 6 (Sept. 16, 1876): 302; Ohio State Board of Agriculture *Annual Report* (1863), 93; *UWH*, July 1, Jan. 14, 1873; *BRH*, Apr. 16, 1863. Bartlett's address was reprinted in the *Ohio Farmer* for 1871: Feb. 18, pp. 100–101, Mar. 18, pp. 162–63, and Mar. 25, p. 178.

55. *UWH*, Apr. 15, July 1, 1873; see also *Moore's* 37 (Aug. 31, 1878): 553; *CG* 28 (Aug. 16, 1866): 114; *BRH*, Apr. 16, 1863, July 15, 1869; Flint, *Milch Cows* (1858 ed.), 141; *CG* 21 (Jan. 22, 1863): 61; U.S. Department of Agriculture *Annual Report* (1866), 365.

56. See also *UWH*, Nov. 14, Aug. 29, 1865; Willard, *Practical Dairy Husbandry*, 50–54.

57. *UWH*, Jan. 14, 1873; *CG* 16 (Sept. 6, 1860): 159.

58. *UWH*, July 4, Oct. 10, 1865; *Rome Sentinel*, Sept. 6, 1870.

59. Arnold, *American Dairying*, 78; see also *UWH*, Aug. 29, 1865, and *American Rural Home* 7 (June 9, 1877): 181, which said that this practice had been discovered by dairy farmers.

60. Allen, *American Cattle*, 311.

61. Total corn acreage dropped slightly, from 19,911 to 18,206 acres, and yields were very poor in 1875. New York State census (1865, 1875); U.S. census (1900).

62. *UWH*, May 23, 1876, Nov. 12, 1872; Allen, *American Cattle*, 311; *Rural Affairs* 1864:99; *UWH*, May 30, 1876; see also *Moore's* 23 (Feb. 11, 1871): 94; *NEF* 16 (Feb. 1864): 50–51; *UWH*, Aug. 25, 1865, Feb. 4, 1873, Mar. 21, 1876; New York State Cheesemakers' Association *Annual Report* (1865), 83; *Boonville Herald*, Oct. 21, 1875; Sheldon, *Dairy Farming*, 400. William H. Storrs, mentioned at the beginning of the chapter, mixed corn and barley for his grade cattle; see *UWH*, Jan. 9, 1865.

63. *GF* 22 (June 1861): 159–60; *Rural Affairs* 1864:99; Willard, *Practical Dairy Husbandry*, 77; *UWH*, July 4, 1865; see also *Rural American*, Aug. 1, 1866, p. 225; *AA* 27 (Aug. 1868): 292; "Butter and Cheese," *Harper's*, 817.

64. *UWH*, Nov. 5, 1872; account books (see, for example, Mar. 29, 1882), Barnes family papers; account book (see, for example, May 27, 1871, p. 22), Nye family papers, New York State Historical Association. See also Lampard, *Rise*, 149.

65. *UWH*, Mar. 21, 1876; *BRH*, July 15, 1869; *UWH*, Apr. 8, 1873, July 4, 1865; *American Rural Home* 7 (Mar. 24, 1877): 62; *NYSAS* 1867:857; see also *American Rural Home* 9 (Sept. 6, 1879): 297.

66. *UWH*, Oct. 29, 1872; Margaret Rossiter, *The Emergence of Agricultural Science* (New Haven: Yale Univ. Press, 1975).

67. Wines, *Fertilizer*, argues for a transition from "closed cycle" fertilizer systems to purchased ones, an escalating spiral of dependency and energy subsidy.

68. *Utica Morning Herald* for Sept. 2, 1864, quoted in New York State Cheesemakers Association *Annual Report* (1864), 158–59; *UWH*, Sept. 2, 1864; see also U.S. Patent Office *Annual Report* (1861), 293.

69. Susan Nye account book, Mar. 6, Nov. 11, 1871, July 17, 1872, ;; B. F. Cloyes diary, Jan. 23, 28, and 30, 1878, Cornell Univ. Archives; *Rome Sentinel*, Apr. 12, 1870; *UWH*, Mar. 4 and 25, 1873, Mar. 28, 1878, Feb. 24, 1874.

70. *UWH*, May 30, 1876.

71. Compare with the tobacco raised in the mid-nineteenth-century Connecticut Valley; see Clark, *Roots*, 86.

Chapter Nine: Fragmentation and Reorientation

1. The milk check came to the men; in this regard it is possible that women experienced a loss of internal household authority from the prefactory era. But since it is not clear that women had enjoyed significant control over cheese income in the

household era, it is difficult to know if the postfactory method of payment made a significant difference. See Jeremiah Sweet account books, New York State Historical Association; account book, Barnes family papers; Susan Nye account book.

2. *American Rural Home* 7 (Mar. 17, 1877): 86. In 1879 an *American Agriculturist* columnist criticized dairying families who consumed few milk products in order to turn every possible drop into money. *AA* 38 (Feb. 1879): 57. Bellesiles, "Account Book," argues that the "communal capitalist" system finally bowed out during the war, and others, including Merrill ("Cash Is Good to Eat") and Clark (*Roots*), also find that the old systems of exchange began to wither in the 1850s and disappeared with the war.

3. *BRH*, July 15, 1869.

4. *American Rural Home* 10 (June 5, 1880): 183.

5. U.S. census, manuscript schedules, Agriculture and Population (1870).

6. In age structure, milk-selling families were not so different from their cheese-making predecessors. Children under the age of ten were still scarce. Although milk dairying did not entail the same high skill level as making cheese had demanded, it still required a good deal of labor in milking, caring for stock, and tending crops, and so teenagers were still valuable and younger children comparatively less so. Moreover, while it was possible to send milk from just a few cows to the factory, it took time to accumulate the money to enter milk selling in a substantial way, and so most factory-patron families tended to be at the peak of the life cycle.

7. Figures on wages are from the surveys published with the New York State census for 1875 ($35 per month) and from *NEF* n.s. 2 (Mar. 1867): 122 ($30 per month for eight months).

8. *American Rural Home* 12 (July 15, 1882): 222; "Farmers' Families and 'Help,'" *CG* 61 (Feb. 1885): 156; see also McMurry, *Families and Farmhouses*, chap. 4.

9. Dairy farmers still relied upon wage labor less than farmers in other branches of agriculture, because it was not as easily rationalized; for example, there were long intervals between milkings. See Osterud, *Bonds of Community*, 156 n. 17, and Mark Kramer, *Three Farms: Making Milk, Meat, and Money from the American Soil* (Cambridge: Harvard Univ. Press, 1987), 40–41.

10. *UWH*, Dec. 31, 1872, p. 3.

11. Hamlin Garland, "Up the Coulee," in *Main-Travelled Roads* (Boston, 1891).

12. Henry Stewart, *The Dairyman's Manual* (New York, 1888), chap. 5.

13. Cyril Tyler, "The Development of Feeding Standards for Livestock," *Agricultural History Review* 4 (1956): 97–108.

14. *AA* 27 (Aug. 1868): 292; *AA* 29 (Feb. 1870): 51.

15. Bateman, "Labor Inputs," implicitly criticizes farm families for not realizing that their labor productivity was on the decline. But the literature of the period—for example, Henry Herbert's poem, and the work of Hamlin Garland—suggests that even if they did not have a precise idea of labor inputs, farmers knew intuitively that they were working harder for smaller returns but did not possess the resources to increase productivity in other ways.

16. Stewart, *Dairyman's Manual*, 9–16.

17. Some farmers, unable to afford the thousands of poles required, planted corn

and let the vines run up the cornstalks. See *Utica Morning Herald*, Aug. 26, 1864; *Rural American* 11 (Feb. 15, 1867): 51.

18. Ibid.

19. B. F. Cloyes diary, May 17, 1864, Jan. 14, May 28, 1878.

20. Osterud, *Bonds of Community*; Phelan, *White Vest*; Cornelia and Peleg Babcock diary. Jane Adams, "The Decoupling of Farm and Household: Differential Consequences of Capitalist Development on Southern Illinois and Third World Family Farms," *Comparative Studies in Society and History* 30 (1988): 453–82, illustrates this trend in New Deal–era Illinois.

21. *Moore's* 23 (Apr. 22, 1871): 254.

22. U.S. census, manuscript schedules, Agriculture (1870). See also "Butter and Cheese," *Harper's*, 817; *GF* 24 (Aug. 1863): 246; *GF* 25 (Apr. 1864): 108.

23. *Ohio Farmer* 18 (Jan. 23, 1869): 49; *AA* 37 (Feb. 1878): 53; "Butter and Cheese," *Harper's*, 818; "American at Work," *Appletons'*, 301; see also Arnold, *American Dairying*, 160; *GF* 26 (Oct. 1865): 312.

24. *Moore's* 37 (Dec. 14, 1878): 796; *UWH*, May 14, 1878.

25. Lamont, "Agricultural Production in New York," 26–27; *Boonville Herald*, Mar. 28, 1878.

26. A piece in *UWH*, Aug. 19, 1873, was addressed to women. See also *Boonville Herald*, Feb. 27, 1873; *UWH*, Mar. 25, Apr. 1, Aug. 19, 1873.

27. Susan Nye account book, see Oct. 26, 1871; B. F. Cloyes diary, July 2, 1872; see also account books, Barnes family papers.

28. Account books, 1882, Barnes family papers; see also "Butter and Cheese," *Harper's*, 827.

29. Virginia McCormick, "Butter and Egg Business: Implications from the Records of a Nineteenth Century Farm Wife," *Ohio History* 100 (winter/spring 1991): 51–61.

30. See the retrospective in *UWH*, Nov. 14, 1876.

31. *BRH*, Sept. 30, 1869.

32. *BRH*, Nov. 18, 1869.

33. *UWH*, Sept. 23, 1873; *Boonville Herald*, Oct. 5, 1871, Sept. 16, 1875; *UWH*, Sept. 30, Oct. 7, 1873; *Boonville Herald*, Sept. 11, 1873.

34. *Boonville Herald*, Jan. 22, 1880, Sept. 16 and 23, 1875.

35. H. J. Goodwin's agricultural warehouse, Oneida County directory (1863); Utica city directory (1869).

36. This was not without ambiguity and conflict. As always, it was hard to establish a line between the useful and the superfluous. A complaint registered in the 1868 *New England Farmer* illustrates. A farmer's wife raised the problem of "losing caste" by making and mending the family's garments. "As my daughters see it—and, I must confess, as I have observed it—what is its effect on our 'respectability,' or 'standing in society'? Is not another class of 'accomplishments' at a premium, even in agricultural communities? Is not a white hand, especially if it is supposed to hold a few greenbacks preferred to one a little browned by efforts to save the same greenbacks?" *NEF* n.s. 2 (July 1868): 327.

37. Diary, Christmas 1871, Nov. 23, 1872, Nye family papers.

Notes to Pages 206–10

38. Faragher, *Sugar Creek*, 190.

39. Elaine Goodale, *Journal of a Farmer's Daughter* (New York, 1881), 1, 2, 74, 111.

40. See Pederson, *Between Memory and Reality*, and Joanne Meyerowitz, "Women and Migration: Autonomous Female Migrants to Chicago, 1880–1930," *Journal of Urban History* 13 (Feb. 1987): 147–68.

41. U.S. census, published summary (1880), vol. 18, pt. 1, Social Statistics of Cities, p. 657.

42. *Camden [Oneida County] Advance*, June 28, 1876.

43. *UWH*, Mar. 17, 1874; see also "What Shall We Do with the Girls?" *Michigan Farmer*, Feb. 14, 1882, p. 7. The importance of cash again appears in a piece from *American Rural Home* 7 (Feb. 15, 1879): 54, which argued that "the daughter invariably takes the place of a hired servant, excepting the wages, which she invariably receives. . . . How many daughters of farmers receive over one-third of that amount ($200 per year) for service rendered far more faithfully?"

44. There is an interesting modern parallel here. Ida Harper Simpson, Pennsylvania State University colloquium, Apr. 29, 1986, noted in a contemporary study of farm women that off-farm employment tended to erode women's commitment to the farm, but not men's.

45. This runs counter to the interpretation Thomas Dublin offers in explaining rural-to-urban migration in "Rural-Urban Migrants in Industrial New England: The Case of Lynn, Massachusetts, in the Mid-Nineteenth Century," *Journal of American History* 73 (Dec. 1986): 623–45. Dublin found that migrants to Lynn were primarily young people who married either just before or just after migration, and he therefore concludes that "migration . . . was fundamentally a family phenomenon. . . . Even when individuals migrated by themselves, they clearly served familial as well as individual purposes" (pp. 628–29). He continues: "For the vast majority of Lynn migrants . . . migration severed links to their past. . . . Migration . . . constituted a new strategy within the lineal family system described by James Henretta. Whereas fathers' choices dominated strategies during the late 18th c. and the first decades of the 19th, by mid-century decisions to migrate by sons and daughters represented equally important elements in the shaping of family residence and wealth patterns across generations" (pp. 638–39). Dublin argues that even though migrants severed ties with their pasts, "it is not so much a question of changing values as it is of adaptation of earlier family strategies in the face of changing economic circumstances" (p. 639). But especially among women, the evidence strongly suggests that migration did represent changing values. They often rejected the salient aspect of the lineal strategies Henretta describes in "Families and Farms": the maintenance of the family farm. Migrants did not simply replace patriarchs in the decision-making process, they invested the decision with very different content.

46. *Rome Sentinel*, Oct. 11, 1870.

47. *Boonville Herald*, Sept. 2, 1880.

48. *American Rural Home* 7 (Sept. 8, 1877): 285; *UWH*, Sept. 3 and 10, 1872, Sept. 9, 1873; *Boonville Herald*, July 29, 1880; Joan Jensen, "Women in the Hop Harvest," in *Labor in the West*, ed. Hugh Lovin (Boulder: Westview, 1986), 97–109.

Notes to Pages 210–16

49. Dora Walker diary, Aug. 28–Sept. 30, no year given, probably late 1860s, New York State Historical Association.

50. Flavia Canfield, *The Hop Pickers: Girl Life in the Sixties* (New York: Harcourt, Brace, 1922).

51. Fink, *Agrarian Women,* 129. In a redefined agrarian ideology, these women believed that "the daughters of pioneer women would reap the fruits of their mothers' sacrifice in the form of lives that were not tied to farms."

Chapter Ten: Rural Communities Transformed

1. Gates, "Agricultural Change in New York State," 115, 118; see also U. P. Hedrick, *History of Agriculture in New York State* (1933; rpt. New York: Hill & Wang, 1966), 437; U.S. census, published schedules, Population and Agriculture (1870, 1880, 1890, 1900).

2. This argument has been most explicitly stated by Allan Kulikoff, but earlier Earl Hayter sought to show that "troubled" farmers resisted agricultural industrialization out of superstition and fear of modernity, while more recently David Danbom, who portrays farmers more sympathetically, nonetheless emphasizes conflict. Kulikoff, "Transition to Capitalism"; Hayter, *Troubled Farmer;* Danbom, *Resisted Revolution.*

3. Barron, *Those Who Stayed Behind,* 14; Osterud, *Bonds of Community;* Baker, *Moral Frameworks of Public Life.*

4. I have excluded some obvious elements of community life from the discussion here, most notably politics, because this study is conceived as an examination of an industry rather than of any specific political entity. A full study of how the dairying economy and the village political scene were related would take up another monograph. For a suggestive account of late-century rural New York politics, see Baker, *Moral Frameworks of Public Life.*

5. Neth, "Preserving the Family Farm," 379.

6. To keep this in perspective, however, it is important to note that compared with large-scale business of the day, cheese factory owners and makers lacked working capital. See Wilde, "Industrialization of Food Processing," chaps. 3 and 4.

7. These are only a few examples from a survey in which I collected 249 names of cheese factory principals and sought biographical information on them in local histories. The biographical information on Brill is from *History of Oneida County,* 596; Wager, *Our County,* 293, 598, 600. On Weaver, from Wager, *Our County,* 200. On Leland, from Rome city directory (1850), 161, 188; Utica city directory (1869), 369; *History of Oneida County,* 442; U.S. census, Population (1860). On Shearman, from Rome city directory (1859), 121. See also CG 21 (Jan. 1, 1863): 18; NEF n.s. 5 (Mar. 1871): 114.

8. Clark, *Roots,* chaps. 6 and 7; Dun and Co. ledgers, 471:551, 472:708.

9. *NEF* n.s. 2 (Mar. 1868): 145.

10. Quoted in *CG* 38 (July 17, 1873): 459.

11. "Butter and Cheese," *Harper's,* 818.

12. *UWH,* Sept. 3, 1872.

13. Nathan Huntington letter, Mar. 25, 1914.

14. *BRH*, Feb. 5, 1863; *Boonville Herald*, Aug. 26, 1880; *UWH*, Aug. 29, 1875.

15. "A Milk Suit at Herkimer," *Moore's* 18 (June 1, 1867): 174.

16. Personal correspondence of Raymond Ernewein, Verona, N.Y., June 22, 1987; see also *BRH*, Feb. 24, 1870.

17. *NYSAS* 1866: 456–61.

18. *CG* 25 (June 15, 1865): 381; American Dairymen's Association *Annual Report* (1870), 143; *Ohio Farmer* 21 (Apr. 22, 1871): 242; *UWH*, Sept. 2, 1873.

19. *CG* 39 (July 30, 1874): 491; *CG* 39 (Nov. 12, 1874): 731; see also *Ohio Farmer* 19 (Feb. 12, 1870): 98; *Boonville Herald*, May 27, 1880; *NEF* n.s. 1 (Aug. 1867): 381.

20. *UWH*, Jan. 9, 1866; see also *CG* 38 (Jan. 23, 1873): 59–60.

21. *AA* 33 (Sept. 1874): 328; *AA* 27 (Mar. 1868): 97; Wager, *Our County*, 514; "Butter and Cheese," *Harper's*, 818.

22. *BRH*, Feb. 24, 1870; *Moore's* 22 (Apr. 9, 1870): 238.

23. See *Moore's* 19 (Feb. 1, 1868): 37.

24. Hahn, *Roots of Southern Populism*.

25. *Ohio Farmer* 16 (Aug. 10, 1869): 254; *Moore's* 24 (May 25, 1872): 346.

26. Cross, *Burned-over District*, 134; Ronald E. Shaw, *Erie Water West: A History of the Erie Canal, 1792–1854* (Lexington: Univ. of Kentucky Press, 1966), 225. See also Curtis Johnson, *Islands of Holiness: Rural Religion in Upstate New York, 1790–1860* (Ithaca: Cornell Univ. Press, 1989). Geib, "Changing Works," 203, notes that in Brookfield, Mass., conflict over farm work on the Sabbath emerged early in the nineteenth century.

27. Anne C. Rose also finds that the Civil War era contributed to secularization in the wider society; see *Victorian America and the Civil War* (Cambridge: Cambridge Univ. Press, 1992).

28. *CG* 23 (June 23, 1864): 397.

29. *CG* 24 (July 7, 1864): 34.

30. *UWH*, Sept. 9, 1873; see also *Moore's* 34 (Dec. 30, 1876): 423; *Ohio Farmer* 20 (Feb. 19, 1870): 118–19.

31. *GF* 18 (Mar. 1857): 90.

32. Okada, "Squires' Diary," 25.

33. "American at Work," *Appletons'*, 302.

34. James Mickel Williams, *An American Town* (New York: James Kempster, 1906), 71.

35. David Russo, "The Origins of Local News in the U.S. Country Press, 1840s–1870s," *Journalism Monographs*, no. 65 (Feb. 1980), argues that local news did not appear in the small-town press until the 1860s, and he suggests several possible reasons. One was that competition forced small-town papers to distinguish themselves from urban dailies; another was that migrants wanted news from home. He further maintains that the makers and writers of local news lacked an analytical or conceptual understanding of nationalizing trends. The evidence from Oneida County suggests, however, that this thesis may require modification. There is no way to assess the argument about urban competition, but it does seem as if migrants hungered for hometown news, because the Utica paper carried a regular column with news from most surrounding rural townships. But another factor should be considered: it seems likely that

Notes to Pages 226–29

there was simply more local news to report in the postwar period, as transportation and economic diversification made possible new cultural forms. Russo's contention about the nature of local news is also open to question. In Oneida County local papers frequently took the opportunity to reflect upon the wonders and perils of modernity; perhaps they were not rigorously analytical, but they did try to assess the implications for their communities of new developments.

36. X. A. Willard thought daughters were actively spurning the dairy work in favor of leisure accomplishments and pursuits: "There is not the careful watchfulness and family interest in [dairying] that there was formerly. Farmers' sons and daughters are of a different stamp than they were twenty, or even ten years ago. We have more pianos, and more fast horses, and less real interest in farm work." American Dairymen's Association *Annual Report* (1870), 113.

37. *UWH*, Mar. 4, 1873.

38. *Rome Sentinel*, Jan. 11 and 18, Feb. 1, 15, and 22, 1870; *BRH*, Feb. 25, 1869; *Boonville Herald*, Jan. 23, 1873, Mar. 8, 1875; *UWH*, Jan. 27, 1874.

39. Donald Scott, "Print and the Public Lecture System, 1840–1860," in *Printing and Society in Early America*, ed. William Joyce (Worcester: American Antiquarian Society, 1983), 278–99. Scott sees the public lecture system in particular as a response to the "atrophy" induced by high population turnover in local, rural communities. The public lecture system, with its nonpartisan, inspirational messages, created a new "cultural core."

40. Jensen, *Loosening the Bonds*; Clifford, "Home and School"; Wayne Fuller, *The Old Country School: The Story of Rural Education in the Middle West* (Chicago: Univ. of Chicago Press, 1982).

41. *BRH*, Jan. 26 and 5, 1871; Utica city directory (1869), 53.

42. *Rome Sentinel*, Mar. 29, 1870; *UWH*, Nov. 26, 1872, Feb. 18, Nov. 25, 1873, Feb. 24, 1874.

43. Faragher, "History from the Inside-Out," 556.

44. Ruth Bordin, *Women and Temperance: The Quest for Power and Liberty, 1873–1900* (Philadelphia: Temple Univ. Press, 1981), p. xviii. See also Jed Dannenbaum, *Drink and Disorder: Temperance Reform in Cincinnati from the Washington Revival to the WCTU* (Urbana: Univ. of Illinois Press, 1984), 192. Later the Grange offered similar organizational opportunities to well-off rural women. See Marti, *Women of the Grange*.

45. Ian Tyrrell, *Sobering Up: From Temperance to Prohibition in Antebellum America, 1800–1860* (Westport, Conn.: Greenwood, 1979), 6–7.

46. Ginzberg, *Women and the Work of Benevolence*.

47. Williams, *American Town*, 33; Faragher, *Sugar Creek*, chap. 19. Fink, *Agrarian Women*, 69, argues that women's experience with violent, drunken husbands propelled them into temperance activity, but it is very difficult to know if this was the case with Oneida County individuals. See also Osterud, *Bonds of Community*.

48. See, for example, Richard D. Brown, "The Emergence of Urban Society in Rural Massachusetts, 1760–1820," *Journal of American History* 61 (June 1974): 29–52; Danbom, *Resisted Revolution*; T. D. S. Bassett, "A Case Study of Urban Impact on Rural Society, Vermont, 1840–1880," *Ag History* 30 (January 1956): 28–35.

49. Neth, "Preserving the Family Farm," introduction.

50. Ryan, *Cradle*. The problematic aspect of this study is that it shifts focus from a rural locale—the Utica suburb of Whitestown—to the city, thus possibly obscuring changes that occurred simultaneously in countryside and city. It is also quite likely that Ryan has overdrawn the demise of "production" in urban households. See Boydston, *Home and Work*.

51. Richard Brown, "Modernization: A Victorian Climax," in *Victorian America*, ed. Daniel Walker Howe (Philadelphia: Univ. of Pennsylvania Press, 1976), 29.

52. Daniel Walker Howe, "Victorian Culture in America," in *Victorian America*, ed. Howe, 1–28.

53. See, for example, Bernard Herman, *The Stolen House* (Charlottesville: University Press of Virginia, 1992).

54. Cott, "Young Women in the Second Great Awakening."

55. Regina Morantz, "Making Women Modern: Middle Class Women and Health Reform in Nineteenth-Century America," *Journal of Social History* 10 (1976–77): 498.

Conclusion

1. See Michael Williams, "Products of the Forest: Mapping the Census of 1840," *Journal of Forest History* 24 (1980): 4–23; Williams, *Americans and Their Forests* (Cambridge: Cambridge Univ. Press, 1989).

Essay on Sources

This reconstruction of nineteenth-century dairying depends upon several types of primary sources. The manuscript schedules of the United States census for the years 1850, 1860, and 1870 offer the basis for reconstructing patterns of household labor and production. The shortcomings of the census are too well known to belabor here. The important point is that despite its flaws, the census offers a rich source of basic information about farming. The same is true of the New York State census for the years 1855, 1865, and 1875, which in some respects—for example, in listing housing construction materials and family relationships—is even more detailed, though unfortunately less accessible.

Manuscript diaries and correspondence also furnish important data. Relationships among farmers, factors, and hired hands are captured particularly well in the Ferris family papers (Knox College, Galesburg, Illinois), the Stephen Moulton Babcock papers (State Historical Society of Wisconsin, Madison), and the Andrew Hurlburt diary (Rome Historical Society, Rome, New York). Materials in the collection of the New York State Historical Association (Cooperstown), especially the Rosetta Hammond Bushnell diary and the John Hughes diary, illuminate day-to-day work. The Lavinia Mary Johnson memoir and the Barnes family papers, both in the Oneida Historical Society, Utica, both provide a long-term perspective. The Mary Root diary, Utica College, offers a young girl's viewpoint on education. The Harvey Day papers, Bentley Library, University of Michigan, Ann Arbor, are unsurpassed for the light they shed on the transition to factory production.

For detailed discussions of farming practices and values, agricultural journals and specialized dairy manuals offer a wealth of viewpoints. Especially useful are the *American Agriculturist, Genesee Farmer, Albany Cultivator* (and later *Country Gentleman*), *Moore's Rural New Yorker, Transactions* of the New York State Agricultural Society, *New England Farmer, Ohio Cultivator,* and *Annual Register of Rural Affairs*. Among published works specifically concerned with dairying, the best are William Townsend, *The Dairyman's Manual* (Vergennes, Vt., 1839); X. A. Willard, *Practical Dairy Husbandry* (New York, 1877); Lauren B. Arnold, *American Dairying: A Manual for Butter and Cheese Makers* (Rochester, 1876); Gurdon Evans, *The Dairyman's Manual* (Utica, 1851); Charles L. Flint, *Milch Cows and Dairy Farming,* new ed. (Boston, 1858); J. P. Sheldon, ed., *Dairy Farming* (London, c. 1880); and Henry Stewart, *The Dairyman's Manual* (New York, 1888). The *Transactions* of the New York State Cheese Manufacturers' Association describe factory practice in detail. Though they were not agricultural journals, *Harper's New Monthly Magazine* and *Appletons' Journal* both

Essay on Sources

published articles about cheesemaking in the 1870s: "Butter and Cheese," *Harper's*, Nov. 1875, and "The American at Work," *Appleton's* 4 (Apr. 1878).

For building a picture of community relationships, local newspapers are indispensable. In Oneida County the most useful are the *Rome Sentinel, Roman Citizen, Black River Herald* (later *Boonville Herald*), *Utica Weekly Herald,* and *Utica Morning Herald.* Local directories for the city of Utica and the county as a whole are also available.

Finally, museum collections afford insights through artifacts. At Greenfield Village, Dearborn, Michigan, there is an especially strong collection of presses, while among the holdings of the New York State Farmers' Museum in Cooperstown is a representative range of vats, cutters, strainers, and ladders. The Frisbie House Museum of the Salisbury Historical Society, Salisbury Center, New York, contains a notable collection of implements that were all used by the same cheesemaker, J. F. Hopson.

Index

Academies, in Oneida County, 115–19, 120. See also Education
Accounting systems, 194
Age, in cheesemaking households, 196. See also Generations
Agriculture: capital-intensive, 195; commercial, 186; "defeminization" of, 1, 145–46; early Mohawk Valley, 10; North vs. South, 44–45
Albany Cultivator, 15, 21, 40, 52, 90, 107
Alfred University, 121
Allen, Lewis Falley, 18, 21, 96, 135, 177, 189, 201
Alvord, H. E., 2
American Agriculturist, 30–31, 68, 71, 89, 109, 131, 132, 143, 168, 173, 177, 178, 182–83, 184, 187, 199, 202, 219
American Dairymen's Association, 187, 218
American Institute Farmers Club, 136
American Rural Home, 141, 195, 210
Angkarloo, Bengt, 145
Annsville, Oneida County, 39
Appleby, Joyce, 44
Appletons' Journal, 175, 202, 205, 225
Archbald, Mary, 9
Arnold, Lauren B., 20, 189
Arr, E. H., 53, 84, 85, 86, 93, 102, 114, 122
Ayrshire cattle, 16, 18

Babcock, Cornelia and Peleg, 27, 41, 47–48, 51, 55, 57, 60, 65, 68–69, 70, 76, 81, 118, 166, 195, 197, 201
Babcock test, 2, 215
Bacteria, in cheesemaking, 83–84
Banking, rapid expansion in, 100. See also Cash system; Exchange
Barn-building boom, 178–79
Barnes family, 195, 204
Barns: commercial designs for, 185; dairy, 180–83; and dairy culture, 32; with dual-purpose basement, 179–80; "English," 10, 32–33, 180, 182; as factories, 184–85; innovations developed for, 33–34; log, 9–10; machinery in, 183–85; milking, 34, 35; multilevel pattern for, 36–38, 183; northern basement, 180–83, 186; various uses of, 38
Barter economy, 98. See also Market
Bartlett, Anson, 187–88
Bateham, Mrs. M. B., 109
Bateman, Fred, 178, 199
Bidwell, Percy Wells, 101
Black River Herald, 131, 133, 157, 165, 190, 195, 204, 205, 219
Black River valley, 10
Boilers, cheese factory, 158
Boonville, Oneida County, 39, 40
Boonville Herald, 176, 179, 203, 210, 216, 218
Boonville's Opera House, 226
Bordin, Ruth, 228
Breeding: acceptance of, 19–20; growing interest in, 177–78; limitations of, 20; and quantity vs. quality, 18
Brokerage system, 46–51, 55, 59
Brown, Richard, 230
Brucellosis, 186
Buel, Jesse, 11
Buildings: for cheesemaking, 92–96; of cheesemaking factories, 152–56, 159–61
"Burned-over District," 221
Burrell, Harry, 59
Burrells and Company, 57
Burritt, Elihu, 119
Buttermaking, 4; centralization of, 3; decline in, 202; of Welsh immigrants, 12; women's control over, 2–3

Calves, raising of, 76
Canada: competition from, 174; dairy industry in, 134
Canfield, Flavia, 210–11

Index

Capitalism: and cheesemaking vs. textile manufacture, 128–29; and conflict, 212–213; in dairying, 211; farmers' attitudes toward, 51–52; and gender organization of labor, 170–71; resistance to, 43, 98; and rise of cheese factories, 124–25; "rural," 125; rural women's relationship to, 146; transition to, 45, 61, 116. *See also* Cash system

Cash crops, 196; growing dependence upon, 193; historical role of, 43; hops, 191–92

Cash economy, competency and, 60

Cash exchange, in postwar period, 195

Cash system: and cheese factories, 124; expansion of, 174; and subsistence production, 72; shift to, 57–61. *See also* Capitalism

Cattle, 11. *See also* Cows

Cattle feed. *See* Feeds

Census, New York State (1865), 162

Census, U.S.: (1870), 161–62; (1900), 170

Centralization, dairy, 125, 234; acceleration of, 133–37, 142; arguments for, 139; and conflict, 213; and farming system, 193; and gender division of labor, 220; impact of Civil War on, 129–33; precedents for, 126–29; and reorganization of household work, 224; and shift to male control, 167; and social changes, 137–45; and women's choices, 146–47. *See also* Factory system

Channel Islands cattle, 16, 177

"Cheddaring" method, 169

Cheese: Cheshire, 14; demand for, 84; factory vs. homemade, 169; Goshen, 127, 173; Herkimer County, 11, 173; Limburgh, 126; regional identities of, 173; "two-meal," 82

Cheese dairying: as commodity-producing business, 81; farming system of, 12; in U.S., 81; withdrawal of women from, 145–47. *See also* Dairying

Cheese houses, 93, 94–95, 129, 203

Cheesemakers: factory, 150; women, 75–76. *See also* Labor; Work

Cheesemaking: as art, 82; buildings for, 92–96; vs. buttermaking, 2–3; collective, 126; "defeminization" of, 152; factory vs. home, 149, 168, 175; household economy of, 62–71, 96–99; as men's work, 167; process of, 82–84; product of, 84; scientific knowledge of, 92; skills needed for, 91–92; substitution of milk selling for, 201; Sunday, 213, 220–24; tools of, 85–88; women's exclusion from, 169–70; women's overwork in, 108–10

Cheesemongers, 51

Cheshire cheese, 14

Children, in cheesemaking households, 62–63

"Cinque," 87

Civil War, 2; historical impact of, 174; impact on cheesemaking of, 129–33

Clarke, Dr. Edward, 22

Cleanliness, problem of, 215–20

Climate: in Dairy Zone, 13; of Oneida County, 7–8

Clinton Liberal Institute, 115

Clover pasture, 28

Colman, Henry, 52

"Commission houses," 46–47

Communities, rural: conflict in, 213–24; declining population of, 212, 233; in factory era, 211; and national culture, 225, 226. *See also* Fairs; Neighborhood system

Competency, 52, 53, 220; and cash economy, 60; elastic nature of, 206; ethos of, 75; and family organization, 62; farmhouses and, 205–6; meaning of, 223–24, 231; and profits, 53, 99; redefining, 194; and superfluity, 110

Conflict: capitalism and, 212–13; and change, 234; family, 101, 106, 107; social basis for, 213–14

Connecticut Valley, 125

Conservatism, 73; and class divisions, 228; vs. dynamic innovativeness, 122

Consumer goods, increased emphasis upon, 206

Cooper, William, 8

Cooperation, vs. competition, 45. *See also* "Mutuality"

Cooperatives, farmers', 129

Corn, 188; cultivation of, 199; disadvantages of raising, 193; increased dependency on, 190

Cott, Nancy, 231

Country Gentleman, 83, 140, 214, 218, 221

Countryside, capitalist transformation of, 170. *See also* Capitalism; Communities, rural

"Cow-barn," 34. *See also* Barns

Cows: acres needed per cow, 187; in Dairy

Zone, 13; new and improved, 175–78; Oneida County, 175–76. *See also* Milch cows
Cradle of the Middle Class: The Family in Oneida County, New York, 1790–1865 (Ryan), 229
"Cream gauge," 217
Cream separator, 2, 3
Cronon, William, 43
Crop rotation, 24, 233
Crops: "catch-crop," 26; of cheese–producing farms, 40; flax, 11; grass, 25; of Oneida County, 10; seasonality of, 73. *See also* Agriculture; Cash crops
"Cultivation," 206
Cultural values, 12; Anglo-American, 42; and dairy centralization, 125; education as threat to, 111; and gendering of work, 78, 145; rejection of rural, 122; traditional and popular, 96–97. *See also* Values
Culture, dairying, emergence of, 3
Curd, in cheesemaking, 82
Curd cutters, 89, 133
Curd-cutting knives, 90, 159, 161
Curd sinks, 159, 161
Curing: houses, 157, 160; process of, 83

Dairy barns, 184
Dairy belt, 172
Dairy communities, development of, 61. *See also* Communities, rural
Dairy cows, 21. *See also* Milch cows
Dairy farms, diversification of, 192. *See also* Farms
Dairy industry, 1851 survey of, 46. *See also* Factory system
Dairying: capitalism in, 211; in crisis, 233; labor productivity in, 199; masculinization of, 208–9; 19th-century, 1; shift to, 73; winter, 185. *See also* Farms; Households, cheesemaking
Dairymaids, 67; in agricultural press, 108; cultural perceptions of, 77–78; idyllic view of, 111. *See also* Women; Work, women's
Dairyman, 19th-century ideal notions of, 200
Dairy supply companies, 157
Dairywomen, wage negotiation of, 68. *See also* Women; Work, women's
Dairy Zone: erosion of, 172–73, 185; ideology of, 12–14; politics of, 15; rhetoric of, 52; soils of, 14
Day, Hannah, 148–51, 161
Day, Harvey, 148–51, 162, 164, 165, 166, 167, 169, 217
Day, Lenore, 151
"Deaconing," 77
Dean, Judge James, 39, 79
"Defeminization": of agriculture, 1; of cheesemaking, 152
Delancey Institute, 115
Delivery window, of early factories, 153–54, 156
Disease, bovine, 186
Domesticity, concept of, 142. *See also* Families; Households
Domestic Seminary, 115
Drought, 134
Drudgery: contradictions surrounding, 139, 140; denunciations of, 144; language of, 143; notion of, 138. *See also* Labor; Work
Dun, R., and Co., 214

Eagle factory, Otsego County, 165
East Springfield Academy, 210
Economy, emotional life and, 105–6
Education, 230; academies, 115–19, 120; advanced, 110–13, 114–15; expanding opportunities for, 100–101; and feminization of teaching, 120; gender organization of, 118; and generational differences, 119–20; professionalizing culture of, 209–10; "select schools," 114, 115; and transition to capitalism, 116; of young women, 110–13, 115–18, 206, 231
Eggs, 203–4. *See also* Poultry raising
Elite, agrarian, 213–14
Ellis, David Maldwyn, 10
Emmons, Ebenezer, 10
Emotional life, economy and, 105–6
Employees: of cheese factories, 161–62; and dairying households, 196–98. *See also* Labor; Work
Employment: of immigrants, 70; women's, 68. *See also* Labor; Work
Emulation, 55
Energy inputs: on cheesemaking farms, 41; shift to steam power, 157
Environment: and Anglo-American cul-

Index

Environment (*continued*)
tural values, 42; cheesemaking, 94–95; ecological degradation, 233
Equipment: cheesemaking, 85–88, 89; of early cheesemaking factories, 156–61; farming, 131
Erie Canal, 2, 11, 23, 39, 100
"Escutcheon," 21, 22
Essex County, Massachusetts, 18–19
Ethnicity: and factory system, 145; of hired labor, 69–70
Evans, Gurdon, 32, 37, 94, 161
Exchange: local systems of, 56; neighborhood networks of, 55–57; work, 66. *See also* Cash system
Export activity, intensification of, 59

Factor system, 46–51, 56, 194
Factories, cheesemaking: buildings of, 152–56; and capitalism, 124–25; conditions in, 136; and cooperative cheesemaking, 127; crossroads, 124, 137, 153; early, 127–28; early equipment of, 156–61; employees of, 161–62; expansion of, 133–34; and feed sales, 190; first modern-style, 123–24, 127; masculinization of, 168; and role for capital, 128; siting of, 153, 154–55; as social solvent, 195; women in, 162–64; working at, 149, 150, 151–52
Factory committees, 194
Factory system: advantages of, 128; changes initiated by, 174; and class divisions, 213; consolidation of, 177; and household reorientation, 194–97; impact of Civil War on, 129–33; in Midwest, 172; and milk adulteration, 215; precedents for, 126–27; problems in, 135; and social organization, 147, 152, 161–67; social roots of, 137–45; unanticipated changes brought about by, 223; women's view of, 140–41. *See also* Cash system; Centralization
Fairfield Academy, Herkimer County, 118
Fairfield factory, Herkimer County, 153
Fairs: and community ties, 225; increased participation in, 234; participation in, 75; proliferation of, 204; women's participation in, 213
Families: in feminist study, 102; negotiation of labor within, 113–14

Families, cheesemaking: and factory system, 136; impact of Civil War on, 130
Families, farm, 5, 119. *See also* Households
Faragher, John Mack, 96, 227
Farmers: conflict among, 213; identification with middle class of, 206; and wage system, 70. *See also* Families; Households
Farmers' daughters, 111, 207. *See also* Dairymaids
Farmhouses, 205–6
Farming: Euro-American, 39; of Native Americans, 7; pastoral vs. arable, 109
Farming systems: cheese-producing farms, 38–42; components of, 12; and dairy centralization, 193; dairying, 4; 19th-century alterations in, 178; role of barns in, 32; wool in, 73
Farms, cheesemaking: auxiliary products of, 192; butter production of, 202; diversity of, 38–42, 175; hay yields of, 29; size of, 39–40; value of, 60, 115–16; work on, 73–85
Feed-cooking devices, 30
Feeds: artificial grasses, 26; corn, 190; grass mixes, 27–28, 30; grasses, 24; and length of milking season, 32; processing of, 183–84, 199; root crops, 190; timothy, 189; traditional mainstays of, 25
Feminist study, families in, 102. *See also* Women; Work, women's
Ferris, Silvanus, 46, 47, 48, 49–50
Fertilizers: "commercial," 191; purchase of, 190; trade in, 29. *See also* Manure
Fish, Alonzo L., 12, 14, 26, 32, 34, 127, 185–86
Flax, 11
Flint, Charles, 19–20, 27, 78
Forest products, as cash crop, 192, 196
Forests: of Oneida County, 8–9, 40; productivity of, 74

Gang Mills factory, 137
"Gang presses," 157, 160
Garland, Hamlin, 198
Gates, Paul, 212
Gender: in cheesemaking households, 101; and family conflict, 107; and negotiating labor, 114. *See also* "Defeminization"; Masculinization
Generations: in cheesemaking households, 101; conservatism and, 122; education and,

Index

111, 119–20; and negotiating labor, 114; views on milking of, 105–6
Genesee County, 11
Genesee Farmer, 17, 23, 24, 74, 78, 79, 103–4, 113, 118, 119, 168, 202
Glassie, Henry, 180, 182
Goodale, Elaine, 207
Goshen, Connecticut, 14
Gowdy, Norman, 65
Grange, the, rise of, 225
Grass seed, trade in, 27–28
Grasses: artificial, 26, 188–89; scarcity of, 186–87; sources of, 26; varieties of, 24, 27; yield of, 28
Grasslands: concern about, 187–90; expansion of, 174; practices, 28
Great Britain: cattle from, 16; cheese exports to, 129
"Great Transformation," 4, 124. *See also* Capitalism
Guenon, François, 21

"Half soiling," 190
Hamilton College, 115
Handcrafts, 204–95, 206. *See also* Fairs
Harper's, 168, 169, 178, 202
Harper's New Monthly Magazine, 215
Hatmaking households, 98
Hay: market for, 193; scarcity of, 186–87. *See also* Feeds
Haying: decreased production, 188; labor involved in, 74; mechanization of, 131, 199; new approaches to, 189; old-time, 195
Help, hired. *See* Labor, hired
Herbert, Henry, 197–98, 202
Herkimer County, 11, 14
Hiring, formal practices of, 67–68
History, women's, 125. *See also* Women
Hogs, disadvantages of raising, 193
Holderness cattle, 16
Holland Land Company, 8
Holstein-Friesian Association, 151, 177
Holsteins, 17, 177, 178, 191
Home Cottage Seminary, 115
Home system, vs. factory system, 178. *See also* Factory system
"Hoop," 87
Hop Pickers, The: Girl Life in the Sixties (Canfield), 210–11
Hops, 191–92; as cash crop, 196; cultivation of, 200, 201; disadvantages of raising, 193; harvesting of, 210; and temperance activity, 229
Horse rake, revolving, 74
Houghton Seminary, 115
Households, analysis of, 4. *See also* Families
Households, cheesemaking: as centers of negotiation, 100–102; composition of, 162–65; and dairy centralization, 125; economic base of, 96–99; postwar changes in, 196; social organization of, 62–71
Households, dairying: in factory era, 211; reorientation of, 194–97. *See also* Centralization
Households, milk-selling: and community ties, 226; daughters in, 207–11
Howe, Daniel Walker, 230
Hubka, Thomas, 33
"Huffing," 150
Huntington, Lydia, 165, 166
Huntington, Nathan, 165, 166, 216
"Hydrometer," 217

Illinois, dairying in, 1
Immigrants, and wage labor, 69. *See also* Ethnicity; Migration
Independent Order of Good Templars, 227
Individualism, 118, 230–31. *See also* Competency
Industrialization: cultural dimension of, 145; farmers' participation in, 126; transformation to, 4
Inspection, provision for, 152. *See also* Milk adulteration
Iroquois confederacy, 7

Jefferson, Thomas, inaugural cheese for, 126
Johnson, Lavinia Mary, 65, 92, 106, 120
Joint-stock companies, 128
Journal of a Farmer's Daughter (Goodale), 207
Judson wives, 120

Kinship: and brokerage business, 57; and factory system, 138. *See also* Families; Households
Knives, dairy, 90, 159, 161
Kulikoff, Allan, 146

Index

Labor: and cow productivity, 199; devaluation of, 96; family exchange of, 66; gender division of, 77–78, 125, 138, 170–71; impact of Civil War on, 130–32; "mixed," 64; predominance of wage, 197; seasonality of, 75; sexual division of, 63, 72, 73–74, 75, 85, 102–8; unfair division of, 108, 109. *See also* Work

Labor, hired, 62, 63; decline in numbers of, 196; ethnicity of, 69–70; in household organization, 64–65; segregation of, 69; women as, 65–67

Lactic acid, 82

Lactometer, 216

Legumes, 28, 29. *See also* Feeds

Legislation, on adulterated milk, 215

Lewis, Dio, 144

Life cycle, family's, 2

Life-cycle phenomenon, cheesemaking as, 63

Livestock: of Oneida County, 10; rising interest in, 11; women's work with, 76. *See also* Cows

Log houses, 9

Lyceums, 226. *See also* Education

Machinery, farm, 183. *See also* Equipment

McLane Report (1832), 55, 128

Manheim Turnpike Association Cheese Factory, 138

Manufacture, cheese, large-scale, 123. *See also* Factory system

Manufacturing, rapid expansion in, 100. *See also* Factories

Manure, uses made of, 28

Manure cellars, 31–32, 179

Market: and dairy centralization, 125; impact on evolving dairying system, 233; and local exchange networks, 55–56; national, 1; for New York–made cheese, 45–46; and production strategies, 51; rural views of, 43; wage-labor, 68

Market economy: and barn design, 183; and social systems, 46; transition to, 124–25. *See also* Capitalism

Markham factory, 137

Marriage, of dairywomen, 68

Marshall cheese factory, 214

Masculinization: in cheesemaking, 169–70; of dairying, 208–9; of dairymen, 200

"Maw," 77

Meadows, makeup of, 26. *See also* Grasslands; Pastures

Mechanization: and barn design, 183; and demand for capital, 174

Media. *See* Press, agricultural

Men: in American cheese dairies, 81; education of, 118, 119–20; and modernization, 231. *See also* Work

Merry, Gottlieb, 160, 162, 217

Michigan Farmer, 96

Middle class: and factory system, 214; and farmers' daughters, 209–10; farmers' identification with, 206; nuclear family and, 229–30; role in community building of, 224; women's alliance with, 213

Middlemen, cheese-selling, 45, 49

Midwest: competition from, 172–73; dairy industry in, 134; decline of rural population in, 212

Migration: New York State, 6; out-migration, 234; rural-to-urban, 207; of young single women, 207–9

Milch cows: in factory system, 174; as machines, 23–24; native, 17–18; 19th-century, 16; selection of, 20–21, 22; shelter for, 30; superannuated, 18. *See also* Milking

Milk adulteration, 213, 215–220

"Milk fever," 186

"Milk mirrors," 21, 22

"Milk suits," 216

Milking: agricultural press on, 78; confining cows for, 31–32; division of labor in, 102; indoor vs. outdoor, 31; sexual differentiation of, 79; social context of, 77–78; women's decreasing involvement in, 203

Milking shed, 34, 35

Minnesota, dairying in, 1

Modernity, agricultural tradition and, 23

Modernization, 230. *See also* Industrialization

Mohawk Valley, 6, 8, 10

Monocrop mentality, Anglo-American, 189

Moore's Rural New Yorker, 126, 127, 132, 134, 135, 139, 155, 173, 182, 184, 202, 216–17

Morantz, Regina, 231

Motherhood, 230. *See also* Women

Mott's Agricultural Furnace, 30

Mount Holyoke, 117, 121

Mowing, mechanization of, 131

Mt. Vernon Boarding School, 115
"Mutuality": division of labor and, 97; and early cheese factories, 163; ethos of, 142–43; and factory system, 219; language of, 144; limitations of, 170; and social organizations, 224; strategies of, 138

National culture, and rural communities, 225, 226
Native Americans, in Oneida County, 6–7
Neighborhood system, 45; and exchange networks, 55–57; in factory era, 211, 224. *See also* Communities, rural; "Mutuality"
New England: dairying in, 1; decline of rural population of, 212; impact on New York of, 119, 233; milking practices in, 79
New England Farmer, 31, 78, 102, 106, 111, 131, 135, 137–38, 140, 214
New York State: cheese production of, 172–73; decline of rural population in, 212; migration to, 6; milking practices in, 79
New York State Cheese Manufacturers' Association, 175, 215, 218
North, decline of rural population in, 212. *See also* South
Nye family, 190, 195

Oats, market for, 193
Ohio, dairying in, 1
Ohio Cultivator, 34, 37, 61, 68, 71, 93, 106, 109, 110–11
Ohio Farmer, 135, 185, 202, 218
Oneida County, 3; climate of, 7–8; creation of, 6; decline in population of, 212; forests of, 8, 40; topology of, 7; water supply in, 9
Oneida County Teachers' Institute, 209
Oneida Weekly Herald, 140
Oneidas, 6, 7, 39
Orchard products, 192
Oriskany Creek valley, 9
Osterud, Nancy Grey, 97, 138
Out-migration, 234
Outwork, 98. *See also* Work
Oversupply, problem of, 134
Overwork, concept of, 108–10. *See also* Drudgery; Labor; Work
Oxford Academy, 121

Pastures: declining productivity of, 187; makeup of, 26, 27; and quality of cheese, 76

Pathogens, in cheesemaking process, 83–84
Patriarchal authority, 102, 113, 114
Pearlash, 8
Pennsylvania, milking practices in, 79
Perkins, Ephraim, 12, 14, 46, 55, 74
Peters, T. C., 89, 109, 112
Plymouth factory, 165
Politics, and women's work, 109
Poultry raising, 192, 201, 203, 234
Prairie Farmer, 112, 168
Press, agricultural: on cheese factories, 140–42; on collective cheesemaking, 126; dairymaids in, 108; diminishing role of women in, 167; on farm machinery, 184; on gendered organization of work, 104–5; on home vs. factory cheesemaking, 168; on masculinization of milking, 202–3; on milk adulteration, 216–19; on Sunday cheesemaking, 221–23; on "winter dairying," 185. *See also specific newspapers*
Presses, 161; "gang," 157, 160; patent, 90, 91; "self-acting," 90
Pressing: equipment for, 88; process of, 83
Prices, cheese, 50, 51–53; and Civil War, 129–30; and family strategies, 114
Private dairy, 165
Productivity: and household organization, 64; 19th-century increase in, 2; and shift to cash terms, 56
Profits: and competency, 53; and competition, 54–55; and Sunday cheesemaking, 223–24
"Pumpkin rumps," 19
Purebreds, 17; attitudes toward, 21; compared with native cows, 176. *See also* Breeding; Milch cows

Quincy, Josiah, Jr., 23

Railroad transportation, 1, 58
Randall, Henry, 16, 19
Rawsonville factory, 151
Reciprocity: argument for, 144; and household hiring, 68; language of, 143. *See also* "Mutuality"
Reformers, agricultural: and accounting procedures, 195; and competency, 52; on economic rationalism, 137; on milking barns, 34, 37; Thoreau's view of, 54; and temperance movement, 228
Refrigeration, 1

Index

Rennet: in cheesemaking process, 82, 83; preparation of, 76–77
Republicanism, 15
Rind, formation of, 83
Rinderpest, 134, 186
Risk: of cheesemaking, 50–51; in cheese-selling market, 45; conservative aversion to, 52
Roberts sisters, 132
Robinson, Solon, 85
Rollins, Ellen Chapman Hobbs. *See* Arr, E. H.
Rome Academy, 115
Rome Cheese Manufacturing Association, 124
Rome Sentinel, 108, 166, 174, 176, 209
Root crops, 190
Root culture, 199
Ross, E. C., 184
Rothenberg, Winifred, 44
Rural Architecture, 96
Rural history, "new," 4. *See also* Communities, rural
Rural New Yorker, 80
Ryan, Mary, 229

Sanborn factory, 155, 160
Sanitation, 1. *See also* Milk adulteration
Scandinavia, cattle from, 16
Scotland, cattle from, 16
Schoolteachers, female, 209
Seasonality, of cheese dairying, 66–67
"Select schools," 114, 115
Self-sufficiency, 43, 44, 45. *See also* Competency
Seminary education, 120
Separate spheres, concept of, 142
Sex ratio, rural-urban, 207–8
Sheldon, J. P., 176
Sheldon, Urania, 117
Shelter, 30; improved, 178–86; and length of milking season, 32; winter, 30, 31. *See also* Barns
Shoemaking, household, 97
Shoemaking industry, farmers' daughters in, 208
Shorthorn cattle, 16, 17, 177
Silos, 2
Skippers, 150
Social relationships, 12. *See also* Families; Households; "Mutuality"

Social structure: of cheesemaking households, 62; conflict and, 213–15; and dairy centralization, 125; and factory system, 137–45
Social systems, and market economy, 46
Soils: and cheese quality, 14, 173; fertility of, 29; overuse of, 10–11; varieties of, 10
Sorghum, 188
South, the: agriculture of, 45; as cheese market, 45–46; resistance to capitalism in, 98
South Plymouth factory, 165
Speculators, 51
Stanchions, 31, 33, 179, 180, 182
Steam power, shift to, 157
Stockbridge Indians, 39
Storage facilities, inadequacy of, 48–49
Subsistence, elements of, 73
Subsistence base, of cheesemaking households, 71
Subsistence exchange, and market participation, 44
Suffrage movement, 228. *See also* Women
Sunday cheesemaking, 213, 220–24

Taberg Good Templars, 227
Taxation, 174, 195
Taxonomies, of dairy cattle, 16, 22
Teaching, feminization of, 120, 130, 209
Technology, of cheesemaking: changes in, 88–92; early tools of, 85–88
Temperance cause, 224, 227–29, 234
Temperature, in cheesemaking, 82
"Tete-a-Tete of the Milkmaids," 108
Textile industry, farmers' daughters in, 208
Textile manufacture: capital requirements of, 128; on cheesemaking farms, 74–75; cheesemaking as substitute for, 103; shift to dairying from, 72–73
Thomas, Didymus, 205, 219
Thoreau, Henry David, 53–54
Timber, as cash crop, 192, 196
Timmerman, M. P., 138
Timothy, excessive reliance on, 189, 193
Tools, cheesemaking, 85–88
Townsend, William, 13, 14
Trade, rapid expansion in, 100
Trading: kinship and, 57; local exchange, 56. *See also* Cash system
Transportation: and dairy centralization, 125; impact on dairying of, 134; railroad, 1, 58;

and shift to cash terms, 58; unreliability of 48–49
Troy Seminary, 117
Tuberculosis, bovine, 186
Turners, cheese, 90
Tyrrell, Ian, 228

Urbanization, 1, 134, 229, 230
Utica, 229
Utica and Schenectady Railroad, 58
Utica Weekly Herald, 165, 168, 176, 177, 183, 189, 191, 197–98, 203, 208, 216

Values: of gentility, 203; middle-class, 138, 139; and Sunday cheesemaking, 221. *See also* Cultural values
Van Eps Institute, 115
Vats, cheese, 88, 133, 157–59
Veterans, Civil War, 130–31
Veterinary medicine, 186
Voelcker, Augustus, 191

Wage system: and cheese factories, 124; farmers and, 70
Wages: in cheesemaking households, 65; of hired labor, 68
Waring, George, 182–183
Water supply, in Oneida County, 9
Watering, problem of, 215–17
Watson, Elkanah, 6, 8
Weeks, Gardner, 135, 167, 169
Welsh: butter dairying of, 12; in Oneida County, 69–70
Western Reserve, 80, 97, 202
Westmoreland, Oneida County, 39
Wheat, as cash crop, 10
Wheat culture, 53
Whey: in cheese factories, 152; in cheesemaking, 82; factory appropriation of, 136; as feed supplement, 30; sour, 169; as waste, 155, 193
Whitesboro factory, Oneida County, 153, 154, 157, 161
Whitestown, Oneida County, 8
Whitford, Maria, 76, 88, 109, 113, 119, 201
Whitford, Samuel, 201
Willard, X. A., 25, 31, 41, 52, 58, 68, 128, 134, 135, 137, 140, 143, 155, 172, 176, 179, 181, 186, 188–89, 218
Williams, James Mickel, 225

Williams, Jesse, 123, 127
Willow Grove factory, Oneida County, 153, 161, 219
"Winter dairying," 185
Winter shelter, development of, 30–31
Wisconsin: cheese production of, 172–73; competition from, 172–73; dairying in, 1, 134
Wives, new occupations for, 201–7
Women: cheesemaking of, 80–81, 234; in cheesemaking households, 63, 64–65, 101; and dairy centralization, 125; education of, 110–13, 115–18, 119, 206, 231; exploitation of, 106; in factory cheesemaking, 162–64; health of, 143–44; as hired labor, 65–67; impact of Civil War on, 129–30, 131–33; impact of farm technology on, 92; increased participation in fairs of, 204; increased social activity of, 227; lack of inheritance for, 112; and modernization, 231; and shift to dairying, 73; and subsistence production, 72; and technology, 90; and temperance reform, 227–28; textile production of, 74–75; wage negotiations of, 68; withdrawal from dairying of, 146–47. *See also* Dairymaids; Work
Wool, in farming economy, 73
Work, gendered organization of, 103–6, 139. *See also* Labor
Work, men's: redefinition of cheesemaking as, 167–71; and rural change, 197–201
Work, women's: on cheese-producing farms, 117, 120–21, 138; during Civil War, 131–32; and dairy centralization, 139, 224; and daughters' choices, 207–11; divesity of, 208; as drudgery, 138, 139, 143–44, in factory system, 152; in farming cultures, 97; as hired labor, 65–67; during hop harvest, 210; marginalization of, 91; and neighborly customs, 166–67; overwork in cheesemaking, 108–10; and political participation, 109; redefinition of, 3; rejection of rural, 122; and rural change, 201–6

"Yankee" dealers, 56
Yields, and cattle feed, 24
Young Ladies Domestic Seminary, 116
Young Ladies School, 115

Library of Congress Cataloging-in-Publication Data

McMurry, Sally Ann, 1954–
Transforming rural life : dairying families and agricultural change, 1820–1885 / Sally McMurry.
   p. cm. — (Revisiting rural America)
Includes bibliographical references (p.) and index.
ISBN 0-8018-4889-X (hc : alk. paper)
   1. Cheesemakers—United States—History—19th century. 2. Dairy farmers—United States—History—19th century. 3. Cheese industry—United States—History—19th century. 4. Dairying—United States—Technological innovations—History—19th century. 5. Rural conditions—United States—History—19th century. I. Title.
II. Series.
SF274.U6M38 1995
306.3′64—dc20
                                                                                             94-12093